Frontiers in Fungal Ecology, Diversity and Metabolites

Published by

I.K. International Publishing House Pvt. Ltd.
S-25, Green Park Extension
Uphaar Cinema Market
New Delhi - 110 016 (India)
E-mail : ik_in@vsnl.net

ISBN: 978-81-89866-91-4

Published by Krishan Makhijani for I.K. International Publishing House Pvt. Ltd., S-25, Green Park Extension, Uphaar Cinema Market, New Delhi - 110 016 and Printed by Rekha Printers Pvt. Ltd., Okhla Industrial Area, Phase II, New Delhi - 110 020.

Dedication

I am grateful to Prof. K.M. Kaveriappa, who impressed me as teacher of microbiology during my M.Sc. Biosciences course (1975-77) and introduced me to the field of mycology in 1978. Although my interest on *mycology* and *ecology* was scanty, I chose the topic '*Ecology of Aquatic Hyphomycetes*' as my research career mainly due to motivation and encouraging words of Prof. Kaveriappa. He was instrumental to initiate and establish the basic facilities for mycological research at Mangalore University. His enthusiasm, involvement and motivation of students to carryout research in mycology at Mangalore University wiped out the difficulties of lack of sophisticated instrumentation. As a student, I was lucky to spend maximum time with Prof. Kavariappa in the laboratory as well Western Ghats during my doctoral studies to understand the basic concepts of aquatic mycology and its relevance to the environment.

Prof. Khandrathanda Monnappa Kaveriappa hails from Virajpet, Kodagu District of Karnataka. He did his early schooling at Virajpet and obtained B.Sc. (Botany, Zoology and Chemistry) and M.Sc. (Botany) degrees from the University of Mysore. Later, he earned Ph.D. in Botany (Plant Pathology) under the guidance of plant pathologist, Late Prof. K.M. Safeeulla at Manasagangotri, University of Mysore and made a substantial contribution to the field of Downy Mildew disease of *Sorghum*. He served as Lecturer and Reader at the Department of Biosciences, Postgraduate Centre, University of Mysore and Applied Botany, Mangalore University since 1971. In 1987 he became professor and in his leadership, the Department of Applied Botany at Mangalore University was established. Besides teaching, he has guided M.Phil. and Ph.D. students in the disciplines of microbiology, forest ecology, forest pathology, plant tissue culture and micropropagation. He has published about 100 research papers in national and international journals. He has co-authered three books '*Plant Resources of Dakshina Kannada and Udupi Districts*' '*Biodiversity Evaluation and Conservation Strategy using Satellite Remote Sensing and Geographic Information System*' and '*Romancing the Elephant: A Strategy for Mitigating Man-Elephant Conflict*'.

Prof. Kaveriappa was a Fulbright Fellow and visited the University of Tennessee in 1994 and successfully carried out micorporpagation of flowering dogwood (*Cornus florida*). He has organized International Paul Neergard Commemoration Symposium on Seed Science and Technology (1988) and Colloquium on Post Harvest Technology of Coffee (1999). He is a life member of Indian Phytopathological Society, National Institute of Ecology, Society of Ethnobotanists, International Society of Plant Morphologists, Society of Indian Foresters and Bombay Natural History Society. He has executed several research projects on plant biodiversity, forest ecology and conservation, diseases of commercial/plantation crops, plant-microbe interactions and micropropagation of endemic and threatened plants funded by several Indian and foreign agencies (Department of Environment, Nuclear Power Corporation of India Ltd., Karnataka Power Corporation, Kudremukh Iron Ore Company Ltd., Indo-Norwegian Environment Programme, John D. and Catherine J. MacArthur Foundation, United States and International Plant Genetic Resources Institute, Malaysia). He has established a model arboretum of endemic, rare and threatened plant species of the Western Ghats as a measure of *ex situ* conservation in 3 ha at Mangalagangotri Campus in association with Late Sri B.V. Shetty. He was also instrumental in establishing another arboretum of Western Ghat plants at Pilikula Nisarga Dhama near Mangalore under Indo-Norwegian Environment Project in 35 ha area. This unique arboretum consists of a repository of 250 plants consists of several endemic and endangered plant species grow in Western Ghats. Pilikula arboretum also consists of green house, orchidarium and vermicompost units facilitating environmental education and training for school children and public.

He has a long administrative experience as Registrar of Mangalore University (1982-1994). He was awarded Nuffield Foundation Administrative Fellowship to visit United Kingdom in 1984 to study the pattern of administration in British Universities to adapt at Mangalore University. He was actively involved in framing statutes, ordinances, rules and regulations pertaining to academic and administrative matters.

Besides academic activities, Prof. Kaveriappa was a member of senate and syndicate; Dean, Faculty of Science and Technology; Director, Planning, Monitoring and Evaluation; Chairman, Academic and Administrative Audit Committee of Mangalore University. He was the member of the peer team of National Assessment and Accreditation Council (NAAC) to evaluate several Indian universities and colleges; Expert to evaluate University Grants Commission sponsored projects; Member of Board of Appointments of various universities. He was also an expert nominee of the Indian Council of Agricultural Research to monitor promotion of Agricultural Research Scientists. He has also served as the Convener of the Government of Karnataka Coastal Agenda Task Force (CATF) to frame a policy report on agriculture and allied sector development of coastal districts of Karnataka.

Prof. Kaveriappa's long academic and administrative experience earned him the coveted position of Vice-Chancellor of Mangalore University in January 2006. He has ample opportunities to nurture higher education for value-addition at the juncture of globalization with his vast academic and administrative experience.

I am dedicating this book to prof. K.M. Kaveriappa, my mentor in the fields of academic and research, whose association has given me immense inspiration to be a teacher and from whom I imbibed qualities of friendship and dedication.

K.R. SRIDHAR

Preface

The diversity, uniqueness and ubiquitous nature fungi elevated their status to the level of '*fifth kingdom*' and represent one of the three major evolutionary segments along with plants and animals. Fungi are placed next to arthropods and angiosperms in their richness. They are known to interact with plants, animals and microbes and lead symbiotic, mutualistic and saprophytic or pathogenic life. The global conservative fungal estimate, 1.5 million is mainly based on angiosperm/fungus ratio (1:0.5-1:33). But only a fraction of this is currently known (about 100,000) and a few have been harnessed. The discipline, *mycology* has multidimensional facets with basic and applied values and the recent investigations have gone beyond routine systematics, diversity and environmental issues. Fungi serve as novel source of food, fuel, biopesticides, metabolites and pharmaceuticals. Being eukaryotes, fungi became important basic tool in biotechnology and bioremediation.

To reflect the recent aspects and views on fungal ecology, diversity and metabolites, I have invited contributions from several active scientists. My concern was to avoid presentation like usual text books. Each chapter has been viewed independently to avoid overlapping and errors to the maximum extent. The main aim was to provide a reasonable coherence between chapters and provide substantial information on the major issues of fungal ecology, diversity and metabolites. Several chapters of this book deal with mycology in less researched or neglected ecosystem and their ecological services. Certainly some gaps remain obviously and I trust such gaps will be filled by additional or novel information to link fungal diversity, ecology and metabolites. Avenues for advancing our knowledge on fungal ecology, diversity and metabolites remain bright if mycologists continue to explore and exploit the diverse mycoflora in critical or less explored habitats and understand or hypothesize their ecosystem services.

In view of tremendous developments in mycology in the recent past, this contribution emphasizes different aspects of contemporary research in mycology with a view to project the issues on ecology, diversity and metabolites. The major areas projected include the ecology, diversity, phylogeny, mutualism, metabolites, pathology and toxins. This volume serves as a valuable source of information on frontier areas of mycology: ecological amplitudes, conservation, novel methods of assessment and bioprospecting. I presume that the readers of this book gain momentum to pursue studies on mycology more precisely by virtue of scope and diverse topics dealt in different fields. The chapters dealt in this book will certainly stimulate mycologists or those who respect mycology to initiate research on various aspects and formulate their own hypothesis to persuade the granting agencies or potential employers.

I am indebted to Prof. Felix Baerlocher (Department of Biology, Mount Allison University, Canada) for his continued support, encouragement and friendliness during my several visits to his laboratory. My repeated visits to Germany were possible due to generosity of Prof. Gerd-Joachim Krauss (Department of Biochemistry and Biotechnology, Martin Luther University, Halle) and Dr. Gudrun Krauss (Department of Environmental Microbiology, Helmholtz Centre for Environmental Research-UFZ, Halle), which stimulated me to investigate fungi in hyper-polluted aquatic habitats of Central Germany. I am grateful to my students, colleagues, friends and well wishers who contributed enthusiastically to this volume and delivered the manuscripts on time. I have special appreciation to my family, past and present Ph.D. students with whom I used to often share, discuss and formulate several ideas in mycology particularly during coffee break.

K.R. SRIDHAR

Contributors

A. Bhanumathi
Department of Microbiology, University of Mysore, Manasagangotri, Mysore 570 006, Karnataka, India

Anuradha N. Naik
National Institute of Oceanography, Dona Paula, Goa 403 004, India

A.H. Rajasab
Mycology and Plant Pathology Laboratory, Department of Botany, Gulbarga University, Gulbarga, 585 106, Karnataka, India

B.F. Rodrigues
Department of Botany, Goa University, Goa 403 206, India

B.S. Girisha
Department of Skin and Venereal Diseases, K.S. Hedge Medical Academy, Deralakatte 574 160, Mangalore, Karnataka, India

C. Leão
Life and Health Sciences Research Institute (ICVS), School of Health Sciences, University of Minho, 4710-057 Braga, Portugal

D.K. Upreti
Lichenology Laboratory, National Botanical Research Institute, Rana Pratap Marg, Lucknow 226 001, Uttar Pradesh, India

D. Shobha
Department of Botany, Mangalore University College, Mangalore 575 001, Karnataka, India

F. Cássio
Centre of Biology, Department of Biology, University of Minho, 4710-057 Braga, Portugal

F. Baltazar
Life and Health Sciences Research Institute (ICVS), School of Health Sciences, University of Minho, 4710-057 Braga, Portugal

G.R. Janardhana
Mycology and Plant Pathology Laboratory, Department of Studies in Botany and Microbiology, University of Mysore, Manasagangotri, Mysore 570 006, Karnataka, India

H. Shekar Shetty
Department of Biotechnology, University of Mysore, Manasagangotri, Mysore 570 006, Karnataka, India

H.S. Prakash
Department of Applied Botany and Biotechnology, University of Mysore, Manasagangotri, Mysore 570 006, Karnataka, India

J.K. Misra
Mycological Research Unit, Department of Botany, Sri Jai Narain Postgraduate College, Lucknow 226 001, India

Jerome Pinto
Department of Skin and Venereal Diseases, K.S. Hedge Medical Academy, Deralakatte 574 160, Mangalore, Karnataka, India

K.R. Sridhar
Microbiology and Biotechnology, Department of Biosciences, Mangalore University, Mangalagangotri, Mangalore 574 199, Karnataka, India

K.R. Chandrashekar
Department of Applied Botany, Mangalore University, Mangalagangotri 574 199, Karnataka, India

K. Natarajan
Centre for Advanced Studies in Botany, University of Madras, Guindy Campus, Chennai 600 025, Tamil Nadu, India

K. Velmourougane
Coffee Research Sub-Station, Coffee Board, Chettalli Post, North Coorg 571 248, Karnataka, India

M. Rajashekhar
Department of Biosciences, Mangalore University, Mangalagangothri 574 199, Karnataka, India

Malathi Srinivasan
Centre for Advanced Studies in Botany, University of Madras, Guindy Campus, Chennai 600 025, Tamil Nadu, India

M.Y. Sreenivasa
Mycology and Plant Pathology Laboratory, Department of Studies in Botany and Microbiology, University of Mysore, Manasagangotri, Mysore 570 006, Karnataka, India

M. Manjunath Shenoy
Department of Skin and Venereal Diseases, K.S. Hedge Medical Academy, Deralakatte 574 160, Mangalore, Karnataka, India

M. Suchitra Shenoy
Department of Microbiology, Kasturba Medical College, Mangalore 575 001, Karnataka, India

M.V. Tejesvi
Department of Applied Botany and Biotechnology, University of Mysore, Manasagangotri, Mysore 570 006, Karnataka, India

Nityanand D. Maddodi
Department of Dermatology, School of Medicine and Public Health, University of Wisconsin, Madison, USA, 53706

N.S. Raviraja
Department of Anatomy and Structural Biology, Albert Einstein College of Medicine, Bronx, New York, USA, 10461

P. Rama Bhat
Department of Biotechnology, Alva's College, Moodbidri 574 227, Karnataka, India

Regina Sharmila Dass
Mycology and Plant Pathology Laboratory, Department of Studies in Botany and Microbiology, University of Mysore, Manasagangotri, Mysore 570 006, Karnataka, India

Rajeev Bhat
Microbiology and Biotechnology, Department of Biosciences, Mangalore University, Mangalagangotri, Mangalore 574 199, Karnataka, India

R. Bajpai
Lichenology Laboratory, National Botanical Research Institute, Rana Pratap Marg, Lucknow 226 001, Uttar Pradesh, India

Rajendra Upadhya
Department of Pathology, Albert Einstein College of Medicine, Bronx, New York 10461, USA

S. Nayaka
Lichenology Laboratory, National Botanical Research Institute, Rana Pratap Marg, Lucknow 226 001, Uttar Pradesh, India

S. Niranjan Raj
Department of Biotechnology, University of Mysore, Manasagangotri, Mysore 570 006, Karnataka, India

S. Mohammed Shabbir
Mycology and Plant Pathology Laboratory, Department of Botany, Gulbarga University, Gulbarga, 585 106, Karnataka, India

Seena Sahadevan
Department of Biology, Mount Allison University, 63B York Street, Sackville, New Brunswick, Canada E4L 1G7
Present address: Centre of Biology, Department of Biology, University of Minho, 4710–057 Braga, Portugal

S. Shishupala
Department of Microbiology, Kuvempu University, Shivagangotri Campus, Davangere 570 002, Karnataka, India

V. Ravishankar Rai
Department of Microbiology, University of Mysore, Manasagangotri, Mysore 570 006, Karnataka, India

Contents

Frontiers in Fungal Ecology, Diversity and Metabolites (edited by K.R. Sridhar), I.K. International Publishing House Pvt. Ltd., New Delhi, 2009.

1

Freshwater Hyphomycetes

K.R. Sridhar
Microbiology and Biotechnology, Department of Biosciences, Mangalore University, Mangalagangotri, Mangalore 574 199, Karnataka, India (E-mail: sirikr@yahoo.com)

Introduction

Aquatic hyphomycetes constitute poylphyletic mitosporic fungal group with mainly ascomycetes and a few basidiomycete representatives. Depending on the conidial state, about 30 species have been connected with teleomorphs (Marvanová, 1997; Webster, 1992). Persisting wood and twigs compared to short-lived leaf litter in streams constitute the sites of sexual reproduction and development of teleomorphs. Several terms have been proposed from time to time to address freshwater hyphomycetes (e.g., aquatic hyphomycetes, water-borne hyphomycetes, Ingoldian fungi) (Descals and Moralejo, 2001). Ingold (1942) preferred the term 'aquatic hyphomycetes' considering underwater existence of fungi to produce, liberate and disperse conidia. He hypothesized that this group of fungi was an evolutionary line that had adapted from a terrestrial existence to flowing fresh waters. Nilsson (1964) theorized that aquatic hyphomycetes might have evolved to produce conventional to well-developed sigmoid or tetraradiate or multiradiate conidia in running waters (Fig. 1, 2). In fact, some freshwater hyphomycetes are known to sporulate in an air-water interface (e.g., *Clavatospora longibrachiata, Filosporella versimorpha, Triscelophorus konajensis*) (Ingold and Cox, 1957; Marvanová *et al.*, 1992; Sridhar and Kaveriappa, 1987a). Freshwater hyphomycetes have worldwide distribution in typical fast flowing, turbulent and well-aerated pristine waters. Besides lotic habitats, they are also found in lentic waters, water droplets or dew on the leaves, rain waters, canopies and stem flow (e.g., Gönczöl and

Fig. 1. Typical conventional (A, B), sigmoid (C, D) and multiradiate (E, F) conidia of freshwater hyphomycetes: *Tumularia aquatica* (A), *Heliscus lugdunensis* (B), *Synnematophora constricta* (C), *Condylospora spumigena* (D), *Trinacrium indica* (E) and *Triscelophorus konajensis* (F) (Scale, 20 μm).

Révey, 2004, 2006; Sridhar *et al.*, 2006). They are also inhabitants of submerged macrophytes and as endophytes in roots of riparian vegetation.

One of the principle challenges of contemporary ecology is to link species and ecosystem functioning (Cardinale *et al.*, 2000). Decomposition of plant litter is an important ecosystem process in terrestrial and benthic habitats (streams and littoral zones of fresh, brackish and marine waters) (Webster *et al.*, 1999). Woodland streams are primarily detritus-based and leaf litter as coarse particulate organic matter (CPOM) constitutes a major source of energy for aquatic heterotrophic microbes for mineralization and transformation into palatable foodstuff for the shredder invertebrate communities (Baerlocher, 1985; Fisher and Likens, 1973). Fungi are most dominant among microbial communities associated with submerged decaying leaf litter in streams and contribute 63-100% of total microbial biomass (Baldy *et al.*, 2002; Gulis and Suberkropp, 2003). Even though bacteria exhibit higher turnover rates of CPOM in streams, fungal production is equivalent or surpasses bacterial production to a factor of 0.9-627 (Baldy *et al.*, 2002; Weyers and Suberkropp, 1996). Freshwater hyphomycetes have been recognized as important links or intermediaries of energy flow between dead plant materials to higher trophic levels (Baerlocher and Kendrick, 1981). Their biomass will build up to 18% of total detritus mass and 80% will be allocated to conidial production (Suberkropp, 1991, 1995). About 300 species of freshwater hyphomycetes are known throughout the world and many more awaiting discoveries (Gulis *et al.*, 2005; Shearer *et al.* 2007; Webster, 1992). About 100 species are known from hotspots of biodiversity in the Indian Subcontinent (Western Ghats and Himalayas) (Rajashekhar and Kaveriappa, 2003; Sati and Tiwari, 1989; Sridhar *et al.*, 1992; Subramanian and Bhat, 1981). Freshwater hyphomycete community is a model to evaluate the community structure and function in freshwater streams. The objectives of the current chapter are to project diversity, structural and functional significance of freshwater hyphomycetes based on some recent developments.

Conidia

Aquatic hyphomycetes produce conidia with a variety of shapes and sizes. Foam accumulated in streams consists of conidia of several species and roughly depicts the fungal population of the catchments. Conidia from fungal biomass of submerged leaf litter can be generated in the laboratory on aeration in water. On aeration in water, pieces of agar culture also generate enormous conidia. The shapes of conidia include conventional (spherical, oval), sigmoid (worm-shaped, scolecoform) and multiradiate (stauroform) measuring about 50-100 μm. Representative conidia of different shapes are given in Figure 1 and 2. Conidia of freshwater hyphomycetes lead planktonic life in flowing waters and are capable of capturing resource and enriching the substrate by absorbing nutrients in water. Conidial adaptation is the result of convergent evolution and the shape of conidia (particularly scolecoform and stauroform) designed for transfer as

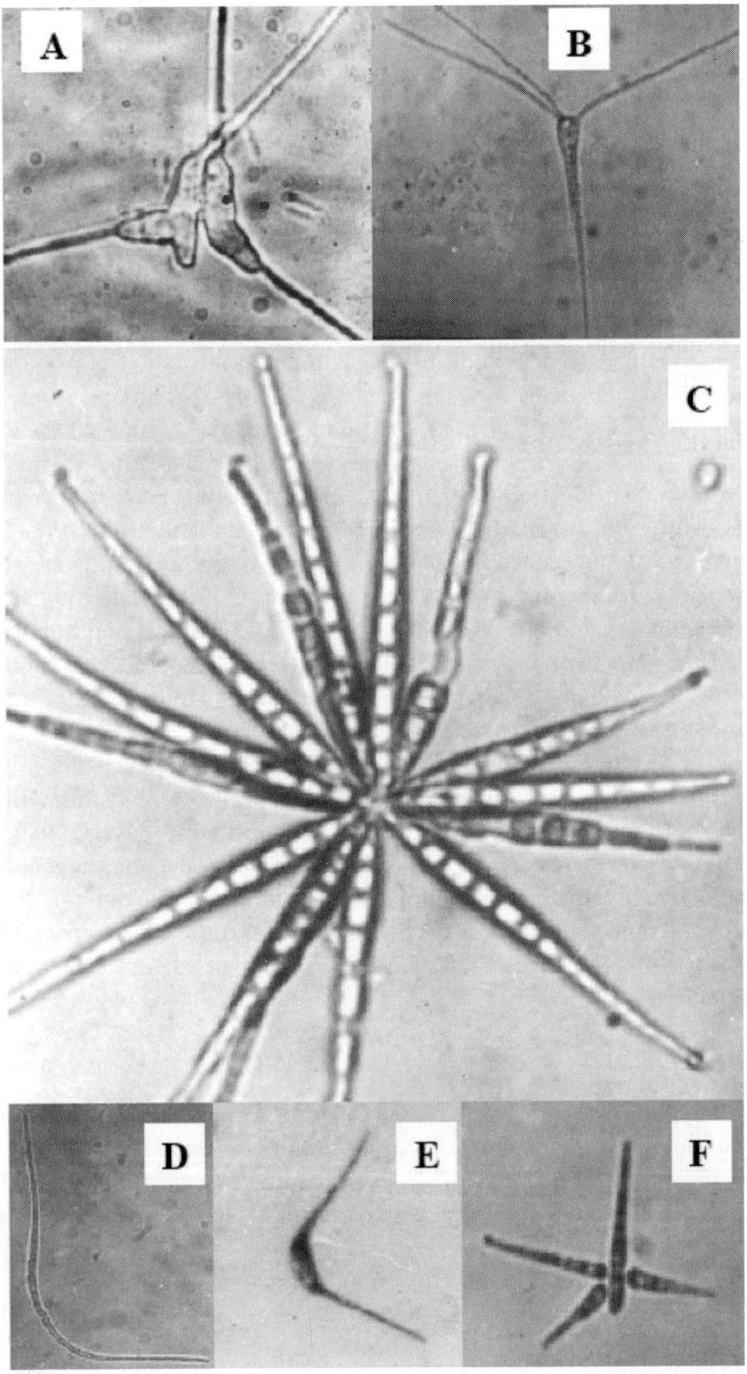

Fig. 2. Some aquatic hyphomycete conidia isolated from the streams of Western Ghats, India: *Campylospora chaetocladia* (A), *Clavatospora tentacula* (B), *Flabellospora multiradiata* (C), *Synnematophora constricta* (D), *Lunulospora cymbiformis* (E) and *Triscelophorus konajensis* (F)

well as trapping on surfaces of organic matter in flowing waters (Webster, 1987). The sigmoid or multiradiate conidial shape minimizes the downstream movement and sigmoid conidia rest with two contact points (one at the tip and another along the length), while multiradiate conidia like tripods on the substrate (Webster and Descals, 1981). Spore attachment and entrapment mechanisms are the adaptations of freshwater hyphomycetes that confer advantages over other fungi (Read, 1990). The mucilaginous material at the tip of each arm gives a strong adherence and prevents spore detachment by strong water currents (Read *et al.*, 1992). The cyclic events in streams include conidial entrapment followed by attachment, germination, production of mucilage, germ tubes, formation of appressoria and colonization.

Distribution and Niches

Most species of freshwater hyphomycetes have been reported from temperate regions. The pattern of distribution of these fungi is biased due to the uneven sampling in different geographical areas. Global report on known and unknown freshwater hyphomycete species (in parenthesis: known/unknown) from different geographic locations include: Arctic (37/2), Western hemisphere: Boreal (130/13), North temperate: Tropics (102/20), Eastern hemisphere: Boreal (75/ 8), European temperate (180/36), Asian temperate (160/48), Northern Africa (45/5), Tropical Africa (37/7), Temperate Africa (90/27), Mid East (20), Tropical Asia (110/11) and Austral Asia (70) (Shearer *et al.*, 2007). The Temperate North America, Europe, United Kingdom, Chile, Japan and Malaysia are the relatively well-sampled locations (Shearer *et al.*, 2007). This feature may change, if intense surveys concentrate on tropical and subtropical regions. However, to achieve a holistic knowledge on fungal resource, a global biodiversity initiative needs to be implemented.

Freshwater hyphomycetes are usually abundant in leaf litter of streams and stream foams, while moderate in rivers, boundaries of stream or river, hyporheos and tree holes. Low population was seen in lentic waters, temporary streams and ponds, estuarine and marine habitats, ground waters, aero-aquatic habitats, riparian roots, tree canopies and terrestrial habitats (e.g., soil, root surface, dead leaf litter). For freshwater hyphomycetes, the stream boundaries become a refuge and reservoir for recolonization on the onset of appropriate conditions (Baerlocher, 1992a). Glass-beads buried at 10 cm depth in a stream with sediment trapped conidia of freshwater hyphomycetes indicated conidial penetration and survival in hyporheic conditions (Baerlocher and Murdoch, 1989). A recent study revealed that they readily disperse within hyporheic zones (Baerlocher *et al.*, 2006). The roots of riparian vegetation are a potential source of freshwater hyphomycetes as endophytes (e.g., Fisher *et al.*, 1991; Marvanová and Fisher, 1991; Nair and Bhat, 2002; Raviraja *et al.*, 1996; Sati and Belwal, 2005; Sridhar and Baerlocher, 1992). Freshwater hyphomycetes are also known from brackish or seawater (Kirk, 1969; Müller-Haecel and Marvanová, 1979; Shearer and Crane, 1971; Sridhar and Kaveriappa, 1988). Gönczöl (1976),

Gönczöl and Révey (2003), and Sridhar *et al.* (2006) recovered freshwater hyphomycetes from tree holes and on the oak leaf basket ferns attached to tree trunks. Many freshwater hyphomycetes have been reported from soil and they are known to survive in soils (Bandoni, 1981; Baerlocher, 1992a). Submerged macrophytes (e.g., *Potamogeton, Ranunculus*, *Apium*) also constitute major substrate in the absence of submerged leaf litter (Baerlocher, 1992b). Woody debris in streams becomes a reservoir, a site of reproduction, competition and dispersal of freshwater hyphomycetes. It is believed that there is no substrate specificity for the occurrence of freshwater hyphomycetes and they are not fastidious as they grow and sporulate even on latex smear (e.g., rubber, banyan) (Sridhar and Kaveriappa, 1987b).

Techniques

The authentic identification of fungi is most important in biodiversity assessment. Some of the monographs facilitate identification of freshwater hyphomycetes (Gulis *et al.*, 2005; Ingold, 1942, 1975; Marvanová, 1997; Nawawi, 1985; Webster and Descals, 1981). Descals (1997) described a variety of basic techniques necessary to evaluate freshwater hyphomycetes. The direct observation of leaf pieces incubated for 24-48 h in Petri dishes reveals colonized fungi. Moist chamber incubation of leaf pieces also helps in conidial development on the surface or cut edges. To generate sufficient conidia, an aeration of leaf pieces suspended in sterile distilled water with a jet of filtered air through aquarium aerator or Pasteur pipette up to 48 h is recommended (Iqbal and Webster, 1973a). Mini glass aeration chamber designed by Suberkropp (1991) to generate conidia from the substrates or cultures facilitates carrying out laboratory microcosm studies more precisely. Aeration facilitates conversion of mycelial biomass into conidial biomass. Conidial output will be up to 8/µg dry mass of leaf, which is close to one million per leaf (Baerlocher, 1992b).

An assessment of foam accumulated in turbulent streams gives a fair estimate of fungal population of the region. Fresh foam can be scanned immediately or fixed in formalin-acetic-alcohol for further assessment. Usually, stream foam concentrates more multiradiate than sigmoid and conventional conidia and the estimation that is entirely based on foam samples may be biased (Iqbal and Webster, 1973a). Foam can be induced with the addition of a small quantity of sodium-lauryl-sulphate into water incubated with leaf followed by aeration. Foam or incubated water can be filtered through membrane filters (5 or 8 µm) for quantification of conidial output (Iqbal and Webster, 1973b). Filters stained by cotton blue in lactophenol, subsequently mounted on glass slide with a few drops of lactic acid helps to enumerate conidia. As there is a direct relation between fungal biomass, ATP and conidial output (Rossett *et al.*, 1982; Suberkropp *et al.*, 1976), conidial estimation becomes an easy and authentic method of assessment of fungal activity in streams. Conidia germinate readily on agar media and isolation from foam or leaf litter will be relatively easy. Cultures of monoconidial isolates can be established through lifting the conidia with the help of a mounted eyelash or an incinerated fine needle and by placing it on malt extract agar (1%) amended with streptopenicillin (Webster and Descals, 1981). Cultures can produce huge amount of conidia on aerating pieces of agar blocks in sterile distilled water. Particle plating (212-700 μm) method developed by Kirby *et al.* (1990) also facilitates isolation of fungi colonized on various substrates in streams. Unlike terrestrial fungi, the growth of freshwater hyphomycetes is fairly slow. Although isolation of freshwater hyphomycetes is easy in temperate environments, in tropical weather (particularly in coastal regions) culture establishment and maintenance is a difficult task.

A variety of techniques can be implemented to assess freshwater hyphomycetes based on the methods outlined by Graça *et al.* (2005). One of the preferred methods for estimating biomass of freshwater hyphomycetes in plant detritus is the extraction and measurement of ergosterol (a filamentous fungal-specific sterol of membranes). However, it does not allow differentiation between different species. Besides solvent extraction of ergosterol (Gessner, 2005), microwave-assisted extraction is also available (Young, 1995). The ergosterol values of decomposing detritus can be converted into fungal biomass assuming that the average of 5.5 mg ergosterol is equivalent to one gram fungal biomass (Gessner and Chauvet, 1993). The estimation of fungal biomass through monoclonal antibodies has been developed by Bermingham *et al.* (1996). Other interesting approaches of fungal biomass estimation include, direct epifluorescence microscopy and glucosamine estimation (Newell, 2000). The most recent approaches for the estimation of freshwater hyphomycete biomass are by modified terminal restriction fragment length polymorphism (T-RFLP) and by measuring band intensities on gels of denaturing gradient gel electrophoresis (DGGE) (Nickolcheva and Baerlocher, 2005a, b).

Ecological Functions

Freshwater hyphomycetes are dominant fungi associated with decaying leaves in streams compared to other fungi (Baerlocher and Kendrick, 1974; Butler and Suberkropp, 1986). The dominance of these fungi on submerged leaves is mainly due to their ability to remain active at low temperatures (Baerlocher and Kendrick, 1974; Godfrey, 1983) and degrade submerged organic matter in a wide range of climatic conditions (Graça and Ferreira, 1995). The production of a rich array of extracellular enzymes is responsible for the degradation of plant polysaccharides (Suberkropp and Krug, 1980). They degrade simple sugars, cellulose and other polymers of plant origin (Chandrashekar and Kaveriappa, 1988; Singh, 1982; Thornton, 1965) and lead to skeletonization and maceration. Experiments revealed that cellulose, xylan and pectin degrading activity by freshwater hyphomycetes are pH-dependent (Chamier and Dixon, 1982; Suberkropp *et al.*, 1983). The rate of leaf decay was correlated to the fungal biomass (Maharning and Baerlocher, 1996). The remaining nitrogen of leaf mass increases during leaf decay by freshwater hyphomycetes (Baerlocher, 1985; Raviraja *et al.*, 1998; Webster and Benfield, 1986) and is partly due to the accumulation of microbial cells (Raviraja *et al.*, 1996). An increase in fungal colonization of leaf results in higher exoenzymatic activity and contributes to mass loss and eventually higher invertebrate feeding as well as susceptibility to mechanical fragmentation (Raviraja *et al.*, 1998).

Triska (1970), and Kaushik and Hynes (1971) demonstrated for the first time the preference of fungal colonized leaves by stream invertebrates and the importance of freshwater hyphomycetes as intermediaries of energy flow between dead leaf litter and invertebrates. Freshwater hyphomycetes maintain territories on submerged leaf litter similar to basidiomycetes on terrestrial wood (Shearer, 1992; Shearer and Lane, 1983) and adapt a ruderal strategy (grow on relatively complex substrata for a short duration coupled with high reproductive potential to escape from predators) than a mass accumulation strategy (Cooke and Rayner, 1984). Suberkropp (1992) further investigated the interaction of freshwater hyphomycetes with stream shredders such as crustaceans (e.g., Amphipoda, Isopoda) and insects (e.g., Diptera, Plecoptera, Trichoptera). Shredders selectively forage patches within the leaf as freshwater hyphomycetes are territorial. The growth and survival of shredders is greatly dependent on fungal biomass in their diet (Arsuffi and Suberkropp, 1986; Baerlocher and Kendrick, 1973). Besides proteins and enzymes, ergosterol of freshwater hyphomycetes plays an important role in the metamorphosis of aquatic insects and palatability to prawns and fishes (Arsuffi and Suberkropp, 1986; Baerlocher and Kendrick, 1973; Chandrashekar *et al.*, 1989; Rao and Manoharachary, 1982).

Two important possibilities on the interactions of freshwater hyphomycetes and stream leaf-shredder communities have been suggested by Baerlocher (1992a): (a) improvement of quality of litter by degradation of polymers by fungi into more digestible forms; (b) fungal extracellular enzymes function in digestive tract of leaf shredders. It is possible to understand precisely the significance of leaf detritus colonized by freshwater hyphomycetes in streams or in the laboratory to shredder invertebrates by the following parameters (Waldbauer, 1968):

$$\textit{Consumption rate (CR)} = \textit{Food ingested per day}$$
$$\textit{Relative consumption rate (RCR)} = [\textit{Food ingested per day}] \div [\textit{Mean animal weight}] \times 100$$
$$\textit{Growth rate (GR)} = \textit{Weight gained per day}$$
$$\textit{Relative growth rate (RGR)} = [\textit{Weight gained per day}] \div [\textit{Mean animal weight}] \times 100$$
$$\textit{Food conversion efficiency (FCE)} = [\textit{Weight gained per day}] \div [\textit{Food ingested per day}] \times 100$$

In addition to follow the interactions of freshwater hyphomycetes and leaf shredder invertebrates, it is possible to understand the extent of leaf processing in streams under different environmental conditions (e.g., physicochemically different or polluted habitats). Many variables can be adapted to address the role of freshwater hyphomycetes in energy transfer to higher trophic levels. Freshwater hyphomycete assemblage and diversity can be linked with the season, succession, leaf litter processing and turnover (dry mass, ash-free dry mass, decay coefficient, leaf chemistry, extracellular enzymes, impacts of stress, human interference, survival and dissemination).

Usually, there will be a small group of fungi called core-group species (frequency of occurrence, >10%) in each survey. The combination and pattern of core-group fungi varies seasonally as well as geographically. Human interferences of stream ecosystem affect the core-group fungal combination and ultimately their ecological services. Studies in Australia revealed that submerged wood consists of rare and a small dominant core-group fungi (Hyde and Goh, 1997, 1998). On testing these core-group fungi, some were unable to produce enzymes capable of degrading lignocellulose and also caused soft-rot cavities. It was hypothesized that no single species is able to produce all enzymes necessary to degrade woody tissue and it is likely a consortium of fungi that involve in the total decomposition process. The core-group species are also regarded as keystone species, and under the threat of human disturbance of freshwater ecosystem, their functions will be severely altered resulting in changes in the rate of detritus decomposition and turnover.

Human Interference

Human interference severely affects the diversity as well as ecological functions of freshwater hyphomycete communities. There is a growing concern on the impact of human influence on watercourses throughout the world (Baerlocher, 1992c). The richness of riparian vegetation was positively correlated with the richness of freshwater hyphomycetes (Fabre, 1996). Deforestation and alterations in riparian vegetation, smoothening of basins and introduction of pollutants into water bodies are the major events. In the interest of improving the drainage, removal of blockages leads to the loss of habitat suitable to freshwater hyphomycetes and shredder communities. The conidial output and colonization of leaf discs exposed in woodland streams was greater than streams with clear-cut riparian vegetation (Metwalli and Shearer, 1989). The addition of pollutants (e.g., sewage, heavy metals, pesticides) and acid precipitation has a major influence on stream fungi and invertebrates (Baerlocher, 1992c). Surprisingly, freshwater hyphomycetes perform their services up to some extent even in polluted or stressed habitats (Baerlocher, 1992c; Bermingham, 1996; Chandrashekar and Kaveriappa, 1989; Krauss *et al.*, 2001, 2003; Raghu *et al.*, 2001; Raviraja *et al.*, 1998; Sridhar and Raviraja, 2001; Sridhar *et al.*, 2005) and thus most likely helps to precisely assess the stream perturbations and recovery.

Studies on the impact of pollutants (organic, heavy metal and pesticide pollution) on freshwater hyphomycetes and their functions are gaining importance (Krauss *et al.*, 2001, 2003; Pascoal and Cássio, 2004; Sridhar *et al.*, 2000, 2001). Organic pollution in a South Indian stream resulted in loss of diversity as well as ecological functions of freshwater hyphomycetes (Raviraja *et al.*, 1998). In an organically polluted low-order stream in Portugal also the fungal diversity affected more than leaf decomposition (Pascoal *et al.*, 2005). Heavy introduction of organics and organic nutrients in the Portuguese stream resulted in a higher contribution of fungi to overall leaf carbon loss than bacteria (29-38.8 vs. 4.2-13.9%) (Pascoal and Cássio, 2004). In the Mansfeld District of Central Germany, due to metallurgical activities, dumping of the spent ore, metalliferous sludge, and mine wastes influenced severely the terrestrial and aquatic habitats. Although freshwater hyphomycete communities in such streams are impoverished, they are diverse (Krauss *et al.*, 2001, Sridhar *et al.*, 2000). They synthesize phytochelatines, sulfur-rich compounds and peptides derived from glutathione for detoxification of heavy metals (Mirsch *et al.*, 2001, 2005). Thus, freshwater hyphomycetes have evolved sophisticated mechanisms to overcome the toxic influence of heavy metals by biosorption and bioaccumulation (Jaeckel *et al.*, 2005; Krauss *et al.*, 2003) and such studies help to employ these fungi in environmental bioremediation.

The growth of *Flagellospora penicillioides, Lunulospora curvula* and *Phalangispora constricta* was not inhibited up to 5 mg/l of herbicides and fungicides (Chandrashekar and Kaveriappa, 1989). Some fungicides and organochlorine insecticides did not inhibit the growth below 10 mg/l, while sporulation was inhibited between 5 and 10 mg/l (Sridhar and Kaveriappa, 1986). This suggests that the reproductive phase will be more affected by herbicides/pesticides than the vegetative phase. Interestingly, Junghanns *et al.* (2005) demonstrated the laccase-catalyzed degradation of xenoestrogen, nonylphenol by *Clavariopsis aquatica* in Central Germany. Thus, the impact of pollution on freshwater hyphomycetes projects a new dimension to understand changes in community structure, ecology and stream ecosystem functioning.

Conclusions and Outlook

There is a vast scope to study the role of freshwater hyphomycetes in detritus turnover, interaction with shredders, pattern of energy flow, stream productivity and impacts of transplants. The study of conidial impaction (availability of surface area), colonization (production of cellulase, xylanase, pectinase and laccase), sporulation (under low nutrients and minerals), spore germination (availability of nutrients), spore viability (under stress) and production of important secondary metabolites (e.g., growth factors and antibiotics) will be useful variables in laboratory microcosm studies. Several ecological principles, pertaining to detritus food web in streams can be addressed through freshwater hyphomycete community organization and functioning. There will be a continuous shift and rearrangement of freshwater hyphomycetes in streams based on the alteration of stream chemistry due to availability of substrates, competition and pollution. It is necessary to understand up to what extent human interference of lotic ecosystem (e.g., multiple stresses) will not upset the diversity and ecological functions of freshwater hyphomycetes. Leaf litter breakdown by freshwater hyphomycetes in streams is an important process as it is governed by a variety of physicochemical and biological factors. The study of the breakdown of leaf litter in streams can be developed as simple, authentic and cost-effective means of assessment of ecosystem functioning. A continuous monitoring of structure (pattern) and function (processes) of freshwater hyphomycetes are required to formulate guidelines or yardsticks in order to control human interference on their ecological functions.

It is necessary to give importance to fungal diversity studies to understand the ecosystem functions that meets the requirement of the Convention on Biological Diversity (CBD). Compared to investigations on other groups of fungi, freshwater hyphomycetes become a neglected group. There are several major gaps in our knowledge on the diversity, ecology and role of freshwater hyphomycetes in the stream ecosystem. Although they are well documented in leaf litter, their occurrence in other habitats (e.g., wood, sediment, aero-aquatic and terrestrial habitats) has been poorly understood as they overlap in freshwater lotic and terrestrial, freshwater lotic and freshwater lentic, freshwater lotic and estuarine (mangrove) or marine (foam) habitats. Techniques need to be employed to establish the adaptations of species as truly aquatic, semi-aquatic or the requirement of the aquatic phase to complete vegetative and reproductive phases. Competition among the freshwater hyphomycetes to exploit the substrate (e.g., leaf litter and wood) is not adequately understood yet. It is not very clear whether single species or consortium of species (e.g., core-group species) is responsible for the decomposition or recycling of substrata in freshwaters. Similarly, competition among fungi to recycle the substrata whether by inhibiting the growth or production of inhibitory compounds is not clear. Tropical regions need special attention in order to fill the gap of knowledge on the occurrence and contributions of freshwater hyphomycetes (see also Chapter 2).

Currently, there are three areas of research on freshwater hyphomycetes drawing serious attention: (a) diversity and ecological functions (Baerlocher, 2005; Pascoal and Cassio, 2004); (b) role in polluted habitats (Krauss *et al.*, 2003; Pascoal *et al.*, 2005); and (c) molecular techniques to address the phylogeny, diversity, ecology and ecosystem services (Baerlocher, 2005, 2007; Baerlocher *et al.*, 2006; Nikolcheva and Baerlocher, 2004; Nikolcheva *et al.*, 2003). Future investigations on freshwater hyphomycetes need the application of traditional and modern approaches to understand the stream ecosystem structure, health and functions more precisely. Special emphasis and techniques are required to explore 'cryptic species' (hidden) of freshwater hyphomycetes. The application of molecular approaches might reveal uncultivable freshwater hyphomycetes and their role in ecological functions. Finally, the rapid identification and report of freshwater hyphomycetes need updated illustrated keys, monographs and databases with accepted names and synonyms.

References

Arsuffi, T.L. and Suberkropp, K. 1986. Growth of two stream caddisflies (Trichoptera) on leaves colonized by different fungal species. *J. N. Am. Benthol. Soc. 5*, 297-305.

Baerlocher, F. 1985. The role of fungi in the nutrition of stream invertebrates. *Bot. J. Linn. Soc. 91*, 83-94.

Baerlocher, 1992a. *The Ecology of Aquatic Hyphomycetes.* Springer-Verlag, Berlin.

Baerlocher, 1992b. Research on aquatic hyphomycetes: Historical background and overview. In: *The Ecology of Aquatic Hyphomycetes* (Ed. Baerlocher, F.). Springer-Verlag, Berlin, 1-15.

Baerlocher, F. 1992c. Human interference. In: *The Ecology of Aquatic Hyphomycetes* (Ed. Baerlocher, F.). Springer-Verlag, Berlin, 173-181

Baerlocher, F. 2005. Freshwater fungal communities. In: *The Fungal Community: Its Organization and Role in the Ecosystem* (Ed. Dighton, J., White, J.F. and Oudemas, P.). Taylor and Frances, Boca Raton, Florida, 39-59.

Baerlocher, F. 2007. Molecular approaches applied to aquatic hyphomycetes. *Fungal Biology Reviews 21*: 19-24.

Baerlocher, F. and Kendrick, B. 1973. Fungi and food preferences of *Gammarus pseudolimnaeus. Arch. Hydrobiol.* 72, 501-516.

Baerlocher, F. and Kendrick, B. 1974. Dynamics of the fungal population on leaves in a stream. *J. Ecol. 62,* 761-791.

Baerlocher, F. and Kendrick, B. 1981. Role of aquatic hyphomycetes in the trophic structure of streams. In: *The Fungal Community: Its Organization and Role in the Ecosystem* (Ed. Wicklow, D.T. and Carroll, G.C.). Marcel Dekker, New York, 743-760.

Baerlocher, F. and Murdoch, J.H. 1989. Hyporheic biofilms - a potential food source for hyporheic animals. *Hydrobiologia 184*, 61-67.

Baerlocher, F., Nikolcheva, L.G., Wilson, K.P. and Williams, D.D. 2006. Fungi in the Hyporheic Zone of a Springbrook. *Microb. Ecol. 52*, 708-715.

Baldy, V., Chauvet, E., Charcosset, J.Y. and Gessner, M.O. 2002. Microbial dynamics associated with leaves decomposing in the mainstream and floodplain pond of a large river. *Aq. Microb. Ecol. 28*, 25-36.

Bandoni, R.J. 1981. Aquatic hyphomycetes from terrestrial litter. In: *The Fungal Community: Its Organization and Role in the Ecosystem* (Ed. Wicklow, D.T. and Carroll, G.C.). Marcel Dekker, New York, 693-708.

Bermingham, S. 1996. Effect of pollutants on aquatic hyphomycetes colonizing leaf material in fresh-waters. In: *Fungi and Environmental Change* (Ed. Frankland, J.C., Magan, N. and Gadd, G.M.). Cambridge University Press, Cambridge, 201-216.

Bermingham, S. L., Maltby, L. and Dewey, F. M. 1996. Monoclonal antibodies as tools to quantify mycelium of aquatic hyphomycetes. *New Phytol. 132*, 593-601.

Butler, S.K. and Suberkropp, K. 1986. Aquatic hyphomycetes on oak leaves: comparison of growth, degradation and palatability. *Mycologia 78*, 922-928.

Cardinale, B.J., Nelson, K., and Palmer, M.A. 2000. Linking species diversity to the functioning of ecosystem: on the importance of environmental context. *Oikos 91*, 175-183.

Chamier, A.-C. and Dixon, P.A. 1982. Pectinases in leaf degradation by aquatic hyphomycetes: the enzymes and leaf maceration. *J. Gen. Microbiol. 128*, 2469-2483.

Chandrashekar, K.R. and Kaveriappa, K.M. 1988. Production of extracellular enzymes by aquatic hyphomycetes. *Folia Microbiol. 33*, 53-58.

Chandrashekar, K.R. and Kaveriappa, K.M. 1989. Effects of pesticides on the growth of aquatic hyphomycetes. *Toxicol. Lett. 48*, 311-315.

Chandrashekar, K.R., Sridhar, K.R. and Kaveriappa, K.M. 1989. Palatability of rubber leaves colonized by aquatic hyphomycetes. *Arch. Hydrobiol. 115*, 361-369.

Descals, E. 1997. Ingoldian fungi: some field and laboratory techniques. *Boll. Soc. Hist. Nat. Beleras 40*, 169-221.

Descals, E. and Moralejo, E. 2001. Water and asexual reproduction in the Ingoldian fungi. *Botanica complutensis* 25, 13-71.

Fabre, E. 1996. Relationships between aquatic hyphomycites communities and riparian vegetation in 3 Pyrenean streams. *C.R. Acak. Sci. Paris, Sciences de la vie / Life Sciences 319*, 107-111.

Fisher, S.G. and Likens, G.P. 1973. Energy flow in Beer Brook, New Hampshire: an integrative approach to stream ecosstem metabolism. *Ecol. Manogr. 43*, 421-439.

Fisher, P.J. Webster, J. and Petrini, 0. 1991. Aquatic hyphomycetes and other fungi in living terrestrial roots of *Alnus glutinosa. Mycol. Res. 95*, 543-547.

Gessner, M.O. 2005. Ergosterol as a measure of fungal biomass. In: *Methods to Study Litter Decomposition - A Practical Guide* (Ed. Graça, M.A.S., Baerlocher, F. and Gessner, M.O.). Springer, The Netherlands, 189-195.

Gessner, M.O. and Chauvet, E. 1993. Ergosterol-to-biomass conversion factors for aquatic hyphomycetes. *Appl. Environ. Microbiol. 59*, 502-507.

Godfrey, B.E.S. 1983. Growth of two terrestrial microfungi on submerged alder leaves. *Trans. Brit. Mycol. Soc. 79*, 418-421.

Gönczöl, J. 1976. Ecological observations on the aquatic hyphomycetes of Hungary II. *Acta Bot. Acad. Sci. Hung. 22*, 51-60.

Gönczöl, J. and RÉVAY, A. 2003. Treehole fungal communities: aquatic, aero-aquatic, and dematiaceous hyphomycetes. *Fungal Diversity 12*, 19-34.

Gönczöl, J. and Révey, A. 2004. Fungal spores in rainwater: stem flow, through fall and gutter conidial assemblages. *Fungal Diversity 16*, 67-86.

Gönczöl, J. and Révey, A. 2006. Species diversity of rain-borne conidia from living trees. *Fungal Diversity 22*, 37-54.

Graça, M.A.S. and Ferreira, C.F. 1995. The ability of selected aquatic hyphomycetes and terrestrial fungi to decompose leaves in freshwater. *Sydowia 47*, 167-179.

Graça, M.A.S., Baerlocher, F. and Gessner, M.O. 2005. *Methods to Study Litter Decomposition – A Practical Guide*. Springer, The Netherlands.

Gulis, V. and Suberkropp, K. 2003. Effects of inorganic nutrients on relative contribution of fungi and bacteria to carbon flow from submerged decomposing leaf litter. *Microb. Ecol. 45*, 11-19.

Gulis, V., Marvanová, L and Descals, E. 2005. An illustrated key to the common temperate species of aquatic hyphomycetes. In: *Methods to Study Litter Decomposition: A Practical Guide* (Ed. Graça, M.A.S., Baerlocher, F. and Gessner, M.O.). Springer, The Netherlands, 153-167.

Hyde, K.D. and Goh, T.K. 1997. Fungi on submerged wood in a small stream on Mt Lewis, North Queensland, Australia. *Muelleria 10*, 145-157.

Hyde, K.D. and Goh, T.K. 1998. Fungi on submerged wood in Lake Barrine, North Queensland, Australia. *Mycol. Res. 102*,739-749.

Ingold, C.T. 1942. Aquatic hyphomycetes of decaying alder leaves. *Trans. Br. Mycol. Soc. 25*, 339-417.

Ingold, C.T. 1975. *An Illustrated Gudie to Aquatic and Waterborne Hyphomycetes (Fungi Imperfecti) with Notes on their Biology*. Freshwater Biol. Assoc. Sci. Publ. # 30.

Ingold, C.T. and Cox, V.J. 1957. *Heliscus stellatus* sp. nov., an aquatic hyphomycete. *Trans. Br. Mycol. Soc. 40*, 155-158.

Iqbal, S.H. and Webster, J. 1973a. The trapping of aquatic hyphomycete spores by air bubbles. *Trans. Br. Mycol. Soc. 60*, 37-48.

Iqbal, S.H. and Webster, J. 1973b. Aquatic hyphomycete spora of the River Exe and its tributaries. *Trans. Br. Mycol. Soc. 61*, 331-346.

Jaeckel, P., Krauss, G., Menge, S., Schierhorn, A., Ruecknagel, P. and Krauss, G.-J. 2005. Cadmium induces a novel metallothionein in phytochelatin 2 in an aquatic fungus. *Biochem. Biophys. Res. Comm. 333*, 150-155.

Junghanns, C., Moeder, M., Krauss, G., Martin, C. and Schlosser, D. 2005. Degradation of xenoestrogen nonylphenol by aquatic fungi and their laccases. *Microbiology 151*, 45-57.

Kaushik, N.K. and Hynes, H.B.N. 1971. The fate of the dead leaves that fall into streams. *Arch. Hydrobiol. 68*, 465-515.

Kirby, J.J.H., Webster, J. and Baker, J.H. 1990. A particle plating method for analysis of fungal community composition and structure. *Mycol. Res. 94*, 621-626.

Kirk, P.W. 1969. Aquatic hyphomycetes on wood in an estuary. *Mycologia 61*, 177-181.

Krauss, G., Baerlocher, F., Shreck, P., Wennrich, R., Glaesser, W. and Krauss, J.-G. 2001. Aquatic hyphomycetes occur in hyper polluted waters in Central Germany. *Nova Hedwigia 72*, 419-428.

Krauss, G., Baerlocher, F. and Krauss, J.-G. 2003. Effect of pollution on aquatic hyphomycetes. In: *Freshwater Mycology* (Ed. Tsui, C.K.M. and Hyde, K.D.). Fungal Diversity Press, Hong Kong, 211-230.

Maharning, A.R. and Baerlocher, F. 1996. Growth and reproduction in aquatic hyphomycetes. *Mycologia 88*, 80-88.

Marvanová, L. 1997. Freshwater hyphomycetes: a survey with remarks on tropical taxa. In: *Tropical Mycology* (Ed. Janardhanan, K.K., Rajendran, C., Natarajan, K. and Hawksworth, D.L.). Science Publishers, New York, 169-226.

Marvanová, L. and Fisher, F. 1991. A new endophytic hyphomycetes from alder roots. *Nova Hedwigia 52*, 33-37.

Marvanová, L., Fisher, P.J., Aimer, R. and Segedin, B. 1992. A new *Filosporella* from alder roots and from water. *Nova Hedwigia 54*, 151-158.

Metwalli, A.A. and Shearer, C.A. 1989. Aquatic hyphomycetes communities in clear-cut and wooded areas of an Illinois stream. *Trans. Illinois Acad. Sci. 82*, 5-16.

Mirsch, J., Tschimedbalshir, M., Baerlocher, F., Grams, Y., Pierau, B., Schierhorn, A. and Krauss, G.-J. 2001. Heavy metal and thiol compounds in *Mucor recemosus* and *Articulospora tetracladia. Mycol. Res. 105*, 883-889.

Mirsch, J., Neumann, D., Menge, S., Baerlocher, F., Baumbach, R. and Lichtenberger, O. 2005. Heavy metals and thiol pool in three strains of *Trtracladium marchalianum. Mycol. Progr. 4*, 185-194.

Müller-haecel, A. and Marvanová, L. 1979. Periodicity of aquatic hyphomycetes in the subarctic. *Trans. Br. Mycol. Soc. 73*, 109-116.

Nair, S.K. and Bhat, D.J. 2002. *Kumbhamaya jalapriya,* a new endophytic hyphomycete from India. *Mycotaxon 84*, 65-68. 2002.

Nawawi, A. 1985. Aquatic hyphomycetes and other water-borne fungi from Malaysia. *Malays. Nat. J. 39*, 75-134.

Newell, S.Y. 2000. Methods for determining biomass and productivity of mycelial marine fungi. In: *Marine Mycology - A Practical Approach* (Ed. Hyde, K.D. and Pointing, S.B.). Fungal Diversity Series 1, Fungal Diversity Press, Hong Kong, 69-91.

Nikolcheva, L.G., Cockshut A. and Baerlocher, F. 2003. Diversity of freshwater fungi on decaying leaves - comparing traditional and molecular approaches. *Appl. Environ. Microbiol. 69*, 2548-2554.

Nikolcheva, L.G. and Baerlocher, F. 2004. Taxon-specific fungal primers reveal unexpectedly high diversity during lead decomposition in a stream. *Mycol. Prog. 3*, 41-49.

Nickolcheva, L.G. and Baerlocher, F. 2005a. Molecular approaches to estimate fungal diversity. I. Terminal restriction fragment length polymorphism (T-RFLP). In: *Methods to Study Litter Decomposition - A Practical Guide* (Ed. Graça, M.A.S., Baerlocher, F. and Gessner, M.O.). Springer, The Netherlands, 169-176.

Nickolcheva, L.G. and Baerlocher, F. 2005b. Molecular approaches to estimate fungal diversity. II. Denaturing gradient gel electrophoresis (DGGE). In: *Methods to Study Litter Decomposition - A Practical Guide* (Ed. Graça, M.A.S., Baerlocher, F. and Gessner, M.O.). Springer, The Netherlands, 177-184.

Nilsson, S. 1964. Freshwater hyphomycetes: Taxonomy, morphology and ecology. *Symb. Bot. Upsal. 18*, 1-130.

Pascoal, C. and Cássio, F. 2004. Contribution of fungi and bacteria to leaf litter decomposition in a polluted river. *Appl. Environ. Microbiol. 70*, 5266-5273.

Pascoal, C. Cássio, F. and Marvanová, L. 2005. Anthropogenic stress may affect aquatic hyphomycetes diversity more than leaf decomposition in a low-order stream. *Arch. Hydrobiol. 162*, 481-496.

Raghu, P.A., Sridhar, K.R. and Kaveriappa, K.M. 2001. Diversity and conidial output of aquatic hyphomycetes in heavy metal polluted river, Southern India. *Sydowia 53*, 236-246.

Rajashekhar, M. and Kaveriappa, K.M. 2003. Diversity of aquatic hyphomycetes in the aquatic ecosystems of the Western Ghats of India. *Hydrobiol. 501*, 167-177.

Rao, M.M. and Manoharachary, C. 1982. Processing of leaves by fungi in an aquatic ecosystem. *Indian Phytopath. 35*, 654-656.

Raviraja, N.S., Sridhar, K.R. and Baerlocher, F. 1996. Endophytic aquatic hyphomycetes of roots of plantation crops and ferns from India. *Sydowia 48*, 152-160.

Raviraja, N.S., Sridhar, K.R. and Baerlocher, F. 1998. Breakdown of *Ficus* and *Eucalyptus* leaves in an organically polluted river in India: fungal diversity and ecological functions. *Freshwat. Biol. 39*, 537-545.

Read, S.J. 1990. *Spore Attachment in Fungi with Special Reference to Freshwater Hyphomycetes.* Ph.D. Thesis, Portsmouth Polytechnic, UK.

Read, S.J., Moss S.T. and Jones, E.B.G. 1992. Attachment and germination of conidia. In: *The Ecology of Aquatic Hyphomycetes* (Ed. Baerlocher, F.). Springer-Verlag, Germany.

Rossett, J., Baerlocher, F. and Oertli, J.J. 1982. Decomposition of conifer needles and deciduous leaves in two Black Forest and two Swiss Jura streams. *Int. Rev. Ges. Hydrobiol. 67*, 695-711.

Sati, A.C. and Belwal, M. 2005. Aquatic hyphomycetes as endophytes of riparian plant roots. *Mycologia 97*, 45-49.

Sati, S.C. and Tiwari, N. 1989. Some aquatic hyphomycetes of Kumaun Himalaya, India. *Mycotaxon 39*, 407-414.

Shearer, C.A. 1992. Role of woody debris. In: *The Ecology of Aquatic Hyphomycetes* (Ed. Baerlocher, F.). Springer-Verlag, Berlin, 77-98.

Shearer, C.A. and Crane, J.L. 1971. Fungi of the Chesapeake Bay and its tributaries. I. Patuxent River. *Mycologia 63*, 237-260.

Shearer, C.A. and Lane, L. 1983. Comparison of three techniques for the study of aqatic hyphomycetes communities. *Mycologia 75*, 498-508.

Shearer, C.A., Descals, E., Kohlmeyer, B., Kohlmeyer, J., Marvanová, L., Padgett, D., Porter, D., Raja, H.A., Schmit, J.P., Thorton, H.A. and Voglymayr, H. 2007 Fungal biodiversity in aquatic habitats. *Biodivers. Conserv. 16,* 49-67.

Singh, N. 1982. Cellulose decomposition by some tropical aquatic hyphomycetes. *Trans. Brit. Mycol. Soc. 79*, 560-561.

Sridhar, K.R. and Baerlocher, F. 1992. Endophytic aquatic hyphomycetes in spruce, birch and maple. *Mycol. Res. 96*, 305-308.

Sridhar, K.R. and Kaveriappa, K.M. 1986. Effect of pesticides on sporulation on spore germination of water-borne hyphomycetes. In: *Environmental Biology-Coastal Ecosystem.* (Ed. Dalella, R.C., Madhyastha, M.N. and Joseph, M.M.). The Academy of Environmental Biology, India, 195-204.

Sridhar, K.R. and Kaveriappa, K.M. 1987a. A new species of *Triscelophorus. Indian Phytopath. 40*, 102-105.

Sridhar, K.R. and Kaveriappa, K.M. 1987b. Culturing water-borne hyphomycetes on plant latex. *J. Indian Bot. Soc. 66*, 232-233.

Sridhar, K.R. and Kaveriappa, K.M. 1988. Occurrence and survival of aquatic hyphomycetes in brackish and seawater. *Arch. Hydrobiol. 113*, 153-160.

Sridhar, K.R. and Raviraja, N.S. 2001. Aquatic hyphomycetes and leaf litter processing in polluted and unpolluted habitats. In: *Trichomycetes and other Fungal Groups* (Ed. Misra, J.K. and Horn, B.W.). Science Publishers, Inc., USA, 293-314.

Sridhar, K.R., Chandrashekar, K.R. and Kaveriappa, K.M. 1992. Research on the Indian Subcontinent. In: *The Ecology of Aquatic Hyphomycetes* (Ed. Baerlocher, F.). Springer-Verlag, Berlin, 182-211.

Sridhar, K.R., Krauss, G., Baerlocher, F., Wennrich, R. and Krauss, G.-J. 2000. Fungal diversity in heavy metal polluted waters in Central Germany. *Fungal Diversity 5*, 119-129.

Sridhar, K.R., Krauss, G., Baerlocher, F., Raviraja, N.S., Wennrich, R., Baumann, R. and Krauss, G.-J. 2001. Decomposition of alder leaves in two heavy metal polluted streams in Central Germany. *Aq. Microb. Ecol. 26*, 73-80.

Sridhar, K.R., Baerlocher, F., Krauss, G.-J. and Krauss, G. 2005. Response of Aquatic Hyphomycete communities to changes in heavy metal exposure. *Int. Rev. Hydrobiol. 90*, 21-32.

Sridhar, K.R., Karamchand, K.S. and Bhat, R. 2006. Arboreal water-borne hyphomycetes with oak-leaf basket fern *Drynaria quercifolia. Sydowia 58*, 309-320.

Suberkropp, K. 1991. Relationships between growth and sporulation of aquatic hyphomycetes on decomposing leaf litter. *Mycol Res. 95*, 843-850.

Suberkropp, K. 1992. Interaction with invertebrates. In: *The Ecology of Aquatic Hyphomycetes* (Ed. Baerlocher, F.). Springer-Verlag, Berlin, 118-134.

Suberkropp, K. 1995. The influence of nutrients on fungal growth, productivity and sporulation during leaf breakdown in streams. *Can. J. Bot. 73*, S1361-S1369.

Suberkropp, K., Arsuffi, T.L. and Anderson, J.P. 1983. Comparisons of degradative ability, enzymatic activity and palatability of aquatic hyphomycetes growing on leaf litter. *Appl. Environ. Microbiol. 46*, 237-244.

Suberkropp, K., Godshalk, G.L. and Klug, M.J. 1976. Changes in the chemical composition of leaves during processing in a woodland stream. *Ecology 57*, 720-727.

Suberkropp, K. and Klug, M.J. 1980. The maceration of deciduous leaf litter by aquatic hyphomycetes. *Can. J. Bot. 58*, 1025-1031.

Subramanian, C.V. and Bhat, D.J. 1981. Conidia from freshwater foam samples from the Western Ghats, Southern India. *Kavaka 9*, 45-62.

Thornton, D.R. 1965. Amino acid analysis of fresh leaf litter and the nitrogen nutrition of some aquatic hyphomycetes. *Can. J. Microbiol. 11*, 657-662.

Triska, F.J. 1970. *Seasonal Distribution of Aquatic Hyphomycetes in Relation to the Disappearance of Leaf Litter from a Woodland Stream.* Thesis, University of Pittsburgh, Pittsburgh.

Waldbauer, G.P. 1968. The composition and utilization of food by insects. *Adv. Insect Physiol. 5*, 229-288.

Webster, J. 1987. Convergent evolution and the functional significance of spore shape in aquatic and semi-aquatic fungi. In: *Evolutionary Biology of the Fungi* (Ed. Rayner, A.D.M., Brasier, C.M. and Moore, D.). Cambridge University Press, Cambridge, 191-201.

Webster, J. 1992. Anamorph-teleomorph relationships. In: *The Ecology of Aquatic Hyphomycetes* (Ed. Baerlocher, F.). Ecological Studies 94, Springer-Verlag, Berlin, 99-117.

Webster, J. and Descals, E. 1981. Morphology, distribution and ecology of conidial fungi in freshwater habitats. In: *Biology of Conidial Fungi*, Volume 1, (Ed. Cole, G.T. and Kendrick, B.). Academic Press, New York, 295-355.

Webster, J. and Benfield, E.F. 1986. Vascular plant breakdown in freshwater ecosystem. *Ann. Rev. Ecol. Syst. 17*, 567-594.

Webster, J.R., Benfield, E.F., Ehrman, T.P., Schaeffer, M.A., Tank, J.L., Hutchens, J.J. and D'angelo, D.J. 1999. What happens to allochthonous material that falls into streams? A synthesis of new and published information from Coweeta. *Freshwat. Biol. 41*, 687-705.

Weyers, H.S. and Suberkropp, K. 1996. Fungal and bacterial production during the breakdown of yellow poplar leaves in two streams. *J. North Am. Benthol. Soc. 15*, 408-420.

Young, J.C. 1995. Microwave-assisted extraction of the fungal metabolite ergosterol and total fatty acids. *J. Agric. Food Chem. 43*, 2904-2910.

Frontiers in Fungal Ecology, Diversity and Metabolites (edited by K.R. Sridhar), I.K. International Publishing House Pvt. Ltd., New Delhi, 2009.

2

Diversity of Aquatic Hyphomycetes of the Western Ghat Rivers

Nityanand D. Maddodi[1]*, N.S. Raviraja[2] and M. Rajashekhar[3]
[1]*Department of Dermatology, School of Medicine and Public Health, University of Wisconsin, Madison, USA, 53706*
[2]*Department of Anatomy and Structural Biology, Albert Einstein College of Medicine, Bronx, New York, USA, 10461*
[3]*Department of Biosciences, Mangalore University, Mangalagangothri 574 199, Karnataka, India*
*(*E-mail: nityamaddodi@gmail.com)*

Introduction

Aquatic hyphomycetes colonize submerged decaying leaves in flowing waters throughout the world (Chauvet, 1991; Sridhar *et al.*, 1992; Suberkropp, 1992; Webster *et al.*, 1994; Webster and Descals, 1981; Wood-Eggenschwiler and Bärlocher, 1985). They play an important role as intermediaries of energy flow between decaying leaves and leaf-eating invertebrates (Bärlocher, 1992). Species assemblage of aquatic hyphomycetes varies in seasons (Bärlocher, 2000; Garnett *et al.*, 2000; Shearer and Webster, 1985b; Suberkropp, 1984, 1992), longitudinal (Shearer and Webster, 1985a) and temporal distribution (Shearer and Webster, 1985b), and with the quality and quantity of riparian vegetation (Gönczöl, 1975, 1989; Iqbal and Webster, 1973; Willoughby and Minshall, 1975; Wood-Eggenschwiler and Bärlocher, 1985). Diversity has been correlated with water chemistry, altitude and riparian vegetation (Bärlocher, 1987; Bärlocher and Rosset, 1981; Chauvet, 1991; Gönczöl, 1989; Raviraja *et al.*, 1998; Shearer and Webster, 1985a, 1985b; Wood-Eggenschwiler and Bärlocher, 1985). Aquatic hyphomycetes have also been reported from terrestrial litter and tree canopies (Bandoni, 1981; Gönczöl and Révey, 2006; Sridhar *et al.*, 2006). Wood-Eggenschwiler and Bärlocher (1985) concluded that the main factors for determining fungal distribution patterns are the temperature and vegetation in different climatic regions worldwide, whereas on a regional scale, stream water chemistry dominates. Major studies on aquatic hyphomycete assemblage and distribution were carried out in temperate streams (Bärlocher, 2000; Descals, 1998; Gulis, 1999; Suberkropp, 1984). More information on distribution, techniques, ecological functions and impact of human interference on aquatic hyphomycetes is available in Chapter 1. In India, aquatic hyphomycetes are mainly reported from the Western Ghats, Himalayan regions and the Eastern Ghats. Raviraja *et al.* (1998) correlated pH, temperature and altitude with fungal richness in a few Western Ghat streams. Studies in Southwestern France suggested that altitude or its associated factors are important in structuring aquatic hyphomycete communities (Chauvet, 1991). A greater emphasis is needed to improve our knowledge about the assemblage and diversity of aquatic hyphomycetes in tropics. Thus, the present chapter reports qualitative and quantitative studies carried out on the aquatic hyphomycete community in leaf litter and foam samples of high and low altitude regions of selected rivers of the Western Ghats with emphasis on temperature and pH.

Rivers and Water Quality

Four major rivers: Aghanashini, Gangavalli, Kali and Sharavathi originates at different peaks of the Western Ghats of Karnataka (13°20' to 16°50' N and 74° 45' to 75°45' E), which meets the Arabian Sea after flowing through the forests and plains. The altitude at mean sea level (MSL) of the sampling locations was determined using an altimeter (BARIGO, Germany). Based on the height from the sea level, Heggarni (HE), Manohikeri (MA), Jog Falls (JO) and Dandeli (DA) sampling locations were classified as high altitude locations, while Hadikop (HA), Hillur (HI), Gersoppa (GE) and Kadra (KA) as low altitude locations. Two sampling locations were selected from each river with one sampling site at high altitude and another at low altitude. Samples were collected in summer (March) and post-monsoon (October) seasons in 1998. Water temperature (mercury thermometer), pH (Systronics, India), and conductivity (Century CK 710, India) were recorded in the field. Dissolved oxygen was detected by Winkler's method (APHA, 1995). Samples and readings were taken in five replicates within one-hour duration (Table 1).

Table 1. *Physicochemical characters of four rivers of the Western Ghats*

River	Location	Altitude (m MSL)	Temperature (°C)	Conductivity (μS/cm)	pH	DO (mg/1)
Aghanashini	Heggarni (HE)	621	28	93.9	7.7	7.95
	Hadikop (HA)	300	29	114.0	7.9	7.54
Gangavalli	Manohikeri (MA)	510	29	192.1	8.3	8.11
	Hillur (HI)	120	28	183.7	8.6	7.70
Kali	Dandeli (DA)	543	28	208.1	7.8	7.95
	Kadra (KA)	222	28	192.3	7.3	7.70
Sharavathi	Jog Falls (JO)	700	29	187.1	7.9	7.58
	Gersoppa (GE)	430	29	209.1	7.1	7.62

Leaf Litter and Foam

The fungal assemblage was assessed in naturally submerged decaying leaf litter and foam samples collected from 50-60m stretch of the river. Submerged decaying leaves were randomly sampled in each sampling location and brought to the laboratory in polythene bags and rinsed in distilled water to remove attached debris and invertebrates. From each location, 4-5 different types of leaf samples at different stage of decomposition were punched into disks (13 mm diameter) with a sterile cork borer and aerated up to 48 h (27±2°C) in 250 ml sterile distilled water in 500 ml Erlenmeyer flasks. Five replicate flasks were prepared from each location (five disks per leaf type). Aerated water was filtered through 5μm membrane filters (Millipore) and the conidia on the filters were immediately stained with lactophenol cotton blue. The stained filters were mounted on slides with lactic acid and scanned for conidia of aquatic hyphomycetes and enumerated. The aerated leaf disks were dried at 100°C for 24 h to determine dry weight to estimate conidia produced per unit dry mass of leaf material.

White (freshly accumulated) foam samples were collected from different locations within 50-60m river stretch of each sapling location. Foam was allowed to settle for 25-30 min in a measuring cylinder and subsequently fixed with an equal volume of formalin-acetic acid-alcohol (FAA) and transported to the laboratory in plastic vials. 5ml each of the foam-FAA mixture was filtered through membrane filters, stained, and scanned for conidia of aquatic hyphomycetes.

Data Analysis

The qualitative and quantitative assessment and diversity (Simpson and Shannon-Wiener) of aquatic hyphomycetes in leaf litter and foam were calculated:

Simpson index: $D' = 11 - \Sigma\ (pi)^2$

(where, pi = proportion of species i in community)

Shannon-Wiener index: $H' = \Sigma\ (pi \log_2 pi)$

(where, pi = proportion of total sample belonging to i^{th} species)

The Evenness: $J' = H'/H'_{max}$

(where, H'_{max} is the maximum value of diversity for the number of species)

One-way ANOVA was employed to find out the difference in fungal diversity between samples of foam and leaf litter (Microsoft, Ecosoft, Inc., 1984). The paired t-test was performed to analyze the difference in Shannon-Wiener diversity index between high and low altitude samples, (Microsoft, Ecosoft Inc., 1984).

Discussion

Altitude, water temperature, pH, conductivity, and dissolved oxygen of the eight sampling sites are summarized (Table 1). A total of 22 species belonging to 14 genera of aquatic hyphomycetes were obtained (Table 2). The location, Dandeli in River Kali possesses the maximum conidia as well as species in foam (114 spores/ml) (19). Leaf litter samples showed a severe decline in mean species richness (4)

Table 2. *Conidial output of aquatic hyphomycetes per ml foam and per mg dry weight of leaf litter (in parenthesis) in high and low altitude locations of River Aghanashini, Gangavalli; Kali and Sharavathi of the Western Ghats (HE, Heggarni; HA, Hadikop; MA, Manohikeri; HI, Hillur; DA, Dandeli; KA, Kadra; JO, Jogfalls; GE, Gersoppa)*

Aquatic hyphomycetes	**River Aghanashini**		**River Gangavalli**		**River Kali**		**River Sharavathi**	
	HE	**HA**	**MA**	**HI**	**DA**	**KA**	**JO**	**GE**
Actinospora megalospora Ingold	0.52	0.06			0.12		0.26	
Alatospora acuminata Ingold					0.12		0.12	0.12
Anguillospora crassa Ingold		2.32	0.26		0.52			2.12
Anguillospora longissima (Sacc. & Syd.) Ingold	25.6 (6.35)	21.9 (7.98)	37.2 (2.51)	6.8 (0.48)	40.46 (6.1)	2.12 (1.04)	21.4 (4.66)	9.92 (1.88)
Articulospora inflata Ingold							0.2	1.12
Campylospora chaetocladia Ranzoni		18.2			20.86 (1.01)		16.06 (1.9)	5.52 (1.24)
C. parvula Kuzuha	0.72	0.2			0.26		0.2	0.2
Clavariopsis aquatica de Wild		0.32		0.2	0.12			
Clavatospora tentacula (Umphlett) Nilsson		0.12		0.06	0.26			
Flabellospora crassa Alasoadura	0.4			0.06	1.26		9.8	
F. multiradiata Nawawi	1.52	0.46	1.26	0.2	0.32	0.06	2.26	0.4
F. verticillata Alasoadura	0.8			0.06	0.72	0.06	1.86	0.8
Flabellospora sp.							0.86	
Flagellospora curvula Ingold	0.32	0.12	0.12	0.46	1.12		0.46	0.6
F. penicillioides *Ingold*		*2.52*						
Isthmotricladia gombakiensis Nawawi					1.12		0.52	
Lunulospora curvula Ingold	(0.38)	14.6	28.7	20.2	20.12 (2.66)	9.2 (0.31)	4.06 (1.17)	21.2 (4.09)
L. cymbiformis Miura	40.26 (1.29)	4.06 (1.92)	21.9 (1.25)	4.92	5.06 (3.96)	3.92 (0.47)	14.6 (4.75)	2.46
Pyramidospora fluminea Miura & Kudo	0.2	0.86			0.8			
Triscelophorus acuminatus Nawawi	8.32 (0.93)	2.46 (0.98)	5.6		16 (0.46)	1.12 (0.14)	6.12 (1.44)	6.12
T. monosporus Ingold	26.06	3.26		2.46	4.86	2.06	26.3	
Tripospermum myrti (Lind.) Hughes		0.12			0.2			0.46
Total conidia/ml foam	104.7	71.3	95.4	35.2	114.4	18.7	105	51
Total species in foam	11	15	8	9	19	8	16	13
Total conidia/mg leaf litter	9	10.9	3.8	0.5	14.2	2	13.9	8.1
Total species in leaf litter	4	3	2	1	5	4	5	4

and mean spore output (8 conidia/mg). Overall, the three most productive species were *Anguillospora longissima*, *Lunulospora curvula* and *L. cymbiformis*. Species richness, Simpson and Shannon-Wiener diversity for aquatic hyphomycetes in leaf litter and foam are given in Table 3. One-way ANOVA showed a significant difference in the diversity of fungi between leaf litter and foam ($p<0.001$). Paired t-test revealed significant differences in the Shannon-Wiener index of fungal diversity between high and low altitude samples ($p = 0.040$).

Many factors in the stream influence the diversity of the fungal community. Aquatic hyphomycete diversity varies in the stream/river from its source to its lower reaches (Raviraja *et al.*, 1998). The change in fungal density and conidial numbers may be due to the change in riparian vegetation, altitude, water flow,

Table 3. *Species richness, diversity, and evenness of aquatic hyphomycetes from four Western Ghat rivers*

Rivers	Sites	Species richness	Diversity Simpson	Shannon-Wiener	Evenness Simpson	Shannon-Wiener
Foam						
Aghanashini	Heggarni	11	0.731	2.120	0.797	0.613
	Hadikop	15	0.800	2.635	0.846	0.675
Gangavalli	Manohikeri	8	0.708	1.924	0.8000	0.641
	Hillur	9	0.627	1.754	0.685	0.553
Kali	Dandeli	19	0.794	2.659	0.830	0.626
	Kadra	8	0.723	2.027	0.790	0.676
Sharavathi	Jog falls	16	0.847	2.935	0.895	0.734
	Gersoppa	13	0.774	2.559	0.822	0.692
Leaf litter						
Aghanashini	Heggarni	4	0.522	1.287	0.627	0.644
	Hadikop	3	0.466	1.082	0.646	0.683
Gangavalli	Manohikeri	2	0.605	0.917	0.984	0.917
	Hillur	1	-	-	-	-
Kali	Dandeli	5	0.749	1.922	0.870	0.828
	Kadra	4	1.288	1.672	1.288	0.836
Sharavathi	Jog falls	5	0.792	2.089	0.926	0.900
	Gersoppa	4	0.747	1.748	0.868	0.874

water chemistry and temperature. Gönozöl *et al*. (2003) studied spatial dynamics of aquatic hyphomycete species and suggested that altitude, water hardness and conductivity were major factors that correlate with the distribution of individual species. Anthropogenic influence increases towards the downstream reaches of the river. This might also be one of the reasons for change in fungal species and conidial density (Allan, 1995; Bärlocher, 1992; Dix and Webster, 1995). Aquatic hyphomycete community is severely reduced in metal polluted sites (Raghu *et al*., 2001; Sridhar *et al*., 2005). In spite of reduction of fungi, the palatability to stream detritivores is improved (Sridhar *et al*., 2001).

Studies on the distribution of aquatic hyphomycetes in relation to altitude are scarce. Chauvet (1991) reported that altitude is more important than pH, temperature or season in 27 streams of Southwestern France in determining the species composition of aquatic hyphomycetes, while Raviraja *et al*. (1998) reported that pH is more important in determining fungal species richness in ten locations of the Western Ghat streams. Gönczöl and Revay (2003) observed more fungal species with a less number of conidia in soft water streams suggesting the abundance of woody substrate in low order stream may have a structuring effect on aquatic hyphomycete communities. The richness of fungal species was high in small streams of the Czech Republic passing through beach forests at an altitude of 500-700m. Fabre (1996) based on 40 collection sites at various distances from the source in the French Pyrenees streams concluded that even though fungi and riparian communities were simultaneously structured, they are independent with respect to environmental conditions. In our study, river locations at high altitude showed the highest spore concentration in leaf litter and foam samples. Conidial output was the maximum except for foam of high altitude locations of River Aghanashini and Gangavalli. Maximum species and spore count may be attributed to thick riparian vegetation. An increase in Shannon-Wiener diversity in the high altitude sampling locations leads to conclusion that these might provide better environment for fungal growth and reproduction.

Aquatic hyphomycete diversity and conidial richness were high in the lower reaches of River Teign in Deven, England, where pH was neutral (7-7.2), riparian trees were abundant in the river source in moorland with acidic pH (5.4-6) and less riparian trees (Shearer and Webster, 1985a, 1985b). Wood-Eggenschwiler and Bärlocher (1983) observed negative correlation between species richness, pH and calcium with the diversity of riparian vegetation that was not significantly correlated, but claimed that the average annual conidial concentration in streams is slightly related to riparian vegetation. High species diversity and conidia were found in streams with a variety of leaf litter and rich riparian vegetation (Sridhar and Kaveriappa, 1989b). Similar observations were made by Chergui (1990) that showed the high spore density in upstream mountain streams and linked to abundance of riparian vegetation in wooded areas than in a clear-cut site in the downstream. Two distinct assemblages of fungi were observed in upstream and downstream with different riparian vegetation by Gönczöl (1975) in Hungary. The difference in quality leaf litter may also play a role on the distribution of species or genotype of aquatic hyphomycetes. Colonization success, fungal-mediated leaf litter decomposition and stimulation of reproduction in aquatic hyphomycetes have been related to nutrient enrichment (e.g., N, P) (Gulis and Suberkroop, 2004). Our study did not show drastic fluctuations in mean pH between high (7.9) and low (7.7) altitude locations. Low species number in high altitude sampling locations of River Aghanashini and Gangavalli might be attributed to less natural vegetation coupled with human inhabitation in the nearby areas.

Sanders and Webster (1980) showed that an increase in the flow rates of streams enhances the release of aquatic hyphomycete spores. They recorded 22 species of aquatic hyphomycetes in randomly collected leaf litter and six species in foam samples of rivers. Conidia in the foam samples were repeatedly found from a particular region and represent its dominant mycoflora. The conidia can be transported to 0.7-0.8 km distances in the stream and this has been considered as half-life of conidial transport in streams (Thomas *et al.*, 1990). Conidial concentration in foam was more in high altitude locations (419.5/ml) than low altitude locations (176.2/ml), which might lead to trapping more conidia in the foam at high altitudes in our study. Naturally deposited decaying leaves in unpolluted streams of the Western Ghats yielded a mean of 15 species per location and liberated about 1,200 conidia/mg leaf dry mass. However, in our study, a severe decline was observed in mean species richness (4 species) and mean spore output (8 conidia/mg leaf). Only six species were recovered from the leaf litter and this might be due to substrate preference by these fungi. Some species might have preferred other substrates like stem or twig, or early colonizers might have altered the chemistry of leaf, which did not support other fungi in streams. This might be due to the sampling of leaves in summer, a few months after the leaf fall in spring. The impacts of abiotic variables and the substrate composition on the structure of aquatic hyphomycte communities are important, but not to the same degree as altitude and water hardness (Gönczöl *et al.*, 2003).

Seasonal occurrence and distribution of aquatic hyphomycetes are influenced by water temperature and pH (Chamier, 1987, 1992; Suberkropp, 1984; Wood-Eggenschwiler and Bärlocher, 1983). Biomass, growth and sporulation of aquatic hyphomycetes are seasonal due to dependence on deciduous leaf litter in temperate streams (Bärlocher, 1992; Suberkropp, 1997). Water temperatures in streams/ rivers of the Western Ghat locations are ideal for the activity of aquatic hyphomycetes. In the locations that were studied, temperature ranged between 28 and 29°C and the difference in temperature between high and low altitude sampling locations was also narrow. This might be the reason for the occurrence of aquatic hyphomycetes almost common to all the sites in high and low altitude sampling locations. In a similar study in the Western Ghat Rivers, Rajashekhar and Kaveriappa (2003) interpreted that pH, low oxygen, high temperature, hardness, iron and sulfides in water were the factors governing the richness of the species. The present study and earlier studies (Chandrashekar *et al.*, 1990; Rajashekhar and Kaveriappa, 1993; Sridhar *et al.*, 1992; Sridhar and Kaveriappa, 1989a,b; 1992) carried out in the Western Ghat locations at altitudes above 500m showed a rich diversity of aquatic hyphomycetes as well as elevated conidial output. The reason for low species richness in other locations at low altitude regions might be due to substrate scarcity, because of sparse vegetation or due to sampling during summer. Abdel-Raheem (2004) used different techniques to study aquatic hyphomycetes in the River Nile and found less number of species in the membrane filtration technique than in the leaf mapping technique. This might also be one of the reasons for the recovery of less number of species in our study.

Mothe-Jean-Louis (1997) observed 46 foam samples for five years in the River Osse, France, and described 65 species and Bärlocher (2000) studied seasonal and yearly variation of spore numbers, the number of individual species and spore taxa for 5 to 12-months period and reported no drastic variation in the total number of species. Interestingly, Descals (1998) in a three-day survey of foam covering 11 mountain streams has encountered 80 taxa of Ingoldian fungi. Input of woody debris can increase detritus retention and can enhance aquatic hyphomycetes diversity and production. The number of species increases during summer and the monsoon/post-monsoon season in the Western Ghats due to the abundance of leaf litter (Chandrashekar *et al.*, 1990). The tropics often pose several substantial logistical problems for fieldwork (Raviraja *et al.*, 1998). Of the eight sites in the current study, several were not accessible for sampling during southwest monsoon and post-monsoon periods due to torrential rains in the Western Ghats (600-700 cm/annum) and sometimes, the access remains hazardous. Thus, conclusions were drawn based on the results of a one-time sampling in each season in the current study.

Based on the altitude, vegetation and temperature, the Western Ghat streams can be divided into four distinct locations: mean species, 10.3 in mountain streams, where riparian vegetation is sparse; mean species, 20.1 at mid altitude streams with rich virgin forests; mean species, 14.2 in foothill streams; mean species, 12.5 in coastal streams where vegetation is sparse and human interference is more with the rise in temperature (Raviraja *et al*., 1998). In our study, high altitude sampling locations possess average species 13.5, while low altitude locations of foothills 11.2. Significant correlation was established between pH and species number, which corroborates with the studies by Raviraja *et al.,* (1998) in the Western Ghats. The maximum number of species was recovered in the post-monsoon season and the minimum in the summer/pre-monsoon season and thus, the availability of oxygen and substrate might play an important role in fungal assemblage (Chandrashekar *et al*., 1990; Sridhar and Kaveriappa, 1984).

The common species found in all the locations are considered as most widespread and considerably tolerant to extreme conditions prevailing in the Western Ghats. *Campylospora chaetocladia* is considered as high temperature preferring, tropical and summer dominant species (Suberkropp, 1984; Chauvet, 1991; Webster and Descals, 1981; Sridhar and Kaveriappa, 1989b) and *Anguillospora longissima* and *Flagellospora curvula* have been considered as species preferring cold climate (Suberkropp, 1984). But, these fungi are known to tolerate extreme conditions (e.g., high pH, low oxygen, and hard water) (Rajashekhar and Kaveriappa, 1993). *Anguillospora longissima, Campylospora chaetocladia, Flabellospora multiradiata, Flgellospora curvula, Lunulospora curvula, L. cymbiformis* and *Triscelophorus acuminatus* are the common aquatic hyphomycetes found in all the locations surveyed. These species are less affected by altitude and pH and have been considered as widespread species occurring in different types of biotypes. However, *Articulospora inflata, Flagellospora penicillioides* and *Isthmotricladia gombakiensis* are known only in the two locations that were studied. This shows that the distribution of some aquatic hyphomycetes species is narrow or restricted.

Conclusions and Future Outlook

The diversity of aquatic hyphomycetes and factors that regulate the community have been understudied or underestimated (Baerlocher, 2005). A number of unresolved issues in the pursuit of understanding the impacts of diversity of aquatic hyphomycetes in ecosystem functioning have received less attention. Greater emphasis needs to be paid to improve our knowledge on the diversity of aquatic hyphomycetes with emphasis on traditional and modern techniques (Baerlocher, 2007). A detailed study, including water chemistry, riparian vegetation, altitude, pH and temperature for an extended period has to be carried out to predict the factors responsible for the loss of species richness, diversity and distribution of aquatic hyphomycetes and to understand their precise role in ecosystem functioning. In addition, it would be valuable to study the abundance and diversity of aquatic hyphomycetes in polluted streams and rivers. These studies will not only document the diversity of fungi in undisturbed streams, but also reveal the impact of pollution on fungal community structure. It might lead to the identification of an indicator or sentinel species. If such is the case, it will be an inexpensive method of explicit assessment for the health of the stream ecosystem.

References

Abdel-Raheem, A.M. 2004. Study of the effect of different techniques on diversity of freshwater hyphomycetes in the River Nile (Upper Egypt). *Mycopathol. 157*, 59-72.

Allan, J.D. 1995. *Stream ecology.* Chapman and Hall, London.

APHA, 1995. *Standard Methods for the Examination of Water and Wastewater.* 19th Edition, American Public Health Association, USA.

Bandoni, R.J. 1981. Aquatic hyphomycetes from terrestrial litter. In: *The Fungal Community: Its Organization and Role in the Ecosystem* (Ed. Wicklow, D.T. and Carroll, G.C.). Marcel Dekker, New York, 693-708.

Bärlocher, F. 1987. Aquatic hyphomycete spora of 10 streams of New Brunswick and Nova Scotia. *Can. J. Bot. 65*, 76-79.

Bärlocher, F. 1992. Research on aquatic hyphomycetes: Historical background and overview. In: *The Ecology of Aquatic Hyphomycetes* (Ed. Bärlocher, F.). Ecological Studies, Volume 94, Springer-Verlag, Berlin, 1-15.

Bärlocher, F. 2000. Water-borne conidia of aquatic hyphomycetes: seasonal and yearly patterns in Catamaran Brook, New Brunswick, Canada. *Can. J. Bot. 78*, 157-167.

Bärlocher, F. 2005. Freshwater fungal communities. In: *The Fungal Community: Its Organization and Role in the Ecosystem* (Ed. Dighton, J., White, J.F. and Oudemas, P.). Taylor and Frances, Boca Raton, Florida, 39-59.

Bärlocher, F. 2007. Molecular approaches applied to aquatic hyphomycetes. *Fungal Biology Reviews 21*: 19-24.

Bärlocher, F. and Rosset, J. 1981. Aquatic hyphomycete spora of two Black forest and two Swiss Jura streams. *Trans. Br. Mycol. Soc. 76*, 479-483.

Chamier, A.-C. 1987. Effect of pH on microbial degradation of leaf litter in seven streams of the English Lake District. *Oecologia 71*, 491-500.

Chamier, A.-C. 1992. Water chemistry. In: *The Ecology of Aquatic hyphomycetes* (Ed. Bärlocher, F). Ecological Studies, Volume 94, Springer-Verlag, Berlin, 152-172.

Chandrashekar, K.R., Sridhar, K.R. and Kaveriappa, K.M. 1990. Periodicity of water-borne hyphomycetes in two streams of the forests of the Western Ghats (India). *Acta Hydrochim. Hydrobiol. 18*, 187-204.

Chauvet, E. 1991. Aquatic hyphomycete distribution in Southwestern France. *Biogeogr. 18*, 699-706.

Chergui, H. 1990. The dynamics of aquatic hyphomycetes in an eastern Moroccan stream. *Arch. Hydrobiol. 118*, 341-352.

Descals, E. 1998. Stream-borne fungi from Karrantza (Basque country) and surroundings. *Societat D' Historia Natural De Les Balears 41*, 191-219.

Dix, N.J. and Webster, J. 1995. *Fungal Ecology.* Chapman and Hall, London.

Fabre, E. 1996. Relationships between aquatic hyphomycete communities and riparian vegetation in 3 Pyrenean streams. *Comptes Rendus Hebdomadaires des Seances des l' Academie des Sciences, Paris,* 319, 107-111.

Garnett, H., Bärlocher, F. and Giberson, D. 2000. Aquatic hyphomycetes in Catamaran Brook: Colonization dynamics, seasonal patterns and logging effects. *Mycologia* 92, 29-41.

Gönczöl, J. 1975. Ecological observations on the aquatic hyphomycetes of Hungary I. *Acta Bot. Acad. Sci. Hung. 21*, 243-264.

Gönczöl, J. 1989. Longitudinal distribution patterns of aquatic hyphomycetes in a mountain stream in Hungary: Experiments with leaf packs. *Nova Hedwigia 48*, 391-404.

Gönczöl, J. and Revay, A. 2003. Aquatic hyphomycetes in the Morgo stream, Hungary tributary communities. *Nova Hedwigia 76*, 173-189.

Gönczöl, J. and Révey, A. 2006. Species diversity of rain-borne conidia from living trees. *Fungal Diversity 22*, 37-54.

Gönczöl, J., Csontos, P. and Revay, A. 2003. Catchment scale patterns of aquatic hyphomycetes. The role of physicochemical variables and substrate composition in structuring conidial communities. *Arch. Hydrobiol. 157*, 249-266.

Gulis.V.I. 1999. Preliminary list of aquatic hyphomycetes from central Belarus. *Mycotaxon 72*, 227-230.

Gulis, V. and Suberkropp, K. 2004. Effects of whole-stream nutrient enrichment on the concentration and abundance of aquatic hyphomycetes conidia in transport. *Mycologia 96*, 57-65.

Iqbal, S.H. and Webster, J. 1973. Aquatic hyphomycete spora of the River Exe and its tributaries. *Trans. Brit. Mycol. Soc. 61*, 331-346.

Mothe-Jean-Louis, A.-M. 1997. Dynamique Pluriannuelle des hyphomycetes aquatiques de Osse, rivere realimentee du sudouest de la France. *Can. J. Bot. 75*, 155-169.

Raghu, P. A., Sridhar, K. R. and Kaveriappa. K. M. 2001. Diversity and conidial output of aquatic hyphomycetes in a heavy metal polluted river, Southern India. *Sydowia 53*, 236-246.

Rajashekhar, M. and Kaveriappa, K.M. 1993. Ecological observations on water-borne hyphomycetes of the Cauvery River and its tributaries in India. *Arch. Hydrobiol. 126*, 487-497.

Rajashekhar, M. and Kaveriappa, K.M. 2003. Diversity of aquatic hyphomycetes in the aquatic ecosys*tems of the Western Ghats of India. Hydrobiol. 501, 167-177.*

Raviraja, N.S., Sridhar, K.R. and Bärlocher, F. 1998. Fungal species richness in the Western Ghat streams (southern India): is it related to pH, temperature or altitude? *Fungal Diversity 1*, 179-191.

Sanders, P. F. and Webster, J. 1980. Sporulation responses of some 'aquatic hyphomycetes' in flowing water. Trans. *Brit. Mycol. Soc. 74*, 601-605.

Shearer, C. A. and Webster, J. 1985a. Aquatic hyphomycete communities in the River Teign I. Longitudinal distribution patterns. *Trans. Brit. Mycol. Soc. 84*, 489-501.

Shearer, C.A. and Webster, J. 1985b. Aquatic hyphomycete communities in the River Teign. II. Temporal distribution patterns. *Trans. Brit. Mycol. Soc. 84*, 503-507.

Sridhar, K.R. and Kaveriappa, K.M. 1984. Seasonal occurrence of water-borne fungi in Konaje stream (Mangalore), India. *Hydrobiol.* 119, 101-105.

Sridhar, K.R. and Kaveriappa, K.M. 1989a. Observations on aquatic hyphomycetes of the Western Ghat streams, India. *Nova Hedwigia 49*, 455-467.

Sridhar, K.R. and Kaveriappa, K.M. 1989b. Colonization of leaves by water-borne hyphomycetes in a tropical stream. *Mycol. Res. 92*, 392-396.

Sridhar, K.R. and Kaveriappa, K.M. 1992. Aquatic hyphomycetes of Western Ghat streams, India. *Sydowia 44*, 66-77.

Sridhar, K.R., Chandrashekhar, K.R. and Kaveriappa, K.M. 1992. Research on the Indian subcontinent. In: *The Ecology of aquatic hyphomycetes* (Ed. Bärlocher, F.). Ecological Studies, Volume 94, Springer-Verlag, Berlin, 182-211.

Sridhar, K.R., Bärlocher, F., Krauss, G.-J. and Krauss, G. 2005. Response of Aquatic Hyphomycete communities to changes in heavy metal exposure. *Int. Rev. Hydrobiol. 90*, 21-32.

Sridhar, K.R., Karamchand, K.S. and Bhat, R. 2006. Arboreal water-borne hyphomycetes with oak-leaf basket fern *Drynaria quercifolia. Sydowia 58*, 309-320.

Sridhar, K.R., Krauss, G., Bärlocher, F., Raviraja, N.S., Wennrich, R., Baumbach, R. and Krauss, G.-J. 2001. Decomposition of alder leaves in two heavy metal polluted streams in Central Germany. *Aquat. Microb. Ecol. 26*, 73-80.

Suberkropp, K. 1984. Effect of temperature on seasonal occurrence of aquatic hyphomycetes. *Trans. Brit. Mycol. Soc. 82*, 53-62.

Suberkropp, K. 1992. Aquatic hyphomycete communities. In: *The fungal community: its organization and role in the ecosystem* (Ed. Carroll, G. C. and Wicklow, D. T.). 2nd Edition, Marcel Dekker, New York, 729-747.

Suberkropp, K. 1997. Annual production of leaf-decaying fungi in a woodland stream. *Freshwat. Biol. 38*, 169-178.

Thomas, K., Chilvers, G.A. and Norris, R.H. 1990. A dynamic model of fungal spora in a freshwater stream. *Mycol. Res. 95*, 184-188.

Webster, J. and Descals, E. 1981. Morphology, distribution and ecology of conidial fungi in freshwater habitats. In: *Biology of Conidial fungi* (Ed. Cole, G. T. and Kendrick, B.). Volume I, Academic Press, New York, 295-355.

Webster, J., Marvanova, L. and Eicker, A. 1994. Spores from foam from South African rivers. *Nova Hedwigia 57*, 379-398.

Willoughby, L.G. and Minshall, G.W. 1975. Further observations on Tricladium giganteum. *Trans. Brit. Mycol. Soc. 65*, 77-82.

Wood-Eggenschwiler, S. and Bärlocher, F. 1983. Aquatic hyphomycetes in sixteen streams in France, Germany and Switzerland. *Trans. Brit. Mycol. Soc. 76*, 479-483.

Wood-Eggenschwiler, S. and Bärlocher, F. 1985. Geographical distribution of Ingoldian fungi. *Ver. Int. Verein. Theor. Angewand. Limnol. 22*, 2780-2785.

Frontiers in Fungal Ecology, Diversity and Metabolites (edited by K.R. Sridhar), I.K. International Publishing House Pvt. Ltd., New Delhi, 2009.

3

Mangrove Fungi of the Indian Peninsula

K.R. Sridhar
Microbiology and Biotechnology, Department of Biosciences, Mangalore University, Mangalagangotri, Mangalore 574 199, Karnataka, India
(E-mail: sirikr@yahoo.com)

Introduction

The 'obligate marine fungi' has been defined as the fungi that grow and sporulate in marine or estuarine habitat, while 'facultative marine fungi' are derived from the freshwater or terrestrial substrates but grow and possibly sporulate in marine habitat (Kohlmeyer, 1974). The first facultative (*Phaeosphaeria typharum*) and obligate (*Halotthia posidoniae*) marine fungi have been described in 1849 and 1869 (see Kohlmeyer and Kohlmeyer, 1979). Development of marine mycology was initiated by Barghoorn and Linder (1944), who described 25 species from submerged wood in New England and California. By 1970, about 100 new species were described from Germany, United States, and Great Britain (Kohlmeyer and Kohlmeyer, 1979). In 1961, the first monograph on marine mycology 'A Treatise on Fungi in Oceans and Estuaries' was published by Johnson and Sparrow. Subsequently, Hughes (1975) summarized the contributions on marine mycology since 1961. Taxonomic and ecological monograph by Kohlmeyer and Kohlmeyer (1979) lists 209 filamentous fungi, which was increased up to 321 in the first decade (Kohlmeyer and Volkmann-Kohlmeyer, 1991) and up to 444 species in subsequent decade (Hyde and Pointing 2000). The latest review of literature by Schmit and Shearer (2003) reveals the occurrence of 625 fungal species (meiosporic and mitosporic) from different mangrove habitats. About 200 species have been considered as obligate marine fungi and 131 species are belonged to sediments or mangrove peat is not totally restricted to marine environments. The present chapter provides current status of mangrove mycology of the Indian Subcontinent with emphasis on major gaps in our knowledge on filamentous mangrove fungi. A checklist of higher mangrove fungi of the Indian Subcontinent has been furnished.

Mangroves

Mangroves are wetland forests established at the intertidal zones (estuaries, backwaters, deltas, creeks, lagoons, marshes and mudflats) of tropical and subtropical latitudes. Approximately 25% of the world's coastline is dominated by mangroves distributed in 112 countries and territories encompassing an area of about 181,000 km^2 (Spalding *et al.*, 1997). Up to 41.5, 26.5 and 22.6% mangroves are found in South-East Asia, America (including Caribbean) and Africa respectively. Among marine ecosystems, in terms of productivity and sustained tertiary yield, mangroves constitute the second most important ecosystem after coral reefs (Qasim and Wafar, 1990). Mangrove flora have been classified into three major categories (Field, 1995; Tomlinson, 1986): (i) True mangroves (about 80 tree and shrub species, which are restricted to intertidal areas between high water levels of neap of spring tides); (ii) Minor species (inconspicuous elements of vegetation and rarely form pure communities); (iii) Mangrove associates (salt-tolerant plant species not found exclusively in the vicinity of mangrove and occur only in transitional vegetation landwards and seawards). More than 100 mangrove tree species have been listed by Chapman (1976). Out of 80 species of true mangrove trees and shrubs, 50-60 species make significant contribution to the structure of mangrove forests (Field, 1995; Tomlinson, 1986). Highest mangrove plant species diversity is seen in South-East Asian region (two-thirds of all species) (Field, 1995). Mangrove vegetation has adapted to harsh conditions (high salinity, tidal extremes, high wind velocity, high temperature and anaerobic clayey soils) and possesses pneumatophores, prop roots and viviparous seedlings (Chapman, 1976). Mangrove forests are great traditional, ecological, economic and social significance (Bandaranayake, 1995, 1998; Kathiresan and Bingham, 2001). Mangroves are detritus-based, thus considerable amount of detritus such as leaf litter, twigs, bark, wood, inflorescence and other detritus accumulate on the forest floor (Wafar *et al.*, 1997). The productivity of mangrove waters depends on the extent of mangrove canopy and supply C, N and P (Wafar *et al.*, 1997). In Peninsular India, mangrove forests spread over 7000 km^2 consisting of 70, 18 and 12% at the east coast, Andaman Nicobar Islands and West Coast respectively (Krishnamurthy *et al.*, 1987; Natarajan, 1998). The total area of the Indian mangroves is estimated to be about 7% of the world mangroves (Untawale, 1987).

Mangrove Fungi

Mangroves constitute 'hotspots' of fungal diversity. Mangrove fungi are decomposers of dead organic materials and act as intermediaries of nutrient flow from organic matter to the higher trophic levels. They are second largest group next to fungi on marine intertidal woody litter among the marine fungi (Hyde, 1990). Cribb and Cribb (1955) reported mangrove fungi in Australia for the first time followed by Kohlmeyer and Kohlmeyer (1979). Detritus and live parts of mangrove vegetation (tree species, mangrove associates, ferns and palms) have been surveyed for the occurrence of higher fungi. Considerable fungal population is involved in detritus processing due to their enzyme equipment and hence special attention is necessary to assess them. Several factors contribute to the diversity of fungi in mangroves (e.g., substrate diversity, salinity, intermittent inundation) and daily changes in water level due to tides (provide further niche differentiation). Substrates beyond the high tide level will be exposed to air and direct sunlight and occasionally exposed to brackish water. The substrates deposited below the high tide zone are less exposed to air and sunlight and occasionally or always be inundated. Studies on fungal communities along the tidal gradients have shown differences in species composition and abundance (Hyde, 1988, 1990).

There is a considerable difference in the sexual state of fungi found on different mangrove substrates. Woody materials are commonly colonized by meiosporic ascomycetes and basidiomycetes (Kohlmeyer and Volkmann-Kohlmeyer, 1991), while leaves and fine roots in sediments by mitosporic fungi (Schmit and Shearer, 2003) and oomycetes (*Halophytophthora*) are frequent on freshly fallen leaves (Newell, 1996). Ascomycetes show specific adaptation to the marine ecosystem in the production of appendages and these appendages help in buoyancy of spores, entrapment and adherence to the surfaces (see Fig. 1). Extensive surveys have been carried out at the Caribbean and Pacific Islands (Kohlmeyer 1984; Volkmann-Kohlmeyer and Kohlmeyer, 1993) and in South-East Asia (Sarma and Hyde, 2001). Maximum mangrove-associated fungi have been reported from South-East Asia than other parts of the world (Schmit and Shearer, 2003). It is not clear yet this is due to uneven sampling or high diversity of mangrove plant species of South-East Asia. So far, 72 plant species have been reported as hosts for mangrove fungi (Schmit and Shearer, 2003). The monographs written by Kohlmeyer and Kohlmeyer (1979), Kohlmeyer (1984), Kohlmeyer and Volkmann-Kohlmeyer (1991), Hyde and Sarma (2000) and Sharma and Vittal (2004) are the major source to identify higher marine fungi. A pictorial identification key for marine fungi has been developed by National Institute of Oceanography, Goa, India, is highly useful source (http://www.indian-ocean.org/bioinformatics/fungi/Micro-cd/main.htm) (Sarma *et al.*, 2000)

Diversity and Distribution

The global distribution of mangrove fungi has been dealt recently by Shearer *et al.* (2007). This report reveals the geographic distribution and number of known species of mangrove-associated meiosporic ascomycetes/mitosporic ascomycetes and basidiomycetes/other fungi (Shearer *et al.*, 2007). Mangrove fungi are known from the North Temperate (55/51/11), Tropics (79/60/13), Asian Temperate (25/17/8), Tropical Africa (21/14/3), Madagascar (0/1/1), Temperate Africa (29/11/1), Middle East (20/9/0), Tropical Asia (225/190/33), Australasia (67/11/11) and Pacific Islands (47/56/18). Among different geographic locations, Tropical Asia clearly showed the highest number of fungi. Although studies on mangrove

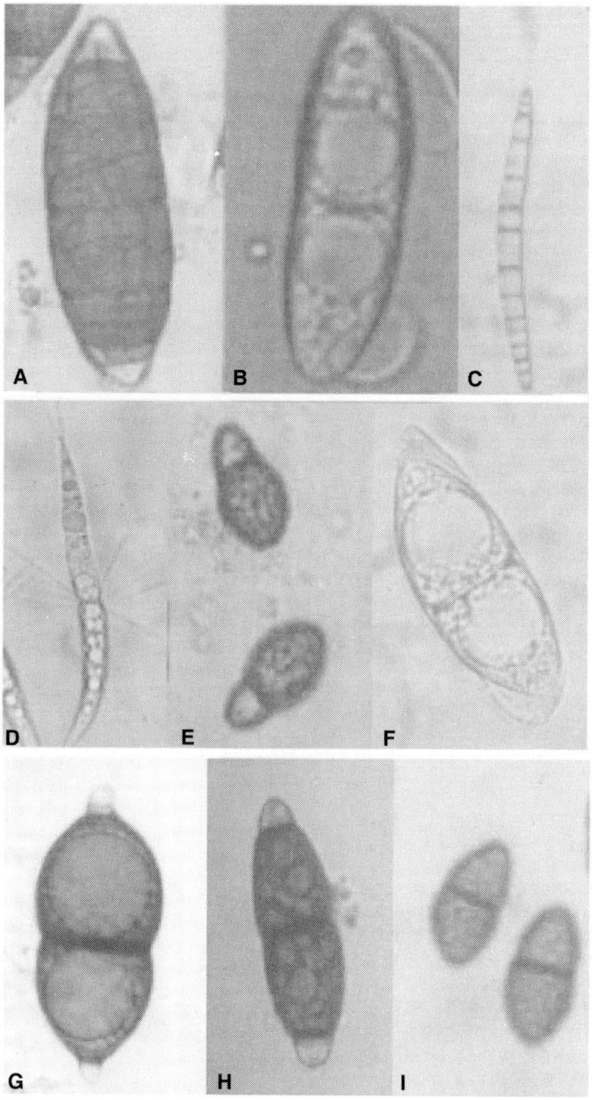

Fig. 1. Asocospores of representative mangrove fungi of the mangroves of Karnataka (west coast of India): A, *Aigialus mangrovei*; B, *Aniptodera mangrovei*; C, *Ascosalsum cincinnatulum*; D, *Corollospora pulchella*; E, *Dactylospora haliotrepha*; F, *Littispora ratnagiriensis*; G, *Passeriniella mangrovei*; H, *Savoryella lignicola*; I, *Verruculina enalia*

fungi have been initiated in India during early 1970s, intense investigations have been performed from 1980 onwards. Considerable floristic studies have been conducted from the mangroves of Maharashtra, Goa, Karnataka, Tamil Nadu and Andhra Pradesh (see Table 1 and 2 for references), mangroves of rest of the coasts (Gujarat, Kerala, Orissa and West Bengal) are less explored and some are virtually unexplored for mangrove fungi. Studies on the mangroves

Table 1. *Check-list of higher mangrove fungi of the Indian Peninsula up to 2006 (core-group fungi: frequency of occurrence, >10% given in bold face)*

Fungus	Location
Ascomycetes	
Acrocordiopsis patilii Borse *et* Hyde	2,9
Aigialus grandis Kohlm. *et* Schatz	1,2,6,7,9
Aigialus mangrovei Borse	2,4,6,7,9
Aigialus parvus Schatz *et* Kohlm.	1,2,6,7,9
Aigialus rhizophorae Borse	2
Aniptodera chesapeakensis Shearer *et* Miller	2,3,4,5,7,9
Aniptodera haispora Vrijmoed, Hyde *et* Jones	7
Aniptodera indica Ananda *et* Sridhar	2,3,4,5
Aniptodera mangrovei Hyde	2,3,4,5,7,8,9
Antennospora quadricornuta (Cribb *et* Cribb) Johnson	1,2,4,6,9
Antennospora salina (Meyers) Yusoff, Jones *et* Moss	1,9
Anthostomella leptospora (Lev. *et* Sacc.) Francis	7
Arenariomyces parvulus Koch	4,8
Arenariomyces trifurcates Hö hnk	4
Ascochyta salicorniae Magnus	2
Ascocratera manglicola Kohlm.	2,4,6,7,9
Ascosalsum cincinnatulum (Shearer *et* Crane) Campb., Anderson *et* Shearer	2,4,5
Asterosphaeriella mangrovis (Kohlm. *et* Vittal) Aptroot *et* Hyde	1,2,3,4,5,6,7,9
Bathyascus avicenniae Kohlm.	6,7,9
Bathyascus grandisporus Hyde and Jones	9
Bathyascus mangrovei Ravikumar *et* Vittal	6
Belizeana tuberculata Kohlm. *et* Volkm.-Kohlm.	6,9
Biatriospora marina Hyde *et* Borse	2,6,9
Biflua physasca Koch *et* Jones	4
Caryosporella rhizophorae Kohlm.	4,8,9
Ceriosporopsis cambrensis Wilson	2
Ceriosporopsis halima Linder	2
Ceriosporopsis sundica Koch *et* Jones	4
Chaetomastia typhicola (Karst.) Barr	7
Corollospora angusta Nakagiri *et* Tokura	4
Corollospora cinnamomea Koch	4
Corollospora intermedia Schmidt	4,6
Corollospora maritima Werdermann	14,5,8
Corollospora pulchella Kohlm., Schmidt *et* Nair	2,3,4,5,7
Coronopapilla mangrovei (Hyde) Kohlm. *et* Volkm.-Kohlm.	4,9
Cryptosphaeria mangrovei Hyde	7
Dactylospora heliotrepha (Kohlm. *et* Kohlm.) Hafellner	1,2,3,4,5,6,7,8,9
Dictyosporium pelagicum (Linder) Hughes	2
Didymella avicenniae Patil *et* Borse	2,4,6
Didymosphaeria lignomaris Strongman *et* Miller	4
Etheirophora blepharospora (Kohlm. *et* Kohlm.) Kohlm. *et* Volkm.-Kohlm.	9
Etheirophora unijubata Kohlm *et* Volkm.-Kohlm.	4
Eutypa bathurstensis Hyde *et* Rappaz	1,3.4,5,7
Halographis runica Kohlm. *et* Volkm.-Kohlm.	4
Halorosellinia oceanica Whalley, Jones, Hyde *et* Laessøe	1,2,3,4,5,7,9
Halosarpheia fibrosa Kohlm. *et* Kohlm.	2,4,5
Halosarpheia hamata (Höhnk) Kohlm.	6
Halosarpheia marina (Cribb *et* Cribb) Kohlm.	2,5,6,7
Halosarpheia minuta Leong	7,9
Halosphaeria cucullata (Kohlm.) Kohlm.	6
Heleococcum japonense Tubaki	6,7
Helicascus kanaloanus Kohlm.	2
Heliscus konaloanus Kohlm.	9
Julella avicenniae (Borse) Hyde	1,2,7,9
Kallichroma tethys (Kohlm. *et* Kohlm.) Kohlm. *et* Volkm.-Kohlm.	1,3,4,5,6,7,8,9
Kirschsteiniothelia maritima (Linder) Hawksw.	7
Lautispora danica (Berl.) Schatz	4
Lautospora gigantea Hyde *et* Jones	5,7

Lecanidion atratum (Hedw. *Ex* Fr.) Endl.	7
Leptosphaeria australiensis (Cribb *et* Cribb) Hughes	1,2,3,4,5,6,7,8,9
Leptosphaeria marina Ellis *et* Everhart	2
Leptosphaeria obiones (Crouan *et* Crouan) Saccardo	2
Leptosphaeria oraemaris Linder	4,6
Leptosphaeria peruviana Speg.	4,6,7
Leptosphaeria salvinii Catt.	4
Lignincola laevis Höhnk	2,3,4,5,7,9
Lignincola tropica Kohlm.	4,5,7
Lindra marinera Meyers	4,8
Lineolata rhizophorae (Kohlm. *et* Kohlm.) Kohlm. *et* Volkm.-Kohlm.	1,4,6,7,9
Littispora abonnis (Kohlm.) Campb., Anderson *et* Shearer	2,5,7,9
Littispora ratnagiriensis Patil *et* Borse	2,4,5,7,9
Lulworthia grandispora Meyers	1,2,3,4,5,6,7,8,9
Lulworthia kniepii Kohlm.	4,5
Lulworthia lindroidea Kohlm.	4
Lulworthia medusa (Ellis *et* Everh.) Cribb *et* Cribb	2
Marinosphaera mangrovei Hyde	7,9
Massarina armatispora Hyde, Vrijmoed, Chinnaraj *et* Jones	3,4,9
Massarina thalassiae Kohlm. *et* Volkm.- Kohlm.	2,6,7,9
Massarina velatospora Hyde *et* Borse	1,2,3,4,6,7,9
Mycosphaerella apophlaeae Kohlm.	4
Mycosphaerella pneumatophorae Kohlm.	6,7
Mycosphaerella salicorniae (Auerswald) Petrak	4
Mycosporella staticicola (Patouillard) Dias	4
Nais glitra Crane *et* Shearer	6,7,9
Natantispora retorquens (Shearer *et* Crane) Campb., Anderson *et* Shearer	2,4,5
Neptunella longirostrias (Cribb *et* Cribb) Pang *et* Jones	2,3,4,5,6,7
Ophiodeira monosemeia Kohlm. *et* Volkm.-Kohlm.	7
Panorbis viscosus (Schmidt) Campb., Anderson *et* Shearer	4,7
Passeriniella mangrovei Maria *et* Sridhar	4,5
Passeriniella obiones (Crouan *et* Crouan) Hyde *et* Mouzouras	7
Passeriniella savoryellopsis Hyde *et* Mouzouras	9
Payosphaeria minuta Leong	9
Pedumispora rhizophorae Hyde *et* Jones	7
Quintaria lignatilis (Kohlm.) Kohlm. *et* Volkm.-Kohlm.	1,6,7,9
Rhizophila marina Hyde *et* Jones	3,4,5,7,9
Saccardoella marinospora Hyde	7
Saccardoella rhizophorae Hyde	7
Salsuginosa ramicola Hyde	9
Savoryella lignicola Jones *et* Eaton	2,3,4,5,7,8,9
Savoryella longispora Jones *et* Hyde	4,5
Savoryella paucispora (Cribb *et* Cribb) Koch	2,3,4,5,9
Splanchnonema britzelmayriana (Rehm.) Boise	7
Swampomyces armeniacus Kohlm. *et* Volkm.-Kohlm.	6,9
Tirispora mandoviana Sarma *et* Hyde	3,4
Torpedospora radiata Meyers	1,2,3,4,8
Trematosphaeria lignatilis Kohlm.	2,6
Trematosphaeria striatispora Hyde	9
Tubeufia setosa Sivanesan *et* Hsieh	7
Verruculina enalia (Kohlm.) Kohlm. *et* Volkm.-Kohlm.	1,2,3,4,5,6,7,8,9
Zopfiella latipes (Lundqvist) Malloch *et* Cain	1,2,4,7
Zopfiella marina Furuya *et* Udagowa	7
Basidiomycete	
Halocyphina villosa Kohlm. *et* Kohlm.	1,2,4,5,6,7,9
Anamorhic fungi	
Alternaria chartarum Preuss	4
Anguillospora marina Nakagiri *et* Tubaki	4
Arthrobotrys oligospora Fresen.	4
Ascochyta salicorniae Magnus	1
Bactrodesmium linderi (Crane *et* Shearer) Palm. *et* Stewart	2, 4,5,7
Brachysporella gayana Batista	4
Camarosporium palliatum Kohlm. *et* Kohlm.	2

Camarosporium roumeguerii Sacc.	1,2,7
Cirrenalia basiminuta Raghukumar *et* Zainal	3,4,6,7
Cirrenalia fusca Schmidt	4
Cirrenalia macrocephala (Kohlm.) Meyers *et* Moore	2,4,6,7
Cirrenalia pseudomacrocephala Kohlm.	4,8,9
Cirrenalia pygmea Kohlm.	1,2,4,6,7,9
Cirrenalia tropicalis Kohlm.	4,5,6,7,9
Cladosporium oxysporum Berk. *et* Curt.	4
Clavatospora bulbosa (Anast.) Nakagiri *et* Tubaki	2,3,4,8,9
Corynespora cassicola (Berk. *et* Curt.) Wei	7
Cumulospora marina Schmidt	4
Curvularia geniculata (Tracy *et* Earle) Boedijn	4
Custingophora olivacia Stock, Hennebert *et* Klopotek	4
Cytospora rhizophorae Kohlm. *et* Kohlm.	1,4,7,9
Delortia palmicola Pat.	4
Dictyopolyschema pirozynskii Ellis	4
Dictyosporium zeylanicum Petch	7
Dicyma pulvinata (Berk. *et* curt.) von Arx	4
Diplocladiella scalaroides Arnaud ex Ellis	4
Ellisembia vagum (Nees *et* Nees) Subram.	7
Emericellopsos terricola van Beyma	4
Endophragmia alternata Tubaki *et* Saito	4
Epicoccum nigrum Link	4
Epicoccum purpurascens Ehrenb. *et* Schlecht.	7
Fusarium oxysporum Schl. *et* Fries	4
Helicoma muelleri Corda	4
Helicoma roseus Link	4
Menispora ciliata (Johnson) Jones	4
Monodictys pelagica (Johnson) Jones	3,4,6,8,9
Monodictys putredinis (Wallr.) Hughes	4
Myxotrichum chartarum Kunze	4
Nigrospora oryzae (Berk. *et* Br.) Petch	4
Periconia prolifica Anastasiou	1,2,3,4,5,6,7,8,9
Phaeoisaria clematides (Fuckel) Hughes	4,5
Phomopsis mangrovei Hyde	1,7,9
Stachybotrys oenanthes Ellis	4
Strachylidium bicolor Link	4
Taeniolella stricta (Corda) Hughes	4
Tetraploa aristata Berk. *et* Br.	4,5
Trichocladium achrasporum (Meyers *et* Moore) Dixon	1,2,4,5,6,7,9
Trichocladium alopallonellum (Meyers *et* Moore) Kohlm. *et* Kolkm.- Kohlm.	1,2,4,6,7,8,9
Trichocladium melhae Jones, Abdel-Wahab *et* Vrijmoid	4
Zalerion maritimum (Linder) Anastasiou	1,2,3,4,5,8
Zalerion varium Anastasiou	1,2,3,4,5,6,7,8,9
Zygosporium gibbum (Sacc., Rouss. *et* Bomm.) Hughes	7
Zygosporium masonii Hughes	4

Location:

1. Gujarat: Borse *et al.*, 2000; Patil and Borse, 2001
2. Maharashtra: Borse, 1984, 1987a-e, 1988; Borse and Hyde, 1989; Borse and Shrivastava, 1988, 1994; Borse *et al.*, 1988; Hyde and Borse, 1986a, b; Kohlmeyer and Vittal, 1986; Maria and Sridhar 2002b; Patil and Borse, 1983, 1985a, b; Shrivastava, 1989, 1994.
3. Goa: Hyde *et al.*, 1992; Chinnaraj and Untawale, 1992; Maria and Sridhar, 2002b; Raghukumar *et al.*, 1988; Sarma and Hyde, 2000.
4. Karnataka: Ananda and Sridhar, 2001a, 2001b, 2003, 2004; Ananda *et al.*, 2007; Chinnaraj and Untawale, 1992; Hyde *et al.*, 1992; Maria and Sridhar, 2002a, b; 2003a; Sridhar and Kaveriappa, 1988; Sridhar and Maria, 2006.
5. Kerala: Maria and Sridhar, 2002b
6. Tamil Nadu: Raghukumar, 1973; Ravikumar and Purushothaman, 1988a, b; Ravikumar and Vittal, 1987, 1991, 1996.
7. Andhra Pradesh: Sarma and Vittal, 2000, 2002, 2004; Vittal and Sarma, 2005
8. Minicoy Island: Ananda and Sridhar, 2003
9. Andaman and Nicobar Islands: Chinnaraj, 1993

of the islands adjacent to Indian coast are also meager. Fourteen new species including three new genera have been described from the mangroves of Peninsular India (Table 2). Most of the new fungi were described from mangroves of Maharashtra.

Table 2. *New genera and new species of mangrove fungi described from the mangroves of Indian Peninsula (*New genera)*

Fungus	Mangrove	Reference
	Ascomycetes	
**Acrocardiopsis patilii* Borse *et* Hyde	Malvan (Maharashtra)	Borse and Hyde, 1989
Aigialus mangrovei Borse	Dabhol (Maharashtra)	Borse, 1987e
Aigialus rhizophorae Borse	Dabhol (Maharashtra)	Borse, 1987e
Aniptodera indica Ananda *et* Sridhar	Udyavara (Karnataka)	Ananda and Sridhar, 2001a
Asterosphaeriella mangrovis (Kohlm. *et* Vittal) Aptroot *et* Hyde	Bombay (Maharashtra)	Kohlmeyer and Vittal, 1986
Bathyascus mangrovei Ravikumar *et* Vittal	Pichavaram (Tamil Nadu)	Ravikumar and Vittal, 1991
**Biatriospora marina* Hyde *et* Borse,	Revadanda (Maharashtra)	Hyde and Borse, 1986a
Didymella avicenniae Patil *et* Borse	Revas (Maharashtra)	Patil and Borse, 1985b
**Julella avicenniae* (Borse) Hyde	Revas (Maharashtra) Pichavaram (Tamil Nadu)	Borse, 1987b; Hyde, 1992 Kohlmeyer and Vittal, 1986
Massarina armatispora Hyde, Vrijmoed, Chinnaraj *et* Jones	Kundapura (Karnataka)	Hyde, Vrijmoed, Chinnaraj and Jones, 1992
	Chorao Island (Goa)	Hyde, Vrijmoed, Chinnaraj and Jones, 1992
Massarina valatospora Hyde *et* Borse	Maharashtra	Hyde and Borse, 1986b
Passeriniella mangrovei Maria *et* Sridhar	Udyavara (Karnataka)	Maria and Sridhar, 2002a
Tirispora mandoviana Sarma *et* Hyde	Chorao (Goa)	Sarma and Hyde, 2000
	Anamorphic fungus	
Cirrenalia basiminuta Raghukumar *et* Zainal	Orda (Goa)	Raghukumar, Zainal and Jones, 1988

In spite of many floristic studies available from the Peninsular India, a few studies provide quantitative data (e.g., Ananda and Sridhar, 2004; Borse, 1988; Borse *et al.*, 2000; Chinnaraj and Untawale, 1992; Maria and Sridhar, 2002b; Patil and Borse, 2001; Ravikumar and Vittal, 1996; Sarma and Hyde, 2000; Sarma and Vittal, 2000; Sridhar and Maria, 2006). Among the 36 species recovered in Maharashtra, the dominant species were *Aigialus grandis, Massarina valatospora* and *Verruculina enalia* (Borse, 1988). Ravikumar and Vittal (1996) from the Pichavaram mangroves, Tamil Nadu, recovered 48 species on the wood, prop roots and seedlings of *Rhizophora apiculata* and *R. mucronata* with a maximum species (44) on proproots. In their study, *Aigialus grandis, Asterosphaeriella mangrovis, Cirrenalia pygmea, Halocyphina villosa* and *Verruculina enalia* were most dominant. From the Krishna and Godavari deltas (Andhra Pradesh), Sarma and Vittal (2000) examined the decaying wood, proproots and seedlings of *Rhizophora apiculata* and recorded 63 species with a maximum on proproots (61) followed by wood (24) and seedlings (21). Sarma and Vittal (2000) also studied the occurrence of mangrove fungi on the wood, roots and pneumatophores of *Avicennia officinalis* and *A. marina* of Krishna and Godavari deltas and recorded 65 species with a maximum of 61 species on wood. On both *Rhizophora* and *Avicennia* spp. *Verruculina enalia* was the most frequent fungus. Detritus of nine host plant species of mangroves of Godavari and Krishna also revealed the highest fungi in *R. apiculata* (Sarma and Vittal, 2001). A seasonal study was carried out at Godavari and Krishna deltas by Sarma and Vittal (1998-1999) and found maximum fungi during the monsoon season, but no definite seasonal trend was evident. There seems to be a drastic difference in the fungal assemblage and their frequency of occurrence between the West Coast and the East Coast of India. The dominant fungi of West Coast (*Aniptodera mangrovei, Cirrenalia pygmea, Lignincola laevis* and *Savoryella lignicola*) (Maria and Sridhar, 2003a) are not common in the mangroves of East Coast (Sarma and Vittal, 2001). Likewise, the dominant fungi on the wood of *Rhizophora* spp. (*Asterosphaeriella mangroves, Dactylospora haliotrepha* and *Verruculina enalia*) of the East Coast (Sarma and Vittal, 2000; Ravikumar and Vittal, 1996) are not dominant on *Rhizophora mucronata* wood of the West Coast (Maria and Sridhar, 2003a). Similarly, dominant *Eutypa bathurstensis* and *Verruculina enalia* on wood of *Avicennia* spp. from the East Coast (Sarma and Vittal, 2000) were not dominant on *Avicennia officinalis* wood of the West Coast (Maria and Sridhar, 2003a). These variations presumably due to the differences in the mangrove habitats between the East (deltaic) and the West Coast (backwater-estuarine). Being deltaic, the East Coast mangroves are more diverse in vegetation than the mangroves of West Coast. Differences in the fungal communities between mangroves of the West Coast of India are also seen. In the mangroves of Maharashtra, *Aigialus grandis, Dactylospora haliotrepha, Halocyphina villosa, Massarina velatospora* and *Verruculina enalia* were dominant (Borse, 1988). Out of them, except for *H. villosa* and *V. enalia* rest of the fungi are not dominant in the mangroves of Karnataka (Maria and Sridhar, 2003a). The dominant fungi of Gujarat mangroves (*Aigialus parvus, Dactylospora haliotrepha, Julella avicenniae, Kallichroma tethys, Lulworthia grandispora* and *Periconia prolifica*) (Borse *et al.*, 2000; Patil and Borse, 2001) are not frequent in mangroves of Karnataka with an exception of *L. grandispora* (Maria and Sridhar, 2003a). These observations clearly indicates the necessity of further investigations to link the differences in fungal assemblage and diversity between East and West Coast mangroves and within mangroves of East and West Coast of India.

A few studies are available from the Islands of the Indian Subcontinent. Andaman and Nicobar consist of about 350 islands and seem to be rich in mangrove mycoflora. The dead and decayed parts of six mangrove plant species yielded 49 ascomycetes, one basidiomycete and 13 anamorphic taxa (Chinnaraj, 1993). The highest number of species was recorded on *Rhizophora mucronata* (53 species). The most dominant fungi include *Verruculina enalia* and *Halocyphina villosa*. On the dead woody litter of mangroves of Minicoy Island (Lakshadweep), 15 fungal species comprising 11 ascomycetes, one basidiomycete and three mitosporic fungi were recorded (Chinnaraj, 1992). Another study at Minicoy Island revealed 20 species (12 ascomycetes, 8 mitosporic fungi) on mangrove woody litter with an average of 1.7 fungi per wood (Ananda and Sridhar, 2003). *Corollospora maritima* was the most frequent species (33.9%) followed by *Torpedospora radiata, Verruculina enalia, Zalerion varium, Kallichroma tethys, Lulworthia grandispora* and *Clavatospora bulbosa* (19.6-10.7%). The updated checklist of mangrove fungi of the Indian Subcontinent reveals 165 species: 111 ascomyctes, 1 basidiomycete and 53 anamorphic taxa (Table 1). Most studies are confined to the mangroves of Gujarat, Maharashtra, Goa, Karnataka, Keraka, Tamil Nadu and Andhra Pradesh. Fungi possess frequency of occurrence of 10% or more have been considered as core-group fungi. These fungi are mainly responsible for driving the ecosystem as they compete efficiently through their growth and sporulation on the substrata. Based on the quantitative data, in Indian mangroves 35 mangrove fungi have been identified as core-group fungi (Table 1). Potential of these fungi can be exploited as they are common and abundant in many Indian mangroves.

Pattern of Colonization

Mangrove detritus (e.g., fallen leaves, twigs, proproots, pneumatophores) support a variety of fungal communities. Senescent mangrove leaves released from the plants reach distinct habitats: (a) Trap in canopy above the water, which is not accessible for tides and reach water due to storm in dried state; (b) Float on the surface as dried or fresh; (c) Reach the mangrove floor under low tide; (d) Entrap at some depth below the water surface; (e) Settle or bury in sediment. Similarly, the dried twigs and branches also detach during monsoon storm and trap in canopy or water with or without bark, settle or trap in sediment. The twigs, stems, proproots and pneumatophores remain attached to plants undergo decay. These specific conditions of substrates influence the diversity and functions of mangrove mycoflora. The lower fungi usually follow the 'substrate-capture' strategy, while higher fungi adapted to 'mass-accumulation' strategy (Newell, 1996). It is suspected that periodical wet and dry regime of mangrove litter leads to higher fungal activity (Newell and Fell, 1997). Some fungi are commonly confined to bark of *Rhizophora apiculata* (e.g., *Hypophloeda rhizospora, Phomopsis mangrovei, Rhizophila marina*), while others confine to woody tissues (e.g., *Caryosporella rhizophorae*) (Hyde *et al.*, 1993). *Eutypa bathurstensis* was confined to the *Avicennia* wood (Sarma and Vittal, 2000). A recent study (Sridhar and Maria, 2006) addressed the pattern of colonization and diversity of filamentous fungi on naturally deposited and introduced *Rhizophora mucronata* wood during monsoon and summer in a southwest mangrove of India and compared overall occurrence with three species co-occurrence. Out of 66 fungi recovered, naturally deposited wood showed higher fungi during monsoon than summer (48 vs. 24), so also among 40 fungi on wood possessing three fungal co-occurrence (21 vs. 18). Percent frequency of occurrence of fungi was not significantly different between wood types and seasons in overall occurrence and three species co-occurrence. Overall, seventeen fungi belonged to the core-group (≥10%). Although *Aigialus mangrovei, Cirrenalia pygmea, C. tropicalis, Halosarpheia cincinnatula, Lulworthia grandispora, Passeriniella mangrovei, Verruculina enalia* and *Zalerion maritimum* are typical mangrove fungi, they belonged to core-group on deliberately introduced wood in monsoon season indicates their high colonization activity on wood even under low salinity. Many terrestrial anamorphic taxa (*Alternaria, Arthrobotrys, Aspergillus, Penicillium, Phoma* and *Tetracrium*) were common during monsoon season. Ananda *et al.* (2007) studied the colonization of fungi on dried and fresh leaves in mangroves of Udyavara, Karnataka. The number of fungi was greatest after 2-4 weeks of exposure and more species were recovered from dried than fresh leaves. Several terrestrial fungi (e.g., *Arthrinium, Aspergillus*) initially common on leaves were subsequently replaced by mangrove fungi.

Animal remains of mangroves also support a variety of fungal flora. Higher fungi have been recovered from a wide variety of animal substrata (corals; shells of snails, balanids, bivalves, foraminifers; hydrozoan exoskeletons; tunicates, shipworm tubes; cuttlefish endoskeletons; feathers; snake skin; crab exoskeletons; beetle wings; bryozoan skeleton; horse hair and teredinid tunnels (e.g., Kohlmeyer and Volkmann-Kohlmeyer, 1990, 1991, 1992; Rees and Jones, 1985; Rosello *et al.*, 1993). Up to 77% of the cuttlefish endoskeletons of the West Coast of India harboured marine fungi (Ananda *et al.*, 1998). Bivalve shells and cuttlefish endoskeletons showed abundant arenicolous fungi (Ananda *et al.*, 1998). *Corollospora intermedia* was the most frequent fungus particularly on calcareous shells and crab exoskeletons (Ananda and Sridhar, 2001b). Prolonged period of incubation was required for the animal remains of the beaches than mangroves to assess colonized fungi (Ananda and Sridhar, 2001b; Ananda *et al.*, 1998).

Litter Decomposition

Litter production in mangroves of the world ranges between 0.01 and 23.69 tons per hectare per annum (Abbey-Kalio, 1992; Bunt, 1995; Chale, 1996; Clarke, 1994; Ghosh *et al.*, 1990; Golley *et al.*, 1962; Gong *et al.*, 1984; Lu and Lin, 1990; Lu *et al.*, 1988; Mall *et al.*, 1991; Slim *et al.*, 1996; Wafer *et al.*, 1997; Woodroffe *et al.*, 1988). The decay coefficient (*k*) per day and half-life (t_{50}, days) of mangrove leaf litter decomposition ranges from 0.00234-0.0516 and 15-70, respectively (Ashton *et al.*, 1999; Boonruang, 1978; Dick and Steever, 2001; Fell *et al.*, 1975; Robertson, 1988; Steinke and Ward, 1987; Tam *et al.*, 1990, 1998). Makey and Smail (1996) showed a perfect linear relationship between mangrove leaf decomposition (t_{50}) versus latitude. Three important phases of leaf litter decomposition of *Rhizophora apiculata* was identified by Raghukumar *et al.* (1995) in Indian mangroves: (i) Rapid loss of detritus mass with reduction of organic constituents (proteins, carbohydrates, reducing sugars, phenolics and cellulose); (ii) Elevated fungal biomass and rapid decline of organic constituents (along with C/N ratio) to the lowest levels within three weeks; (iii) Decline in fungal and bacterial biomass after three and five weeks respectively followed by decline in cellulose and lignin.

Leaf litter of *Rhizophora mucronata* decomposed in the laboratory and field showed a decline in organic carbon and ash, while increase in total nitrogen, organic matter, protein and calorie (Sumitra *et al.*, 1980). The patterns of fungal colonization, mass loss and biochemical changes during the decomposition of dried and fresh leaves of *Rhizophora mucronata* at the mid-tide zone were studied in a southwest mangrove of Karnataka (Udyavara) up to 14 weeks (Ananda *et al.*, 2007). Ergosterol and nitrogen levels peaked between 4 and 8 weeks of exposure, subsequently ergosterol declined, while nitrogen was stable in dried leaves and declined in fresh leaves. The dynamics of leaf mass remaining for the first 8 weeks of exposure were described by a double-exponential decay model. Subsequently, the decay rate accelerated and the second phase was best described by a single exponential decay model. The leaf breakpoint coincided with an increase in the salinity of the mangrove swamp. Further study on the pattern of leaf degradation in mangroves need to be followed on introducing the leaf litter in different seasons.

The richness and diversity of mangrove fungi are dependent on chemical composition of wood, presence or absence of bark and period of incubation (Hyde and Lee, 1995; Jones, 2000; Nakagiri, 1993; Prasannarai and Sridhar, 1997). Maria *et al.* (2006) studied the breakdown of dead twigs of *Avicennia officinalis* and *R. mucronata* up to 18 months in a southern mangrove of Karnataka (Udyavara). The twig decay was slow during the first 10-12 months and accelerated subsequent six months. Nitrogen increased in both substrates with a peak around 120 days and declined thereafter. Phosphorus rapidly declined in the first two months and then gradually elevated. Fluctuations in phenolics of the twigs were best described by an exponential loss of function. Among water parameters (temperature, pH, oxygen and salinity), only salinity fluctuated between 1.1‰ during monsoon (due freshwater inflow) and 34.1‰ during dry season (due to decrease of freshwater inflow). Future wood decomposition studies should distinguish the pattern of wood decomposition on introduction at different seasons.

The crude lipid of *Rhizophora mucronata* leaves incorporated at 0.5, 1 and 2% (w/w) to a basal diet (protein, 42%; lipid, 10%; carbohydrates, 4%; vitamins and minerals, 2%) of prawn (*Penaeus indicus*), promoted the growth as well as efficiency of assimilation (Ramesh and Kathiresan, 1992). Prawns, crabs and fishes were attracted towards the decomposing leaves during the pre-monsoon and post-monsoon when senescent leaves of *Rhizophora apiculata* and *Avicinnia marina* were introduced (Rajendran and Kathiresan, 1999). Incorporation of leaf litter of *Acanthus ilicifolius, Avicennia officinalis* and *Rhizophora mucronata* along with rice bran increased the assimilation efficiency of *Metapenaeus manoceros* up to 86% (Ramadhas and Sumitra- Vijayaraghavan, 1979).

Endophytic Fungi

Endophytic fungi invade the tissues of living plants throughout or part of their life cycle and cause asymptomatic infections without disease symptoms (Wilson, 1995). Foliar endophytes of mangrove plant species of the East Coast of India have been extensively studied (Kumaresan and Suryanarayanan, 2001, 2002; Suryanarayanan *et al.,* 1998; Suryanarayanan and Kumaresan, 2000). Single species dominance in foliar endophytes was seen in many mangrove halophytes: *Avicennia marina* (*Phoma* sp.), *Bruguiera cylindrica* (*Colletotrichum gloeosporioides*), *Rhizophora apiculata* (*Sporormiella minima*), *Rhizophora mucronata* (*Sporormiella minima*) and *Suaeda maritima* (*Camarosporium palliatum*) (Kumaresan and Suryanarayanan, 2001; Suryanarayanan *et al.,* 1998; Suryanarayanan and Kumarsean, 2000). Root endophytes of mangrove plant species of the West Coast of India (*Avicennia officinalis, Rhozophora mucronata* and *Sonneratia caseolaris*) revealed multiple species dominance (Ananda and Sridhar, 2002). Multiple species dominance was also seen in the leaves of palm and deciduous trees (Fisher and Petrini, 1990; Frölich *et al*., 2000). *Acremonium* sp., *Alternaria* sp., *Cladosporium* sp., *Colletotrichum* sp. and *Fusarium* sp. were common foliar endophytes in mangrove plants (Kumaresan and Suryanarayanan, 2001, 2002; Suryanarayanan *et al.,* 1998; Suryanarayanan and Kumaresan, 2000), seagrass (Devarajan *et al.,* 2002) and beach halophyte (Fisher and Petrini, 1987).

Dominant endophytic fungi differed on host plants of mangrove community of the East Coast of India (Kumaresan and Suryanarayanan, 2001). Typical marine fungi were not dominant in roots of mangrove plant species (Ananda and Sridhar, 2002), mangrove associates (Maria and Sridhar, 2003b) and coastal sand dune halophytes (Beena *et al.,* 2000) and legumes (Seena and Sridhar, 2004) of the West Coast of India. *Colletotrichum* spp. is the most frequent endophyte in mangrove plant community of India (*Acanthus ilicifolius, Arthrocnemum indicum, Sesuvium portulacastrum, Avicennia marina, Bruguiera cylindrica, Ceriops decandra, Excoecaria agallocha* and *Lumnitzera racemosa*) (Kumaresan and Suryanarayanan, 2001; Suryanarayanan and Kumaresan, 2000) and tropical plant species of Hong Kong and Australia (Brown *et al.,* 1998). Both *Acremonium* and *Colletotrichum* were dominant in senescent standing herbaceous wood of *Acanthus ilicifolius* and they may switch over to saprophytic life style as endophytes in senescent mangrove leaves (Kumaresan and Suryanarayanan, 2002). *Phomopsis* spp. and *Phyllosticta* spp. were common foliar endophytes of mangrove plants (Kumaresan and Suryanarayanan, 2001, 2002; Suryanarayanan *et al.,* 1998; Suryanarayanan and Kumaresan, 2000). Similarly, *Phomopsis* spp. was also common root endophyte of *Avicennia officinalis* and *Rhizophora mucronata* of the West Coast mangrove (Ananda and Sridhar, 2002).

Twenty-five endophytic fungi comprising three ascomycetes, 20 anamorphs and two sterile fungi were recovered from two mangrove associates (*Acanthus ilicifolius* and *Acrostichum aureum*) of the West Coast mangrove (Maria and Sridhar, 2003b). Overall colonization by endophytes ranged between 74.5%

(*A. ilicifolius*) and 77.5% (*A. aureum*). Out of four tissues screened, species richness and diversity were high in stems of *A. ilicifolius* and roots of *A. aureum. Colletotrichum* sp. was the most dominant endophyte in prop roots of *A. ilicifolius* and yeast sp. in rhizomes of *A. aureum. Acanthus ilicifolius* showed single species dominance by *Colletotrichum* sp., while multiple species dominance in *A. aureum* (*Acremonium* sp., *Penicillium* sp. and yeast). *Cumulospora marina* was the only typical marine anamorphic fungus recovered from the roots of *A. ilicifolius*.

Terrestrial Fungi

Terrestrial fungi are common in waters, mud (Chowdhery *et al.*, 1982; Garg, 1983), leaves (Raghukumar *et al.*, 1995), wood (Sridhar and Maria, 2006), seagrasses (Devarajan *et al.*, 2002), animal remains (Ananda and Sridhar, 2001b) and rhizosphere (Nair *et al.*, 1991; Salique *et al.*, 1985; Venkatesan and Natarajan, 1985) of Indian mangroves. Terrestrial fungi have been isolated from mangrove leaves (Raghukumar *et al.*, 1995) and rhizosphere (*Avicennia officinalis* and *Rhizophora mucronata*). Pichavaram mangrove consists of several terrestrial fungi (Venkatesan and Natarajan, 1985; Salique *et al.*, 1985). Mangrove mud, rhizosphere, rhizoplane and non-rhizosphere zones of Sunderban mangrove swamps also yielded many terrestrial fungi (Garg, 1983; Chowdhary *et al.*, 1982). Rhizosphere and non-rhizosphere soils (*Avicennia officinalis*) of Maharashtra yielded several terrestrial fungi (Nair *et al.*, 1991). Mangrove mud at Vellar Estuary, Cochin backwaters and Andaman Nicobar Islands consists of terrestrial fungi (Misra, 1986; Prabhakaran and Gupta, 1990; Silique *et al.*, 1985). In Cochin backwaters mangrove, leaves, stem, roots and pneumatophores of *Avicennia* and *Acanthus* yielded terrestrial fungi along with mangrove fungi.

Chandralata (1999) and Raghukumar and Raghukumar (1998) interpreted that adaptation and activity of terrestrial fungi under mangrove ecosystem refers to facultative or indwellers or residents. Seasonal sampling of the leaf litter from the Nethravathi estuary also revealed the occurrence of many freshwater hyphomycetes and a typical mangrove fungus, *Clavatospora bulbosa* (Sridhar and Kaveriappa, 1988). It is clear that terrestrial and freshwater fungi involve in plant detritus degradation in mangrove habitats under certain conditions and seasons. For instance, leaf, bark and wood decay in mangrove canopies or decomposition of pneumatophores, prop roots and twigs on mangrove floor during low saline (or rainy season) or air-water interphase by terrestrial fungi cannot be ruled out. Similarly, aquatic hyphomycetes will be active during low saline conditions of mangroves and backwaters (Sridhar and Kaveriappa, 1988).

Metabolites

Mangrove fungi proved to be important sources of new bioactive compounds including enzymes (Grant *et al.*, 1996; Pointing and Hyde, 2000; Pointing *et al.*, 1998). Pointing *et al.* (1998) showed the cellulolytic activity of marine fungi throughout the salinity range (0-34‰). For additional information on marine fungal metabolites and techniques, see Chapter 4. A few studies have been carried out on the enzymatic capabilities of mangrove fungi in India. Extracellular enzyme potential of foliar endophytic fungi of *Rhizophora apiculata* has been studied by Kumaresan and Suryanarayan (2002) and linked their involvement in litter degradation during senescence and death. Several mangrove fungi showed laccase-positive activities (Kumaresan and Suryanarayanan, 2002, Raghukumar *et al.*, 1994, 1999). Cellulase-free xylanase or xylanase with low cellulase activity are effective in bleaching paper pulp, biopolishing of textiles and application of such enzymes considerably reduce the use of chlorine compounds for bleaching purposes. *Flavodon flavus* and *Podospora anserina* of marine origin produced many enzymes those modify lignin (Raghukumar *et al.*, 1999). *Flavodon flavus* decolourizes and degrades a variety of polymeric dyes. Such decolorization of dyes even in the presence of artificial seawater is advantageous since industrial effluents usually contain high quantity of salts (e.g., chlorides and sulphates). Seven endophytic fungi were assessed for extracellular enzymes (amylase, cellulase, chitinase, laccase, lipase, protease and tyrosinase) by culture plate method (Maria *et al.*, 2005). Cellulase and lipase activity was seen in all fungi, amylase and protease activity was confined to a few, while they were devoid of chitinase, laccase and tyrosinase activity. Enzyme production by *Pestalotiopsis* sp. (cellulase by submerged fermentation; xylanase, pectinase and protease by solid substrate fermentation) at pH 7 and pH 9 during 3-15 days of fermentation was assessed. The cellulase activity was low at 9th day at pH 9, while xylanase was highest reveals the potentiality of endophytic *Pestalotiopsis* sp. in the production of high xylanase and low cellulase. Lignocellulolytic activities of mangrove fungi and their ability to degrade the colored paper and pulp industrial effluents have been reviewed by Chandralata (2005). Amount and type of lignin degrading enzymes production by a mangrove basidiomycete (NIOCC #2a) was dependent on the type of nitrogen source used (D'Wouza-Ticlo *et al.*, 2006). The amount of extracellular peroxidases increased many folds in the presence of industrial effluents, while absence of the effluent resulted in negligible production of enzyme. A few new isozymes of laccase were induced in the presence of industrial effluents. Efficiency of decolorization of effluents by the concentrated culture filtrate obtained from the media containing different nitrogen sources proved the importance of the type of nitrogen milieu in decolorization of coloured effluents of industries.

Antimicrobial potential of 14 endophytic fungi isolated from *Acanthus ilicifolius* and *Acrostichum aureum* of Nethravathi mangrove cultivated through solid substrate fermentation were tested against selected bacteria and fungi (Maria *et al.*, 2005). All test bacteria were inhibited by sterile isolate MSI 1 as well as *Aspergillus* spp. *Cumulospora marina* and *Pestalotiopsis* sp. showed considerable antagonistic effect on Gram-positive as well as Gram-negative bacteria. The crude ethyl acetate extracts of four endophytes (grown under submerged fermentation) partially purified by TLC revealed several fluorescent fractions. Two such fractions of *Aspergillus* sp. showed high antimicrobial activity. One of the fractions of *Pestalotiopsis* sp. exhibited high inhibitory activity against *Bacillus subtilis, Staphylococcus aureus* and *Candida albicans*.

Conclusions and Outlook

Conservative estimate of the world's fungal resource based on plant/fungus ratio is about 1.5 million species (Hawksworth, 1991). In temperate locations, plant/fungus ratio has been predicted as 1:6, while 1:26 in tropics (Hyde, 1996). The ratio in tropics has been updated as 1:33 based palm fungi in Australia and Brunei Darussalam (Fröhlich and Hyde, 1999). Over three decades of studies on mangrove/marine fungi in India confined mainly to floristic with a few studies on application of these fungi. Although east coast of India comprises of 70% of mangrove cover of the Indian Subcontinent, mycological studies on these mangroves are inadequate. Several cryptic species might have adapted to the local habitats of mangroves. As the mangrove flora of Indian Subcontinent is diverse, intense survey may reveal many cryptic species. In addition to leaves and wood, sediments and animal remains need to be explored for fungi. Duration of incubation of mangrove substrata and intervals of screening in the laboratory has to be considered as an important aspect of diversity studies. Usually, incubation of leaf and woody litter up to eight weeks results in dominance of terrestrial fungi, sporulating marine fungi reaches a peak at about 16 weeks, while arenicolous (sand-inhabiting) fungi appear after 16 weeks (Ananda and Sridhar, 2004). If the interval of observation is too long, there are chances to miss many sporulating anamorphic taxa. Overlap between terrestrial, freshwater and marine fungi is common in mangroves. As many so-called terrestrial fungi isolated from mangrove habitats are important source of metabolites, their documentation and exploitation should not be ignored. In the new millennium, studies on mangrove fungal diversity should accompany with their exploitation for industrial and health applications.

If the mangrove host plant is endemic, fungi associated may also have a restricted distribution (e.g., endophytic fungi). Loss of such plant species results in total elimination of host-specific fungi from the ecosystem. For instance, *Kandelia candel* (Rhizophoraceae) has been recorded only in two locations of the West Coast, while *Heritiera fomes* (Sterculiaceae) and *Nypa fruticans* (Palmaceae) in one location of the East Coast of India (Blasco and Aizpuru, 1997; Rao and Suresh, 2001; Sarma, 2007). As mangroves are highly threatened by human interference, endemic plants needs conservation and special attention should be focused on mycological survey. Many saprophytic fungi carry out different functions simultaneously and attained important position in the food web possibly serve as keystone species (Hawksworth, 1991). For example, some marine/mangrove fungi also lead endophytic life with mangrove or coastal sand dune plant species (Ananda and Sridhar, 2002; Beena *et al.*, 2000; Seena and Sridhar, 2004). Possibly these fungi are important in defence or nutrition of mangrove or coastal sand dune flora. A large number of endophytic fungi of mangrove plant speciess did not sporulate on the semi defined media (Ananda and Sridhar, 2002; Kumaresan and Suryanarayanan, 2001; Suryanarayanan *et al.*, 1998). Such fungi should not be ignored to test for bioactive metabolites. Mangrove fungi are usually alkaliphilic and possess several metabolites of industrial and environmental importance. The lignin-degrading enzymes of mangroves will be immense value in bioremediation of industrial wastewaters or effluents with high chlorides and sulphides or dye-containing effluents (Chandralata, 2005) or effluents with heavy metals. There is vast scope to immobilize mangrove fungal enzymes or fungal mycelia to enhance their bioremediation potential. Challenging future is ahead in exploration of mangrove fungi of the Indian Subcontinent for nutrition, health and environmental abatement.

References

Abbey-Kalio, N.J. 1992. A pilot study of mangrove litter production in the Bonny Estuary of southern Nigeria. *Discovery and Innovation 4*, 71-78.

Ananda, K. and Sridhar, K.R. 2001a. *Aniptodera indica*, a new species of mangrove inhabiting ascomycete from west coast of India. *J. Environ. Biol. 22*, 283-286.

Ananda, K. and Sridhar, K.R. 2001b. Mycoflora on dead animal materials of mangrove habitats of Karnataka Coast, India. *Sri Lanka J. Aquatic Sci. 6*, 85-93.

Ananda, K. and Sridhar, K.R. 2002. Diversity of endophytic fungi in the roots of mangrove species on west coast of India. *Can. J. Microbiol. 48*, 871-878.

Ananda, K. and Sridhar, K.R. 2003. Filamentous fungal assemblage of two island mangroves of India. In: *Prospects and Problems of Environment across the Millennium* (Ed. Madhyastha, M.N., Sridhar, K.R. and Lakshmi, A.). Daya Publishing House, Delhi, India, India, 35-44.

Ananda, K. and Sridhar, K.R. 2004. Diversity of filamentous fungi on decomposing leaf and woody litter of mangrove forests of southwest coast of India. *Current Science 87*, 1431-1437.

Ananda, K., Prasannarai, K. and Sridhar, K.R. 1998. Occurrence of higher marine fungi on animal substrates along the west coast of India. *Indian J. Mar. Sci. 27*, 233-236.

Ananda K., Sridhar K.R., Raviraja N.S. and Baerlocher, F. 2008. Breakdown of fresh and dried *Rhizophora mucronata* leaves in a mangrove of Southwest India. *Wetlands Ecol. Managem. 16,* 1-9.

Ashton, E.C., Hogarth, P.J. and Ormond, R. 1999: Breakdown of mangrove leaf litter in a managed mangrove forest in peninsular Malaysia. *Hydrobiol. 413*, 77-88.

Bandaranayake, W.M. 1995. Survey of mangrove plants from Northern Australia for phytochemical constituents and UV-absorbing compounds. *Current Topics in Phytochem. 14*, 69-78.

Bandaranayake, W.M. 1998. Traditional and medicinal uses of mangroves. *Mangroves and Salt Marshes 2*, 133-148.

Barghoorn, E.S. and Linder, D.H. 1944. Marine fungi: Their taxonomy and biology. *Farlowia 1*: 395-467.

Beena, K.R., Ananda, K. and Sridhar, K.R. 2000. Fungal endophytes of three sand dune plant species of west coast of India. *Sydowia 52*, 1-9.

Blasco, F. and Aizpuru, M., 1997. Classification and evolution of the mangroves of India. *Tropical Ecol. 38*, 357-374.

Boonruang, P. 1978. The degradation rates of mangrove leaves of *Rhizophora apiculata* (Bl.) and *Avicennia marina* (Forsk.) Vierh. at Phuket Island, Thailand. *Research Bulletin # 26*, Phuket Marine Biological Centre, Ministry of Agriculture and Cooperatives, Thailand.

Borse, B.D. 1984. Marine fungi from India I. *Indian Bot. Reptr. 3*, 156-157.

Borse, B.D. 1987a. Marine fungi from India IV. *Trans. Mycol. Soc. Japan 28*, 55-61.

Borse, B.D. 1987b. Marine fungi from India V. *Current Science 56*, 1109-1111.

Borse, B.D. 1987c. Marine fungi from India VI. *Indian Bot. Reptr. 6*, 108-110.

Borse, B.D. 1987d. Marine fungi from India VII. *Indian Bot. Reptr. 6*, 62-64.

Borse, B.D. 1987e. New species of *Aigialus* from India. *Trans. Br. Mycol. Soc. 88*, 424-426.

Borse, B.D. 1988. Frequency of occurrence of marine fungi from Maharashtra coast, India. *Indian J. Mar. Sci. 17*, 165-167.

Borse, B.D. and Hyde, K.D. 1989. Marine fungi from India III. *Acrocordiopsis patilii* gen. *et* sp. nov. from mangrove wood. *Mycotaxon 30*, 535-540.

Borse, B.D. and Shrivastava, A.D. 1988. Marine fungi from India VIII. *Indian Bot. Reptr. 7*, 26-30.

Borse, B.D. and Shrivastava, A.D. 1994. Marine fungi from India IX: Loculoascomycetes. *Indian Bot. Reptr. 7*, 18-25.

Borse, B.D., Ramesh, Ch. and Shrivastava, A.D. 1988. Marine Fungi from Maharashtra (India). *Indian Bot. Reptr. 7*, 18-25.

Borse, B.D., Kelkar, D.J. and Patil, A.C. 2000. Frequency of occurrence of marine fungi from Pirotan Island (Gujarat), India. *Geobios 27*, 145-148.

Brown, K.B., Hyde, K.D. and Guest, D.I. 1998. Preliminary studies on endophytic fungal communities of *Musa acuminata* species complex in Hong Kong and Australia. *Fungal Diversity 1*, 27-51.

Bunt, J.S. 1995. Continental scale patterns in mangrove litter fall. *Hydrobiol. 259*, 135-140.

Chale, F.M.M. 1996. Litter production in an *Avicennia germinans* (L.) stern forest in Guyana, South America. *Hydrobiol. 330*, 47-53.

Chandralata, R. 1999. What are geofungi doing in the marine environment? *Asian Mycol. Congress*. Chennai, India, 12.

Chandralata, R. 2005. Marine Fungi and their Enzymes for Decolorization of Colored Effluents. In: *Marine Microbiology: Facets and Opportunities* (Ed. Ramaiah, N.). National Institute of Oceanography, Goa, India, 145-158.

Chapman, V.J. 1976. *Mangrove Vegetation*. Cramer Vadyz, Liechtenstein, Germany.

Chinnaraj, S. 1992. Higher marine fungi of Lakshadweep Islands and a note on *Quintaria lignatilis*. *Cryptogamie, Mycol. 13*, 312-319.

Chinnaraj, S. 1993. Higher marine fungi from mangroves of Andaman and Nicobar Islands. *Sydowia 45*, 109-115.

Chinnaraj, S. and Untawale, A. G. 1992. Manglicolous fungi from India. *Mahasagar 25*, 25-29.

Chowdhery, H.J., Garg, K.L. and Jaitly, A.K. 1982. Occurrence of fungi in rhizosphere, rhizosplane and non-rhizosphere zones of some mangroves. *Indian J. Mar. Sci. 11*, 138-142.

Clarke, P.J. 1994. Baseline studies of temperate mangrove growth and reproduction: Demographic and litterfall measures of leafing and flowering. *Aust. J. Bot. 42*, 37-48.

Cribb, A.B. and Cribb, J.W. 1955. Marine fungi from Queensland-I. *Univ. Queensl. Pap., Dep. Bot. 3*, 77-81.

Devarajan, P.T., Suryanarayanan, T.S. and Geetha, V. 2002. Endophytic fungi associated with the tropical seagrass *Halophila ovalis* (Hydrocharitaceae). *Indian J. Mar. Sci. 31*, 73-74.

Dick, T.M. and Steever, W.J. 2001. Decomposition of *Avicennia marina* on an iron-smelting slag substrate. *Austral Ecol. 26*, 127-131.

D'Wouza-Ticlo, D., Verma, A.K., Mathew, M. and Raghukumar, C. 2006. Effect of nutrient on laccace production, its isozymne pattern and effluent decolorization by the fungus NIOCC #2a isolated from mangrove wood. *Indian J. Mar. Sci. 35,* 364-372.

Fell, J.W., Cefalu, R.C., Master, I.M., Tallman, A.S. 1975. Microbial activities in the mangrove (*Rhizophora mangle*) leaf detrital system. In: *Proceedings of International Symposium in Biology and Management of Mangroves* (Ed. Walsh, G.E., Snedaker, S.C. and Teas, H.J.). University of Florida, Gainesville, Florida, 661-667.

Field, C. 1995. *Journey amongst Mangroves*. International Society for Mangrove Ecosystems, Okinawa, Japan, South China Printing Co., Hong Kong.

Fisher, P.J. and Petrini, O. 1987. Location of fungal endophytes in tissues of *Suaeda fruticosa*: a preliminary study. *Trans. Br. Mycol. Soc. 89*, 246-249.

Fröhlich, J. and Hyde, K.D. 1999. Biodiversity of palm of fungi in the tropics: are global fungal diversity estimates realistic? *Biodivers. Conser. 8*, 977-1004.

Fröhlich, J., Hyde, K.D. and Petrini, O. 2000. Endophytic fungi associated with palms. *Mycol. Res. 104*, 1202-1212.

Garg, K.L. 1983. Vertical distribution of fungi in Sunderban Mangrove mud. *Indian J. Mar. Sci. 12*, 48-51.

Ghosh, P.B., Singh, B.N., Chakrabarty, C., Saha, A., Das, R.L. and Choudhury, A. 1990. Mangrove litter production in a tidal creek of Lothian Island of Sunderbans, India. *Indian J. Mar. Sci. 19*, 292-293.

Golley, F.B., Odum, H.T. and Wilson, R.F. 1962. The structure and metabolism of a Puerto Rican red mangrove forest in May. *Ecology 43*, 9-19.

Gong, W.K., Ong, J.E., Wong, C.H. and Dhanarajan, G. 1984. Productivity of mangrove trees and its significance in managed mangrove ecosystems in Malaysia. In. *Proceedings of the Asian Synmposium on Mangrove Environment: Research and Management* (Ed. Seopadmo, E., Rao, A.N. and Macintosh, D.J.), University of Malaysia and UNESCO, 216-225.

Grant, W.D., Atkinson, M., Burke, B. and Molloy, C. 1996. Chitinolysis by the marine ascomycete *Corollospora maritima* Werdermann: Purification and properties of a chitobiosidase. *Bot. Mar. 39*, 177-186.

Hawksworth, D.L. 1991. The fungal dimension of biodiversity: Magnitude, significance and conservation. *Mycol. Res. 95*, 641-655.

Hughes, G.C. 1975. Studies of fungi in oceans and estuaries since 1961. I. Lignicolous, caulicolous and foliicolous species. *Oceanogr. Mar. Biol. Ann. Rev. 13*, 69-180.

Hyde, K.E. 1988. Observations on the vetical distribution of marine fungi on *Rhizophora* spp., at Kampong Danau mangrove, Brunei. *Asian Mar. Biol. 5*, 77-81.

Hyde, K.D. 1990. A study of the vertical zonation of intertidal fungi on *Rhizophora apiculata* at Kampong Kapok mangrove, Brunei. *Aquat. Bot. 36*, 255-262.

Hyde, K.D. 1992. *Julella aviceniae* (Borse) comb. nov. (Thelenellaceae) from intertidal mangrove wood and miscellaneous fungi from the NE coast of Queensland. *Mycol. Res. 96*, 939-942.

Hyde, K.D. 1996. Biodiversity of microfungi in North Queensland. *Aust. System. Bot. 9*, 261-271.

Hyde, K.D. and Borse, B.D. 1986a. Marine fungi from Seychelles V. *Biatriospora marina* gen *et* sp. nov. from mangrove wood. *Mycotaxon 26*, 263-270.

Hyde, K.D. and Borse, B.D. 1986b. Marine fungi from Seychelles VI. *Massarina velatospora,* a new marine Ascomycotina from mangrove wood. *Mycotaxon 27*, 161-167.

Hyde, K.D. and Lee, S.Y. 1995. Ecology of mangrove fungi and their role in the nutrient cycling: what gaps occur in our knowledge? *Hydrobiol. 295*, 107-118.

Hyde, K.D. and Pointing, S.B. 2000. *Marine Mycology - A Practical Approach.* Fungal Diversity Press, Hong Kong.

Hyde, K.D. and Sarma, V.V. 2000. Pictorial key to higher marine fungi. In: *Marine Mycology – A Practical Approach.* Fungal Diversity Press, Hong Kong, 201-264.

Hyde, K.D., Vrijmoed, L.L.P, Chinnaraj, S. and Jones, E.B.G. 1992. *Massarina armatispora* sp. nov., a new intertidal Ascomycete from mangroves. *Bot. Mar. 35*, 325-328.

Hyde, K.D., Chalermongse, A. and Boonthavikoon, T. 1993. The distribution of intertidal fungi on *Rhizophora apiculata.* In: *The Marine Biology of South China Sea.* (Ed. Morton, J.B.). Hong Kong University Press, Hong Kong, 643-652.

Johnson, T.W. and Sparrow, F.K. 1961. *Fungi in Oceans and Estuaries.* J. Cramer, Weinheim, Germany.

Jones, E.B.G. 2000. Marine fungi: some factors influencing biodiversity. *Fungal Diversity 4*, 53-73.

Kathiresan, K. and Bingham, B.L. 2001. Biology of mangrove ecosystem. *Adv. Mar. Biol. 40*, 81-251.

Kohlmeyer, J. 1984. Tropical marine fungi. *P.S.Z.N.I. Marine Ecology 5*, 329-378.

Kohlmeyer, J. and Kohlmeyer, E. 1979. *Marine Mycology: The Higher Fungi.* Academic Press, New York.

Kohlmeyer, J. and Vittal, B.P.R. 1986. *Lophiostoma mangrovei*, a new marine ascomycete from the tropics. *Mycologia 78*, 485-489.

Kohlmeyer, J. and Volkmann-Kohlmeyer, B. 1990. New species of *Koralionastes* (Ascomycotina) from the Caribbean and Australia. *Can. J. Bot. 68*, 1554-1559.

Kohlmeyer, J. and Volkmann-Kohlmeyer, B. 1991. Illustrated Key to the filamentous higher marine fungi. *Bot. Mar. 34*, 1-61.

Kohlmeyer, J. and Volkmann-Kohlmeyer, B. 1992. Two Ascomycotina from coral reefs in the Caribbean and Australia. *Cryptogamic Bot. 2*, 367-374.

Krishnamurthy, K., Choudhury, A. and Untawale, A.G. 1987. *Status Report – Mangroves in India.* Ministry of Environment and Forests, Government of India, New Delhi.

Kumaresan, V. and Suryanarayanan, T.S. 2001. Occurrence and distribution of endophytic fungi in a mangrove community. *Mycol. Res. 105*, 1388-1391.

Kumaresan, V. and Suryanarayanan, T.S. 2002. Endophyte assemblage in young, mature and senescent leaves of *Rhizophora apiculata*: Evidence for the role of endophytes in mangrove litter degradation. *Fungal Diversity 9*, 81-91.

Lu, C.Y. and Lin, P. 1990. Studies on litter fall and decomposition of *Bruguiera sexangular* community in Hainan Island, China. *Bull. Mar. Sci. 47*, 139-148.

Lu, C.Y., Wong, Y.S., Tam, N.F.Y., Ye, Y., Chi, S.H. and Lin, P. 1988. Preliminary studies on methane fluxes in Hainan mangrove communities. *Chinese J. Ocean. Limnol. 16*, 64-71.

Makey, A.P. and Smail, G. 1996. The decomposition of mangrove litter in a subtropical Mangrove forest. *Hydrobiol. 332*, 93-98.

Mall, L.P., Singh, V.P. and Garge, A. 1991. Study of biomass, litter fall, litter forests of Andaman Islands (India). *Trop. Ecol. 32*, 144-152.

Maria, G.L. and Sridhar, K.R. 2002a. A new ascomycete *Passeriniella mangrovei* sp. nov. from the mangrove forest of India. *Indian J. Forestry 25*, 319-322.

Maria, G.L. and Sridhar, K.R. 2002b. Richness and diversity of filamentous fungi on woody litter of mangroves along the west coast of India. *Current Science 83*, 1573-1580.

Maria, G.L. and Sridhar, K.R. 2003a. Diversity of filamentous fungi on woody litter of five mangrove plant species from the southwest coast of India. *Fungal Diversity 14*, 109-126.

Maria, G.L. and Sridhar, K.R. 2003b. Endophytic fungal assemblage of two halophytes from west coast mangrove habitats, India. *Czech Mycol. 55*, 241-251.

Maria, G.L. and Sridhar, K.R. 2004. Fungal colonization of immersed wood in mangroves of the southwest coast of India. *Can. J. Bot. 82*, 1409-1418.

Maria, G.L., Sridhar, K.R. and Raviraja, N.S. 2005. Antimicrobial and enzyme activity of mangrove endophytic fungi of southwest coast of India. *J. Agric. Technol. 1*, 67-80.

Maria, G.L., Sridhar, K.R. and Baerlocher, F. 2006 Decomposition of dead twigs of *Avicennia officinalis* and *Rhizophora mucronata* in a mangrove in southwest India. *Bot. Mar. 49,* 450-455.

Misra, J.K. 1986. Fungi from mangrove muds of Andaman Nicobar Islands. *Indian J. Mar. Sci. 15*, 185-186.

Nair, L.N., Rao, V.P. and Chaudhuri, S. 1991. Micorflora of *Avicennia officinalis* Linn. In: *Proceedings of the Symposium on Significance of Mangroves* (Ed. Agate, A.D., Bonde, S.D. and Kumaran, K.P.N.). Pune, India, 52-55.

Nakagiri, A. 1993. Intertidal fungi from Iriomote Island. *Institute of Fermentation of Osaka Research Communications (Osaka) 16*, 24-62.

Natarajan, R. 1998. The mangroves: A natural renewable economic resource needs conservation. In: *An Anthology of Indian Mangroves.* ENVIS Centre, Annamalai University, Parangipettai, India, 1-6.

Newell, S.Y. 1996. Establishment and potential impacts of eukaryotic mycelial decomposers in marine/terrestrial ecotones. *J. Exp. Mar. Biol. Ecol. 200*, 187-206.

Newell, S.Y. and Fell, J.W. 1997. Competition among mangrove oomycetes and between oomycees and other microbes. *Aquatic Microb. Ecol. 12*, 21-28.

Patil, S.D. and Borse, B.D. 1983. Marine fungi from Maharashtra (India) III. Some fungi from mangroves. *Indian Bot. Reptr. 2,* 56-58.

Patil, S.D. and Borse, B.D. 1985a. Marine fungi from Indian mangroves. In: *The Mangroves: Biology, Utilization and Conservation of Mangroves* (Ed. Bhosale, L.J.). Maharashtra, India, 151-152.

Patil, S.D. and Borse, B.D. 1985b. Marine fungi from Maharashtra (India) IV: Some Loculoascomycetes. *Trans. Mycol. Soc. Japan 26*, 271-276.

Patil, K.B. and Borse, B.D. 2001. Studies on higher marine fungi from Gujarat coast (India). *Geobios 28*, 41-44.

Pointing, S.B. and Hyde, K.D. 2000. Lignocellose-degrading marine fungi. *Biofouling 15*, 221-229.

Pointing, S.B., Vrijmoed, L.L.P. and Jones, E.B.G. 1998. A qualitative assessment of lignocellulose degrading enzyme activity in marine fungi. *Bot. Mar. 41*, 293-298.

Prabhakaran, N. and Gupta, R. 1990. Activity of soil fungi of Mangalvan, the mangrove ecosystem of Cochn backwater. *Fishery Technol. 27*, 157-159.

Prasannarai, K. and Sridhar, K.R. 1997. Effect of incubation period of driftwood on the occurrence of marine fungi. *Indian J. Mar. Sci. 26*, 380-382.

Qasim, S.Z. and Wafar, M.V.M. 1990. Marine resources in the tropics. *Resource Management and Optimization 7*, 141-169.

Raghukumar, S. 1973. Marine lignicolous fungi from India. *Kavaka 1*, 73-85.

Raghukumar, C. and Raghukumar, S. 1998. Barotolerance of fungi isolated from deep-sea sediments of the Indian Ocean. *Aquatic Microb. Ecol. 15*, 153-163.

Raghukumar, S., Zainal, A. and Jones, E.B.G. 1988. *Cirrenalia basiminuta*: A new lignicolous marine Deuteromycete from the tropics. *Mycotaxon 31*, 163-170.

Raghukumar, S., Sathe-Pathak, V., Sharma, S. and Raghukumar, C. 1995. Thraustochytrid and fungal component of marine detritus III. Field studies on decomposition of leaves of the mangrove *Rhizophora apiculata. Aquat. Microb. Ecol. 9*, 117-125.

Raghukumar, C., D'Souza, T.M., Thorn, R.G. and Reddy, C.A. 1999. Lignin-modifying enzymes of *Flavodon flavus*, a basidiomycete isolated from a coastal marine environment. *App. Environ. Microbiol. 65*, 2103-2111.

Rajendran, N. and Kathiresan, K. 1999. Do decomposing leaves of mangroves attract fishes? *Current Science 77*, 972-976.

Ramadhas, V. and Sumitra-Vijayaraghavan, 1979. Efficiency of energy utilization in the shrimp, *Metapenaeus monoceros* fed mangrove leaves. *Indian J. Mar. Sci. 8*, 114-115.

Ramesh, M.X. and Kathiresan, K. 1992. Mangrove cholesterol in the diet of panaeid prawn *Penaeus indicus. Indian J. Mar. Sci. 21*, 164-166.

Rao, T.A. and Suresh, P.V. 2001. *Coastal Ecosystems of the Karnataka State, India – 1. Mangroves.* Karnataka Association for the Advancement of Science, Bangalore, India.

Ravikumar, D.R. and Purushothaman, A. 1988a. *Cirrenalia tropicalis*, an addition to Indian marine fungi. *Current Science 57*, 674.

Ravikumar, D.R. and Purushothaman, A. 1988b. *Corollospora intermedia*, a lignicolous marine fungus from India. *Current Science 57*, 898-899.

Ravikumar, D.R. and Vittal, B.P.R. 1987. Studies on mangrove fungi of India I. Note on some Ascomycetes. *Kavaka 15*, 99-103.

Ravikumar, D.R. and Vittal, B.P.R. 1991. A new species of *Bathyascus* from Indian mangroves. *Mycol. Res. 95*, 370-371.

Ravikumar, D.R. and Vittal, B.P.R. 1996. Fungal diversity on decomposing biomass of mangrove plant *Rhizophora* in Pichavaram estuary, east coast of India. *Indian J. Mar. Sci. 25*, 142-144.

Rees, G. and Jones, E.B.G. 1985. The fungi of coastal sand dune system. *Bot. Mar. 28*, 213-220.

Robertson, A.I. 1988. Decomposition of mangrove leaf litter in Tropical Australia. *J. Exp. Mar. Biol. Ecol, 116*, 235-247.

Rosello, M.A., Descals, E. and Cabrer, B. 1993. *Nia epidermoidea*, a new marine gasteromycete. *Mycol. Res. 97*, 68-70.

Salique, S.M., Krishnamurthy, K. and Palaniappan, R. 1985. Ecology of fungal population in the Vellar Estuary and Pichavaram Mangroves, South India. In: *The Mangroves: Biology, Utilization and Conservation of Mangroves* (Ed. Bhosale, L.J.). Maharashtra, India, 153-162.

Sarma, V.V. 2007. *Nypa fruticans* needs more conservation efforts. *Current Science 92*, 270.

Sarma, V.V. and Hyde, K.D. 2001. A review on frequently occurring fungi in mangroves. *Fungal Diversity 8*, 1-34.

Sarma, V.V. and Hyde, K.D. 2000. *Tirispora mandoviana* sp. nov. from Chorao mangroves, Goa, the west coast of India. *Aust. Mycol. 19*, 52-56.

Sarma, V.V. and Vittal, B.P.R. 1998-1999. Studies on the ecology of manglicolous fungi from east coast of India, observations on the seasonal occurrence. *Kavaka* 26-27, 105-120.

Sarma, V.V. and Vittal, B.P.R. 2000. Biodiversity of mangrove fungi on different substrata of *Rhizophora apiculata* and *Avicennia* spp. from Godavari and Krishna deltas, east coast of India. *Fungal Diversity 5*, 23-41.

Sarma, V.V. and Vittal, B.P.R. 2001. Biodiversity of manglicolous fungi on selected plants in the Godavari and Krishna deltas, east coast of India. *Fungal Diversity 6*, 115-130.

Sarma, V.V. and Vittal, B.P.R. 2002. A preliminary study on verticaldistributionof manglicolous fungi on prop roots of *Rhizophora apilulata* Blume at Krishan delta, eas coast of India. *Kavaka 30*, 21-29.

Sarma, V.V. and Vittal, B.P.R. 2004. Manglicolous fungi recorded from Godavary and Krishna deltas, A.P., East coast of India along with a dichotomous key and notes on some taxa. *Kavaka 32*, 65-111.

Sarma, V.V., Raghukumar, S., Hyde, K.D., Vittal, B.P.R. and Chandramohan, D. (2000). A CD-ROM for documentation and identification of mangrove fungi. *Fungal Diversity 5*, 195-200.

Schmidt, J.P. and Shearer, C.A. 2003. A checklist of mangrove-associated fungi, their geographical distribution and known host plants. *Mycotaxon 53*, 423-477.

Seena, S. and Sridhar, K.R. 2004. Endophytic fungal diversity of 2 sand dune wild legumes from the southwest coast of India. *Can. J. Microbiol. 50*, 1015-1021.

Shearer, C.A., Descals, E., Kohlmeyer, B., Kohlmeyer, J., Marvanova, L. Padgett, D.E. Porter, D., Raja, H.A., Schmit, J.P., Thorton, H.A. and Voglymayr, H. 2007. Fungal biodiversity in aquatic habitats. *Biodivers. Conserv. 16*, 49-67.

Shrivastava, A.D. 1989. Marine fungi from Bombay III. Some loculoascomycetes. *Indian Bot. Reptr. 8*, 16-19.

Shrivastava, A.D. 1994. Marine fungi from Bombay IV. *Indian Bot. Reptr. 13*, 92-93.

Slim, F.J., Gwada, P.M., Kodjo, M. and Hemminga, M.A. 1996. Biomass and litterfall of *Ceriops tagal* and *Rhizophora mucronata* in the mangrove forest of Gazi Bay, Kenya. *Mar. Freshwat. Res. 47*, 999-1007.

Spalding, M., Blasco, F. and Field, C. 1997. *World Mangrove Atlas*. Cambridge Samara Pub. Co., Cambridge, UK.

Sridhar, K.R. and Kaveriappa, K.M. 1988. Occurrence and survival of aquatic hyphomycetes in brackish and sea water. *Arch. Hydrobiol. 113*, 153-160.

Sridhar, K.R. and Mraia, G.L. 2006. Fungal diversity on woody litter of *Rhizophora mucronata* in a southwest Indian mangrove. *Indian J. Mar. Sci. 35,* 318-325.

Steinke, T.D. and Ward, C.J. 1987. Degradation of mangrove leaf litter in the St. Lucia Estuary as influenced by season and exposure. *S. Afr. J. Bot. 53*, 323-328.

Sumitra-Vijayaraghavan, Ramadhas, V., Kumari, K.L. and Royan, J.P. 1980. Biochemical changes and energy content of the mangrove, *Rhizophora mucronata* leaves during decomposition. *Indian J. Mar. Sci. 9*, 120-123.

Suryanarayanan, T.S. and Kumaresan, V. 2000. Endophytic fungi of some halophytes from an estuarine mangrove forest. *Mycol. Res. 104*, 1465-1467.

Suryanarayanan, T.S., Kumaresan, V. and Johnson, J.A. 1998. Foliar fungal endophytes from two species of the mangrove *Rhizophora. Can. J. Microbiol. 44*, 1003-1006.

Tam, N.F.Y., Vrijmoed L.L.P. and Wong Y.S. 1990. Nutrient dynamics associated with leaf decomposition in a small subtropical mangrove community in Hong Kong. *Bull. Marine Sci. 47*, 68-78.

Tam, N.F.Y., Wong, Y.S., Lan, C.Y. and Wang, L.N. 1998. Litter production and decomposition in a subtropical mangrove swamp receiving wastewater. *J. Exp. Mar. Biol. Ecol. 226*, 1-18.

Tomlinson, P.B. 1986. The Botany of Mangroves. Cambridge University Press, Cambridge, UK. Untawale, A.G. 1987. Country Reports: India. In: *Mangroves of Asia and the Pacific: Status and Management*. Technical Report of the UNDP/UNESCO Research and Training Pilot Programme on Mangrove Ecosystems, 51-87.

Untawale, A.G. 1987. *Mangrove of Asia and Pacific: Status and Management.* A Technical Report of the UNDP/UNESCO Research and Training pilot programme on Mangrove Ecosystem in Asia and the Pacific.

Venkatesan, T. and Natarajan, R. 1985. The Mangroves. In: *Biology, Utililzation and Conservation of Mangroves* (Ed. Bhosale, L.J.). Maharashtra, India, 163-169.

Vittal, B.P.R. and Sarma, V.V. 2005. Fungal diversity on mangroves. In: *The Fungi – Diversity and Conservation in India* (Ed. Dargan, J.S., Atri, N.S. and Dhingra, G.S.). Bishen Singh Mahendra Pal Singh, Dehra Dun, Uttaranchal, India, 33-45.

Volkmann-Kohlmeyer, B. and Kohlmeyer, J. 1993. Biogeographic observations on Pacific marine fungi. *Mycologia 85*, 337-346.

Wafar, S., Untawale,A.G. and Wafar, M. 1997. Litterfall and energy flux in a mangrove ecosystem. *Estuar. Coast. Shelf Sci. 44*, 111-124.

Wilson, D. 1995. Endophyte – the evolution of a term, a clarification of its use and definition. *Oikos 73*, 274-276.

Woodroffe, C.D., Bardsley, K.N., Ward, P.J. and Hanley, J.R. 1988. Production of mangrove litter in a macrotidal embayment, Darwin Harbour, NT, Australia. *Estuar. Coast. Shelf Sci. 26*, 581-598.

Frontiers in Fungal Ecology, Diversity and Metabolites (edited by K.R. Sridhar), I.K. International Publishing House Pvt. Ltd., New Delhi, 2009.

4

Marine Fungi and Novel Metabolites

Rajeev Bhat and K.R. Sridhar*
Microbiology and Biotechnology, Department of Biosciences, Mangalore University, Mangalagangotri, Mangalore 574 199, Karnataka, India
*(*E-mail: sirikr@yahoo.com)*

Introduction

The oceans and seas cover 70% of the earth's surface and possess a wide diversity of natural flora and fauna. Among the 36 known living phyla, 34 are found in marine environments with more than 0.3 million known species of flora and fauna (Faulkner, 2001). These oceans and seas are considered as an untapped rich source of potential foods and drugs that are beneficial to humankind. A recent upsurge in scientific inventions in the search of inexpensive and reliable substances of economic value has opened up the possibilities of exploring the bountiful wealth of marine environment. Today, there is a mounting interest in marine products for new and promising bioactive substances, as they are expected to serve as leading compounds for drug development or pharmacological tools.

Several reviews deal with the pharmacological importance of marine natural products (Konig and Wright, 1996; Mayer and Hamann, 2002, 2004, 2005; Mayer and Lehmann, 2000). Marine natural products may be biologically active secondary metabolites, lipids, hetero-polysaccharides and enzymes derived from marine organisms. Studies have pointed out that bioactive compounds isolated from marine invertebrates (sponges, coelenterates and molluscs) originate from symbiotic microorganisms (e.g., bacteria, fungi) (Donia *et al.*, 2003; Haefner, 2003). Available reviews have shown a rising trend in exploring the secondary metabolites of marine microorganisms and many microbes are known to produce pharmaceutically valued compounds (Bernan *et al.*, 2004; Fenical, 1993, Jensen and Fenical, 2000; Kelecom, 1999; Kobayashi and Ishibashi, 1993; Newman *et al.*, 2000; Proksch *et al.*, 2003). To date, over 10,000 metabolites have been isolated from marine organisms (Fusetani, 2000). The biomedicals have shown exceptional pharmacodynamic properties like antibiotic, antifungal, cytotoxic, neurotoxic, antimitotic, antiviral, antineoplastic, immunosuppressive, anti-inflammatory, anti-Alzheimer's, anti-ageing and some prevention of tropical diseases (Kelecom, 1999). The exploration of marine fungi for bioactive metabolites has resulted in a high percentage of positive results. In this chapter, we have highlighted the importance of marine fungi and explored their metabolites to generate industrially valued products. In view of the exploration of novel metabolites of marine fungi, techniques employed for isolation, identification, and the constraints associated with marine fungal research are discussed briefly.

Marine Fungi

Marine mycology was enriched by the first detailed study published by Barghoorn and Linder (1944). Marine fungi have been defined as the species whose optimum growth and reproduction occur in total saline conditions rather than being only saline-tolerant (Gold, 1959). However, Kohlmeyer and Kohlmeyer (1979) have divided marine fungi into two categories: obligate marine fungi (grow and sporulate exclusively in marine or estuarine habitat) and facultative marine fungi (belong to fresh water and terrestrial habitats and able to grow and sporulate in marine habitat). Jones (1988) has provided substantial information on the global distribution of marine fungi including distribution in depth, substrata and adaptation to sea environs. The global distribution of higher marine fungi has been updated recently by Shearer *et al.* (2007) (see details in Chapter 3). Taxonomically, the updated checklist of fungi from European marine waters has been published recently by Landy and Jones (2006). They listed up to 338 species as European marine fungi, which accounts to one third (33.8%) of known obligate and facultative marine fungi worldwide (Jones and Mitchell, 1996). The updated checklist of mangrove fungi of the Indian Subcontinent reveals 165 species: 111 ascomyctes, 1 basidiomycete, and 53 anamorphic taxa (see Chapter 3).

Isolation

Marine fungi are divergent in morphology, physiology and adaptation to aquatic habitats (Meyers, 1996). The isolation of a particular marine fungus is entirely dependent on the substrate (e.g., soft wood or hard wood, seaweeds, animal exoskeleton, hairs or keratinaceous substrates, sediments, sea foam) and place of sampling (e.g., temperate, tropics, arctic regions, mid dune, and hind dune). Marine fungi have also been reported as endophytes of sea animals, seaweeds and sea grasses (Belofsky *et al.*, 1999; Nielsen *et al.*, 1999; Porter, 1986; Sugano *et al.*, 1995). Physicochemical features like pH, temperature, light, salinity, carbon, nitrogen, vitamin and growth factors will also play an important role in fungal growth (Bolton *et al.*, 1988; Booth and Kenkel, 1986; Jennings, 1983;

Jones, 1993; Jones and Byrne, 1976). Spore attachment through appendages also contributes significantly during the isolation of marine fungi (Hyde *et al.*, 1993; Rees and Jones, 1984). For instance, the greatest diversity of appendages like ribbon-shaped, pad-like, filamentous or spines has been observed in marine ascomycetes.

Various methods are available for the isolation of marine fungi. They can be isolated by direct enumeration by floatation methods (Kirk, 1983), single spore isolation (Choi *et al.*, 1999; Goh, 1999; Johnson and Sparrow, 1961), growth on sea water medium (Gray *et al.*, 1963), plating on synthetic media (Hyde *et al.*, 1987), sand grain method (Rees and Jones, 1985), Shearer method (Shearer and Von Bodman 1983), simulation of deep sea conditions (Lorenz and Molitorius, 1997), aeration and collection of spores, and by damp incubation (Kohlmeyer, and Kohlmeyer, 1979). Prioritization of fungal strains cultivated in Petri dish based on antagonism may also be useful in isolation (Yuen *et al.*, 1999).

For the measurement of the fungal biomass, several methods are in practice: direct-epifluorescence microscopy, chromatographic assay, fumigation-extraction, glucosamine (GlcN) and ergosterol estimation (Daniel *et al.*, 1995; Newell, 1992, 1993; Nylund and Wallander, 1992; Paul and Johnson, 1977). Most commonly employed methods are the GlcN and ergosterol estimation techniques (Lee *et al.*, 1980; Newell, 1992, 1996). As the GlcN technique has several inherent drawbacks (Newell, 1992), egrosterol has become the most reliable index of true fungal mass and thus, the most extensively used (Nylund and Wallander, 1992).

Identification

An authentic identification of marine fungi is usually possible based on the experience of a researcher. However, several monographs and reviews with sketches are available from time to time for the authentic identification and confirmation of taxon (Jones 1995; Kohlmeyer and Kohlmeyer, 1979, Kohlmeyer and Volkmann-Kohlmeyer, 1991). Usually, the phenology of the fungal spores (colour, size and appendages) and the morphology of conidia or ascospores have been considered important for identification. Identification and classification of marine fungi that are based on the ultrastructural observations employing light and scanning electron microscopy and histochemical methods are also advocated (Johnson *et al.*, 1991; Jones and Moss, 1980; Moss, 1990, Read *et al.*, 1992). The DNA fingerprinting using molecular hybridization and PCR techniques are the methods of choice that are routinely employed to resolve the genetic diversity among marine fungi. Reconstruction of fungal phylogeny can be established by an analysis of ribosomal DNA nucleotide sequences. Usually, nucleotide sequence analysis of various fungi results in the maintenance of databases, which can be made available electronically for species identification, confirmation, and the discovery of new fungi. In fungal systematics, the most commonly employed technique involves gel electrophoresis of soluble proteins and enzymes (isozymes), and DNA analysis [DNA-DNA hybridization, electrophoretic karyotyping, micro-satellite-primed PCR, Restriction Fragment Length Polymorphism (RFLP), Random Amplified Polypmorphic DNA (RAPD), Amplified Fragment Length Polymorphism finger printing (AFLP), sequencing of ribosomal RNA genes]. Molecular techniques have also been employed to understand the evolution of marine fungi from terrestrial origin (Spatafora *et al.*, 1998). Studies on molecular phylogenetic relationship of various fungi (mainly on Halosphaeriaceae) are available (Chen *et al.*, 1999; Okada *et al.*, 1997; Spatafora *et al.*, 1998). For example, using molecular phylogenetic analysis and morphology, a new family and order, Lulworthiaceae and Lulworthiales have been described by Kohlmeyer *et al.* (2000) (e.g., genera of *Lindra* and *Lulworthia*). Reports are also available on the application of Enzyme-Linked Immunoabsorbent Assay (ELISA) for species differentiation (e.g., *Dendryphiella*) (Mohamad *et al.*, 1989) and the technique can also be applied to distinguish between the taxa of fungi.

Preservation

Intense search for novel secondary metabolites makes it obligatory to carefully handle and maintain the cultures (Monaghan *et al.*, 1995). For the industrial exploitation of fungi of interest, care should be taken to maintain the viability and genetic constituents. Maintenance of marine fungal cultures is rather difficult compared to terrestrial fungi. Lyophilization and liquid drying has been assumed to be the best for sporulating fungi (Nakagiri and Jones, 2000). The usual methods employed for the maintenance of fungal culture includes: lyophilization or preservation in a refrigerator that is maintained below –4°C. Reports are available on long-term maintenance of marine fungal cultures through freezing at ultra low temperatures (Morris *et al.*, 1988, Nishii and Nakagiri, 1991). Storage of cultures under liquid nitrogen (–196°C) has been successful and has been recommended as it ensures minimal genetic change (Smith and Onions, 1994). However, fungal cultures can also be maintained in glass tubes (agar slants) or in Eppendorf micro-centrifuge tubes. The advantages and disadvantages of employing Eppendorf micro-centrifuge tubes for the storage of fungal culture have been described briefly by Suryanarayan and Kumaresan (1998). Even though a number of preservation techniques are available to date, it has been generally accepted that the most important method of conservation of the genetic resource of fungi is to conserve the natural habitat as they are habitat-dependent (Hawksworth, 1991).

Fungal Metabolites

Terrestrial fungi

Any chemical compound obtained from a fungus having a wide range of activity (beneficial or harmful) against other microbes, cells or organisms could be considered to be a bioactive compound. The early discovery of penicillin by

Alexander Fleming, a novel secondary metabolite by *Penicillium* sp. (now mutated strains of *P. chrysogenum*) possessing antibiotic character opened up several new fields for exploration of secondary metabolites of fungal origin. This pioneering discovery was followed by extensive screening of soil fungi for the presence of industrially-valued secondary metabolites like cyclosporin, lovastatin, griseofulvin and streptomysin (Erkel, 1997; Monaghan *et al.*, 1995, Wildman, 1997). Some of the common fungi from which pharmaceutically-valued metabolites are obtained belonged to *Aspergillus* (e.g., *A. terreus*), *Penicillium* (e.g., *P. griseofulvum*) and *Tolypocladium inflatum*.

Research on fungal chemistry has led to the discoveries of new pharmaceutrical products (Masurekar, 1992). Cyclosporins, first isolated from *T. inflatum* in 1976 as antifungal compounds were later shown to possess immunosuppressive properties. Cyclosporin A is currently the most widely used drug for preventing the rejection of human organ transplants. Similarly, *mevinolin*, a cholesterol-lowering compound has been isolated from terrestrial fungi. Other pharmaceutically-valued products derived from fungal metabolites include: asperlicin, echinocandins, papulacandins and squalestatins (Caporale, 1995; Masurekar, 1992). Although fungi are well known producers of a wide range of secondary metabolites, some of these are well known potential toxins. This is true in terrestrial fungi as they are capable of producing aflatoxins, ochratoxin, fumonisin, zearalenone or other carcinogenic compounds that affect humans and livestock. However, an assortment of fungi that produce different compounds useful in food and pharmaceutical industries is well established. Dreyfuss and Chapela (1994) have estimated about 4000 secondary metabolites of fungal origin to be biologically active. It is assumed that around 1.5 million of fungi exist worldwide, but only a few (5-10%) have been identified (Fröhlich and Hyde, 1999) and very few are screened for secondary metabolites.

Marine fungi

Marine fungi have been ascertained to be an important source of new bioactive compounds and enzymes (Grant *et al.*, 1996; Pointing *et al.*, 1998; Pointing and Hyde, 2000; Raghukumar *et al.*, 1999). Cuomo and coworkers (1995) compared a large number of terrestrial (1450) and marine (1500) fungal isolates and found a higher number of antimicrobial activities among marine fungi than terrestrial fungi. It has been anticipated that a total of 1500 marine fungal species exist (Jones and Mitchell, 1996). Christophersen *et al.* (1999) evaluated 227 marine fungi isolated from animals, plants and sediments of Venezuelan waters, mangroves and lagoons. Many endophytic fungi belonging to *Pestalotiopsis* sp. of marine origin are known to produce antifungal, antioxidant and anticancer metabolites (Li *et al.*, 1998; Li and Strobel, 2001; Mayer and Gustafson, 2003; Strobel *et al.*, 2002). A variety of novel metabolites have been generated and evaluated from fungi of marine origin (Table 1). These metabolites possess antibacterial, antifungal, antiviral, antimalarial, anti-inflammatory, cytotoxic, and antineoplastic properties.

***Table 1.** Some novel secondary metabolites isolated from marine and marine-derived fungi*

Fungi	Isolated compound	Activity	Reference
Aigialus parvus	Aigialomycin D	Antimalarial, Cytotoxic (KB, Vero)	Isaka *et al.* (2002)
	Hypothemycin	Antimalarial, Cytotoxic (BC-1, Vero)	
Ascochyta salicorniae	Ascosalipyrrolidinone A	Antimicrobial	Osterhage *et al.* (2000)
Aspergillus fumigatus	Tryprostatins A and B	Mammalian cell inhibitors	Cui *et al.* (1995)
Aspergillus niger	Asperazine	Cytotoxic alkaloid	Varoglu *et al.* (1997)
Corollospora maritima	Corollosporine (see Fig. 1)	Antibacterial	Liberra *et al.* (1998)
Corollospora pulchella	Melinacidins III, IV and Ganacidin W	Antibiotic	Furuya *et al.* (1985)
	Pulchellalactam	Inhibits CD45 (CD45 activates B and T cells)	Alvi *et al.* (1998)
Diaporthe sp.	Diaporthelactone, 7-methoxy-4,6-dimethyl-3H-isobenzofuran-1-one and Mycoepoxydiene	Anticancer and Antimicrobial	Lin *et al.* (2005)
Emericella variecolor	varixanthone	Antimicrobial	Malmstrom *et al.* (2002)
Fusarium sp.	JM 47 (cyclic tetrapeptide)	Antibiotic and Antiprotozoal	Jiang *et al.* (2002)
Gymnasella dankaliensis	Gymnasterones A and B	Cytotoxic	Amagata *et al.* (1998)
Halorosellinia oceanica	2-hexylidene-3-methyl-succinic acid	Cytotoxic (KB, BC-1)	Chinworrungsee *et al.* (2001)
	Cytochalasin Q	Antimalarial, Cytotoxic (KB, BC-1)	
	5-carboxymellein	Antimalarial	
	Halorosellinic acid	Antimycobacterial	
Hypoxylon oceanicum (LL-15 G256)	15G256γ	Antifungal	Albaugh *et al.* (1998)
Keissleriella sp. (YS4108)	EPS2 (exopolysaccharide)	Antioxidant (acting against peroxide hydrogen)	Sun *et al.* (2005)
Keissleriella sp. (YS4108)	YCP (polysaccharide)	Anticoagulant and Antiplatelet aggregation	Han *et al.* (2005)
Kirschsteiniothelia funiculata	Kirschsteinin	Cytotoxic (human tumor cell lines)	Liberra and Lindequist (1995)
Leptosphaeria oraemaris	Culmorin	Antifungal	Liberra and Lindequist (1995)
	Obioninene, Oraemarin	Antibiotic	Miller and Savard (1989)
Lignincola laevis	Thrausochytrosides A-C	Cytotoxic	Abraham *et al.* (1994)
Microascus longirostris	Epoxysuccinates	Cysteine proteases inhibitor	Yu *et al.* (1996a)
Microsphaeropsis *olivacea*	Fatty acid and Glyceride	-	Yu *et al.* (1996b)
Microsphaeropsis sp.	2-betaenone derivatives, 3-tetrahydroxyanthra-quinone congeners	Protein kinases inhibitor	Brauers *et al.* (2000)

Passseriniella obiones	Obionin A	CNS activity	Liberra and Lindequist (1995)
Penicillium sp.	Epolactaene	Neuritogenic	Kakeya *et al.* (1995)
Penicillium terrestre	Polyketides (Penicillones A 1 and B 2)	Anti-tumor	Liu *et al.* (2005)
Periconia sp.	Pericosines A and B, Macrosphelides E-H	Anti-tumor	Numata *et al.* (1997)
Phoma sp.	Phomactin A, B, B_1, B_2, D, E, F, G	Platelet activating factor antagonists	Sugano *et al.* (1991, 1994, 1995)
Phoma herbarum (YS 4108)	YCP - YCP-S1 and YCP-S2, (sulphated polysaccharide)	Antioxidant	Yang *et al.* (2005)
Phomopsis sp.	Phomopsidin	Anti-microtubule	Kobayashi *et al.* (2003)
Pestalotia sp.	Pestalone	Antibiotic	Cueto *et al.* (2001)
Zopfiella latipes	Zopfiellamides A and B	Antimicrobial	Daferner *et al.* (2002)

Structures of six metabolites derived from marine and marine-derived fungi are presented in Figure 1.

A marine fungal strain identified as *Penicillium* sp. (BM 1689-P), produced a neuritogenic compound (epolactaene), which has high commercial and pharmaceutical value to treat various neurodegenerative diseases involving dementia (Kakeya *et al.*, 1995) (Fig. 1). Metabolites from cultures of *Emericella variecolor* collected in the Caribbean Sea, off the coast of Venezuela yielded varitriol, which exhibited increased potency towards some renal, central nervous system and breast cancer cell-lines in the NCI's 60-cell line panel (Malmstrom *et al.*, 2002) (Fig. 1). *Wardomyces anomalus* produced several xanthone derivatives, which showed antioxidant and tyrosine kinase inhibitor activities (Abdel-Lateff *et al.*, 2003a). The xanthone derivatives isolated were: 2,3,6,8-tetrahydroxy-1-methylxanthone and 5-(hydroxymethyl)-2-furanocarboxylic acid, which showed significant antioxidant properties, while 2,3,6,8-tetrahydroxy-1-methylxanthone, 3,6,8-trihydroxy-1-methylxanthone and 5-(hydroxymethyl)-2-furanocarboxylic acid showed inhibition of p56(lck) tyrosine kinase. Chinworrungsee *et al.* (2001) isolated 2-hexylidene-3-methyl-succinic acid, cytochalasin Q and 5-carboxymellein from *Halorosellinia oceanica* possessing moderate cytotoxicity against KB and BC-1 cell lines with IC_{50} values of 1-13μg/ml. Until recently, no antiviral agents were available for the treatment of infection by the pathogenic poxvirus. Molluscum Contagiosum Virus (MCV) infection in healthy person causes only small papules that are easily treated, however, in AIDS patients, MCV causes severe lesions that are essentially untreatable. Recently, Hwang *et al.* (1999) identified and characterized a novel marine fungal product sansalvamide A (a cyclic depsipeptide), which inhibits the topoisomerase of the pathogenic poxvirus MCV. This was a breakthrough that generated more interest to explore and exploit marine fungi.

Among the obligate marine fungi, *Dendryphiella salina* was the first to yield a new metabolite, dendryphiellin A (Guerriero *et al.*, 1988) and subsequently many new compounds including dendryphiellin E 1 were produced (Jensen and Fenical, 2002) (Fig. 1). Abdel-Lateff *et al.* (2002) isolated novel hydroquinone derivatives from a marine fungal isolate, *Acremonium* sp., which showed high antioxidant properties. Abdel-Lateff (2003b) isolated new natural products from *Epicoccum* sp. (4,5,6-trihydroxy-7-methylphthalide and epicoccone), which showed significant antioxidative properties with 95% DPPH radical scavenging effects at 25 μg/ml and also the inhibition of peroxidation of linolenic acid in the TBARS assay (62% inhibition at 37 μg/ml). Epolactaene's

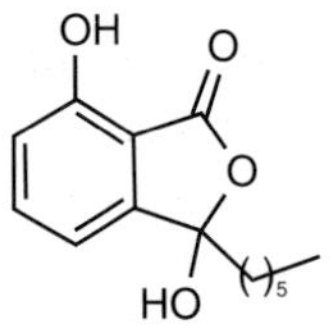

Corollosporine
(Corollospora maritima)

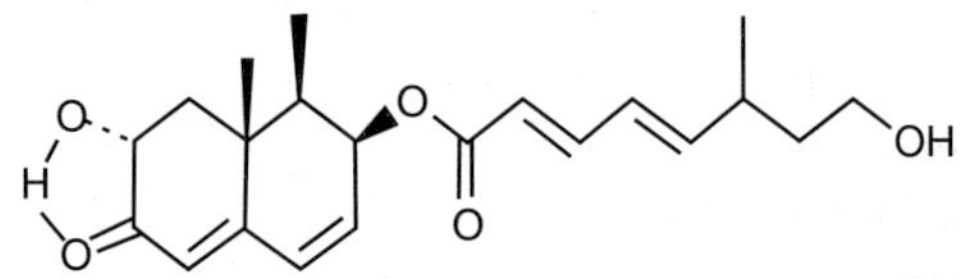

Dendryphiellin A
(Dendryphiella salina)

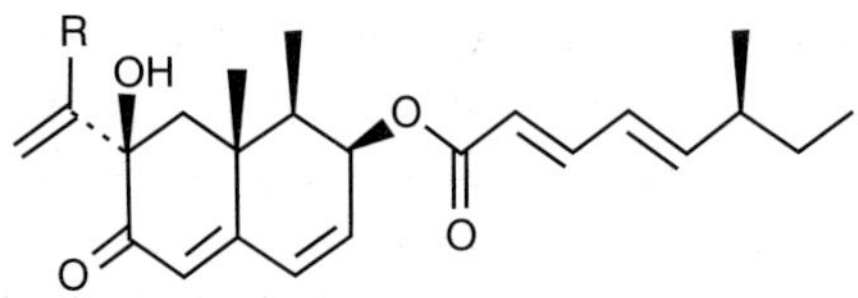

Dendryphiellin E 1
(Dendryphiella salina)

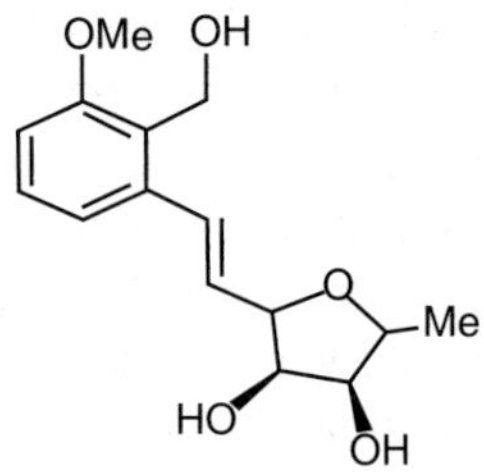

Varitriol
(Emericella variecolor)

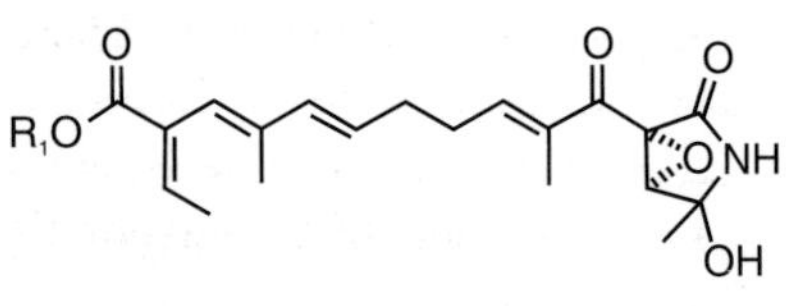

Epolactaene
(Penicillium sp.)

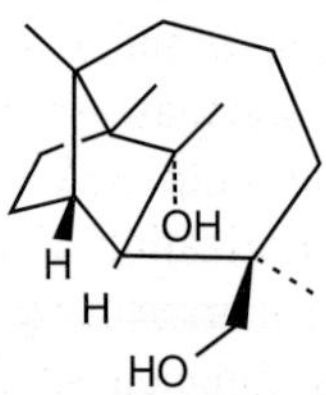

Isoculmorin
(Kallichroma tethys)

Fig. 1. Representative metabolites produced by marine and marine-derived fungi.

lactam moity is responsible for the growth inhibition of human cancer cells through modification of HsP60 (Nagumo *et al*., 2004). Liu *et al*. (2006) isolated new diphenylether dimethyl 2,3'-dimethylosoate and monomethylsulochrin through bioassay-guided fractionations of a marine-derived fungal (*Aspergillus* sp.) culture, B-F-2 showed cytotoxicity against K562 cell line. Huang (2006) isolated a novel isocoumarin from a marine-derived fungus, *Alternaria tenuis* Sg17-1, which on evaluation showed cytotoxic effects *in vitro*. Li *et al*. (2004) isolated an alkaloid, dihydroxyisoechinulin A from a marine-derived *Aspergillus* isolate, which exhibited significant radical scavenging activity against 1,1-diphenyl-2-picrylhydrazyl (IC50, 20 mM) similar to ascorbic acid used for positive control. This compound also showed ultraviolet-A radiation protecting activity with ED50 value of 130 mM, which is more active than the currently used sunscreen agent, oxybenzone (ED50, 350 mM). Yanagihara *et al*. (2005) isolated leptosins (Leps) F and C (indole derivatives) from a marine fungus (*Leptoshaeria* sp.), which showed *in vitro* cytotoxic effects. They showed that Leps might form a group of anticancer chemotherapeutic agents with single or dual catalytic inhibitory activities against topoisomerases (topos I and II). The Lep F inhibited the activity of topos I and II, whereas Lep C topo I *in vitro*. The DNA topo I and II are molecular targets of several potent anticancer agents and the inhibitors of these enzymes are potential candidates or model compounds for anticancer drugs.

Fungal enzymes are important in the food, beverages, confectioneries, textile and leather industries. Most of the enzymes of industrial use have been derived from microbes (bacteria and fungi) of terrestrial origin (Chandrashekaran, 1997). Some studies are available on the enzymatic capabilities of marine fungi (e.g., Pointing *et al*., 1999; Raghukumar *et al*., 1999; Rohrmann and Molitoris, 1992). Pisano *et al*. (1964) reported protease activity in 13 out of 14 marine fungi. Many marine fungi possess laccase-positive properties (Kumaresan and Suryanarayanan, 2002). Marine fungi such as *Flavodon flavus* and *Podospora anserina* produce lignin-modifying enzymes (Liberra and Lindequist, 1995; Raghukumar *et al*., 1999). Raghukumar *et al*. (1994) showed *xylanase* activity in 11 marine fungi and *Aigialus mangrovei, Gongronella* sp., *Halorosellinia oceanica* and *Asterosphaeriella* of mangroves showed high activity. Cellulase-free xylanase or xylanase with low cellulase activity is effective in bleaching paper pulp and bio-polishing textiles (Srinivasan *et al*., 1999). Recently, the decolorization of black liquor, textile dye wastewaters and molasses spent wash by a marine fungal isolate (NIOCC # 2a) producing laccase as a major lignin-degrading enzyme has been reported (D'Souza *et al*., 2006).

There is a growing interest to investigate the occurrence of essential fatty acids in marine microorganisms. Special emphasis has been laid to investigate the possible presence of polyunsaturated fatty acids (PUFA) as they are important in preventing cardiovascular disease and dietary requirements of humans and livestock. Although some reports are available on PUFA in lower marine fungi belonging to thraustochytrids, no detailed studies are available on higher marine fungi (Bergé and Barnathan, 2005; Kamlangdee and Fan, 2003). Cooney *et al.* (1993) reported fatty acid composition of ten strains of marine fungi (2 lignicolous, 8 arenicolous). In each of the fungi isolated, the fatty acids ranged between 14 and 22 carbons. Palmitic (16:0), oleic (18:1n9) and linoleic (18:2n6) acids were the principal fatty acids (89% or more of the total peak). The total unsaturated fatty acids varied considerably ranging from 33 to 79%. No hydroxy- or cyclopropyl fatty acids were detected. The n-3 and n-6 linolenic (18:4n3) and arachidonic (20:4n6) acids of commercial value were detected in small amounts in all the fungi (Kendrick and Ratledge, 1992). Khudiakova *et al.* (2000) have reported the fatty acid composition of a marine fungus (*Chaetomium spiculipilium*) that was isolated from the sediments of the Sea of Japan. See Chapter 3 for details on studies carried out on the mangrove fungal metabolites in India.

Techniques

Currently, Bu'Lock's review (1975) on induction and the principles involved in secondary metabolism in marine fungi is considered as preeminent. Strongman *et al.* (1987) opined that a secondary metabolite like antibiotics might be produced for defensive purposes or for survival in a competitive environment. The induction of secondary metabolite in fungi can be initiated by stress or by altering the physicochemical environment. Nitrogen, carbon and oxygen are the main limiting factors for the induction of secondary metabolites *in vitro*. Reports are available on induction of metabolites by limiting carbon, oxygen and phosphate (Miller *et al.*, 1994; Miller and Blackwell, 1986; Miller and Savard, 1989). Some trace elements (e.g., zinc, copper) also substantially contribute to the induction of secondary metabolites in fungi. Glover (1995) correlates the production of secondary metabolites with the ecological niches. To screen for a novel secondary metabolite having industrial applications, it is necessary to understand the structure of the chemical compound involved. The main rationale behind the chemical investigation is to isolate pure and biologically active metabolites from fungal culture extracts. To achieve this, selected strains have to be cultured in a large-scale, metabolites are extracted and these extracts have to be separated using different chromatographic and spectroscopic methods. Some of the routinely employed techniques for the identification of bioactive compounds involve Thin Layer Chromatography (TLC), High Performance Liquid Chromatography (HPLC), Gas Liquid Chromatography (GLC), Gas Chromatography-Mass Spectrometry (GC-MS), NMR spectroscopy, IR spectroscopy, Liquid Chromatography-Mass Spectrometry (LC-MS) and X-ray crystallography. After extraction of the fungal biomass with an appropriate solvent, the extracted solvent has to be injected into one of the above instruments and the compounds detected have to be matched with the available databases of compounds from marine or terrestrial sources. If new compounds of interest are detected, the reproducibility has to be checked and reported as a new metabolite after characterization and testing biological activities. The chemical investigations will be completed through structural elucidation and physical characterization of the isolated metabolites.

Constraints

Studies on marine microorganisms are facing some expected as well as unexpected difficulties. We need to have a clear-cut idea whether a fungus that is isolated from marine environment is active in marine habitat or transient or tolerant to extreme conditions. Contradictorily, some marine fungi identified as terrestrial (e.g., *Aspergillus* spp.) were found to grow profusely in high saline concentrations due to many reasons. Firstly, variations can be seen in the physiological characteristics between marine and terrestrial isolates although they are morphologically similar (Masuma *et al.* (2001). Secondly, the taxonomical identity of marine fungi is defectively defined, wherein the identifications are based on numbered strains of otherwise partially or totally undefined organisms. This may lead to the unfeasibility of employing chemotaxonomic techniques for research or comparative studies resulting in a duplication of data. Thirdly, the yield obtained by the harvest of marine fungi may be very less making it impossible to produce in large scale and this may be time consuming. Genetic engineering techniques may allow the option of identifying the active gene in a fungus and cloning to another fungus for easy cultivation or handling (Newman *et al*., 2000). Studies are necessary to understand whether these compounds are truly essential or significant in the life history of the host species.

A short period of incubation of woody debris collected from beaches usually results in the isolation of typical terrestrial or weed fungi. Hyde (1992) has addressed the impact of the incubation period on the occurrence of fungi on woody litter collected from maritime habitats. Six months incubation yielded more fungi as well as higher frequency of occurrence than 1-2 weeks of incubation. The incubation period of woody substrata of tropical beaches has a considerable impact on fungal composition, diversity and similarity (Prasannarai and Sridhar, 2003). Incubation and repeated screening up to 18 months would result in a reasonable assessment of fungal assemblage, diversity and frequency of occurrence. It is possible that new or rare fungal species, which are not competitive, might grow on beach-collected woody litter on prolonged incubation in laboratory microcosms (e.g., moist chambers with different regimes of water activity). Panebianco *et al.* (2002) have shown that balsa test blocks pre-colonized with fungi in the laboratory (exposed to sea for 2-6 months) inhibited the sporulation of other fungi. *Kallichroma tethys* grown in an enriched seawater medium yielded isoculmorin (Fig. 1) (Alam *et al*., 1996). The biological activity of isoculmorin is of ecological interest as it is involved in the interference competition among lignicolous fungi. Studies on the co-occurrence of fungal taxa in marine (Prasannarai *et al*., 1999) and mangrove (Sridhar and Maria, 2006) woody litter reveal the nature of competition, importance of rare and core-group fungi in the production of metabolites of industrial and medicinal importance.

Conclusions and Outlook

Some of the most recent molecular techniques can be used to update origin, phylogenetic relationships, and untapped properties. Future emphasis has to be laid on the natural product chemistry so as to broaden the chemotaxonomic knowledge of marine fungi. A detailed study has to be undertaken worldwide to screen the novel metabolites of marine fungi on pilot scale with genuine assistance of government and other funding agencies. Methods to induce early sporulation in marine fungi need immediate attention as they are slow growing. Recent molecular techniques of fungal genome analysis and manipulation will open up several possibilities to explore the biosynthesis of secondary metabolites. The use of biosensors for investigating new and biologically active compounds can also prove to be useful. The occurrence and role of fungi in deep-sea sediments have received little attention (Gonda *et al.*, 2000; Johnson and Sparrow, 1961; Kohlmeyer, 1977; Raghukumar and Raghukumar, 1998). Remote sensing techniques have immensely contributed to the collection of marine samples from waters up to 900m depth and the isolation of over 5000 novel compounds (McCarthy and Pomponi, 2004). Gautschi *et al.* (2004) isolated *Penicillium* from deepwater sediment (1335m) and reported its cytotoxic activity. Future studies should be initiated to explore the possibilities of isolating novel metabolites from marine fungi of deep-sea origin with the aid of remote sensing.

The exploration of metabolites from marine fungi in food and nutraceutical industry is still in its infant stage. The isolation or induction of pigments in marine fungi has wide applications in food, textile and cosmetic industries. There is an increased consumers' preference for natural bioflavours and additives than chemical or synthetic ones as 'chemophobia' persists (Cheetham, 1997). The use of fungi and their enzymes for the production of fruity flavours has been discussed in detail by Vandammae (2003). The same strategy can be applied to explore the possibilities for the production of flavor and precursors of aromatic compounds from marine fungi. Attempts to understand possible degradation of plastic, oil and radioactive pollutants by marine fungi might reveal interesting results. Studies on interactions and antagonistic activities among marine fungi on specific substrata in relation to secondary metabolites might provide their complex role in a marine ecosystem. Increased human interference (e.g., release of sewage, industrial effluents, shelling, sand mining, oil spilling, establishing nuclear/thermal reactors in marine vicinity) of marine habitats has a negative influence on marine fungi. Minimization of anthropogenic stresses on marine habitats may facilitate to preserve the diversity of marine fungi and in turn, exploitation of their unique bioactive metabolites.

References

Abdel-Lateff, A., König, G.M., Fisch, M.K., Höller, U., Jones, P.G. and Wright, A.D. 2002. New antioxidant *hydroquinone* derivatives from the *algicolous* marine fungus *Acremonium* sp. *J. Nat. Prod. 65*, 1605-1611.

Abdel-Lateff, A., Klemke, C., Konig, G.M. and Wright, A.D. 2003a. Two new *xanthone* derivatives from the *algicolous* marine fungus *Wardomyces anomalus. J. Nat. Prod. 66*, 706-708.

Abdel-Lateff, A., Fisch, K.M., Wright, A.D., Konig, G.M. 2003b. A new antioxidant *isobenzofuranone* derivative from the algicolous marine fungus *Epicoccum* sp. *Planta Med. 69*, 831-834.

Abraham, S.P., Hoang, T.D., Alam, M. and Jones, E.B.G. 1994. Chemistry of the cytotoxic principles of the marine fungus *Lignincola laevis. Pure and Appl. Chem. 66*, 2391-2394.

Alam, M., Jones, E.B.G., Hossain, M.B. and van der Helm, D.J. 1996. Isolation and structure of *isoculmorin* from the marine fungus *Kallichroma tethys. Nat. Prod. 59*, 454-456.

Albaugh, D., Albert, G., Bradford, P., Cotter, V., Froyd, J., Gaughran, J., Kirsch, D.R., Lai, M., Rehnig, A., Sieverding, E. and Silverman, S. 1998. Cell wall active antifungal compounds produced by the marine fungus *Hypoxylon oceanicum* LL-15G256. III. Biological properties of 15G256g. *J. Antibiotics 51*, 317-322.

Alvi, K.A., Casey, A. and Nair, B.G. 1998. Pulchellalactam: A CD45 protein *tyrosine phosphatase* inhibitor from the marine fungus *Corollospora pulchella. J. Antibiot. 51*, 515-517.

Amagata, T., Minoura, K., and Numata, A. 1998. Gymnasterones, novel cytotoxic metabolites produced by a fungal strain from a sponge. *Tetrahedron Lett. 39*, 3773-3774.

Barghoorn, E.S. and Linder, D.H.1944. Marine fungi: Their taxonomy and biology. *Farlowia 1*, 395-467.

Belofsky, G.N., Jensen, M.K. and Fenical, W. 1999. Sansalvamide: A new cytotoxic cyclic depsipeptide produced by a marine fungus of the genus *Fusarium. Tetrahedron Lett. 40*, 2913-2916.

Bergé, J.-P. and Barnathan, G. 2005. Fatty acids from lipids of marine organisms: molecular biodiversity, roles as biomarkers, biologically active compounds, and economical aspects. *Adv. Biochem. Eng. Biotechnol. 96*, 49-125.

Bolton, R., Denkmann, D., Langly, D. and Dawson, M.J. 1988. Effects of salinity and pH on antibiotic production by marine fungi. In: *The 31st harden conference.* Wye College, Kent, USA.

Booth, T. and Kenkel, N. 1986. Ecological studies of *lignicolous* marine fungi: a distribution model based on ordination and classification. In: *The Biology of Marine Fungi* (Ed. S.T. Moss). Cambridge University press, UK, 297-310.

Brauers, G., Edrada , R.A., Ebel R., Proksch, P., Wray, V., Berg, A., Grafe U., Schachtele, C., Totzke F., Finkenzeller, G., Marme, D., Kraus, J., Munchbauch, M., Michel, M., Bringmann, G. and Schaumann, K. 2000. *Anthraquinones* and *betaenone* dérivatives from the sponge-associated fungus *Microsphaeropsis* species: Novel inhibitors of protein kinases. *J. Nat. Prod. 63*, 739-745.

Bu'Lock, J.D. 1975. Secondary metabolism in fungi and its relationship to growth and development. In: *The Filamentous Fungi - Industrial Mycology*, Volume 1 (Ed. Smith, J.E., and Berry, D.R.). Wiley, New York, 35-58.

Caporale, L.H. 1995. Chemical ecology: A view from the pharnmaceutical industry. *Proc. Natl. Acad. Sci. USA 92*, 75-82.

Chandrashekaran, M. 1997. Industrial enzymes from marine microorganisms: The Indian Scenario. *J. Mar. Biotech. 5*, 86-89.

Cheetham, .P.S.J. 1997. Combining the technical push and the business pull for natural flavours. *Adv. Biochem. Eng. Biotech. 55*, 1-49.

Chen, W., Shearer, C.A., and Crane, J.L. 1999. Phylogeny of *Ophioceras* spp. based on morphological and molecular data. *Mycologia 91*, 84-94.

Chinworrungsee, M., Kittakoop, P., Isaka, M., Rungrod, A., Tanticharoen, M. and Thebtaranonth, Y. 2001. Antimalarial halorosellinic acid from the marine fungus *Hallorosellinia oceanica. Bioorg. Med. Chem. Lett. 11*, 1965-1969.

Choi, Y.W., Hyde K.D. and Ho W.H., 1999. Single spore isolation of fungi. *Fungal Diversity 3*, 29-38.

Christophersen, C., Crescente, O., Frisvad, J.C., Gram, L., Nielsen, J., Nielsen, P.H. and Rahbaek, L. 1999. Antibacterial activity of marine-derived fungi. *Mycopathol. 143*, 135-138.

Cooney, J.J., Doolittle, M.M., Grahl-Nielsen, O., Haaland, M.I. and Kirk, P.W. Jr. 1993. Comparison of fatty acids of marine fungi using multivariate statistical analysis. *J. Indust. Microbiol. 12*, 373-378.

Cueto, M., Jensen, P.R., Kauffman, C., Fenical, W., Lobkovsky, E., Clardy, J., 2001. Pestalone, a new antibiotic produced by a marine fungus in response to bacterial challenge. *J. Nat. Prod. 64*, 1444-1446.

Cui, C.B., Kakeya H., Okada, G., Onose, R., Ubukata, M., Takahashi,, I., Isono K. and Osada, H. 1995. Tryprostatins A and B, novel mammalian cell inhibitors produced by *Aspergillus fumigatus. J. Antibiot. 48*, 1382-1384.

Cuomo, V., Palomba, I., Perretti, A., Guerriero, A., D'Ambrosio, M. and Pietra, F. 1995. Antimicrobial activities from marine fungi. *J. Mar. Biotechnol. 2*, 199-204.

Daferner, M., Anke, T., and Sterner, O. 2002. Zopfiellamides A and B, antimicrobial *pyrrolidinone* derivatives from the marine fungus *Zopfiella latipes. Tetrahedron 58*, 7781-7784.

Daniel, O., Schönholzer, F. and Zeyer, J. 1995. Quantification of fungal hyphae in leaves of deciduous trees by automated image analysis. *Appl. Environ. Microbiol. 61*, 3910-3918.

Donia, M. and Hamann, M.T. 2003. Marine natural products and their potential applications as anti-infective agents. *Lancet 3*, 338-348.

Dreyfuss, M.M. and Chapela, I.H. 1994. Potential of fungi in the discovery of novel, low-molecular weight pharmaceuticals. In: *The Discovery of Natural Products with Therapeutic Potential* (Ed. Gullo, V.P.). Butterworth-Heinemann, Boston, USA, 49-80.

D'Souza, D.T., Tiwari, R., Sah, A.K., Raghukumar, C. 2006. Enhanced production of laccase by a marine fungus during treatment of colored effluents and synthetic dyes. *Enz. Microb. Technol. 38*, 504-511.

Erkel, G. 1997. *Fusidic acid, griseofulvin* and *pleuromutilin*. In: *Fungal Biotechnology* (Ed. Anke, T.). Chapman and Hall, Weinheim, Germany, 128-135.

Faulkner, D.J. 2001. Marine natural products. *Nat. Prod. Rep. 18*, 1-49.

Fenical, W. 1993. Chemical studies of marine bacteria: developing a new resource. *Chem. Rev. 93*, 1673-1683.

Fröhlich, J. and Hyde, K.D. 1999. Biodiversity of palm fungi in the tropics: Are fungal diversity estimates realistic? *Biodiver. Conser. 8*, 977-1004.

Furuya, K., Okudaira, M., Shindo, T. and Sato, A. 1985. *Corollospora pulchella,* a marine fungus producing antibiotics, melinacidins III, IV and gancidin W. *Sankyo Kenkyusho Nempo 37*, 140-142.

Fusetani, N. 2000. Introduction. In: *Drugs from the Sea* (Ed. Fusetani, N.). Karger, Basel, 1-5.

Gautschi, J.T., Amagata, T., Amagata, A., Valeriote, F.A., Mooberry, S.L. and Crews, P. 2004. Expanding the strategies in natural product studies of marine-derived fungi: A chemical investigation of *Penicillium* obtained from deep water sediment. *J. Nat. Prod. 67*, 362–367.

Glover, J.B. 1995. The chemistry of fungal antagonism and defense. *Can. J. Bot. 73*, S1265-S1274.

Goh, T.K. 1999. Single spore isolation using a hand made glass needle. *Fungal Diversity 2*, 46-63.

Gold, H.S. 1959. Distribution of some lignicolous *Ascomycetes* and fungi imperfecti in an estuary. *J. Elisha Mitchell Sci. Soc., 75*, 25-28.

Gonda, K.E., Jendrossek, D., Molitoris, H.P. 2000. Fungal degradation of the thermoplastic polymer *poly-β-hydroxybutyric* acid (PHB) under simulated deep-sea pressure. *Hydrobiologia 426*, 173-183.

Grant, W.D., Atkinson, M., Burke, B. and Molloy, C. 1996. *Chitinolysis* by the marine *ascomycete Corollospora maritima* Werdermann: Purification and properties of a chitobiosidase. *Bot. Mar. 39*, 177-186.

Gray, W.D., Pinto, P.V.C. and Pathak, S.G. 1963. Growth of fungi in sea water medium. *Appl. Microbiol. 11*, 501-505.

Guerriero, A., D'Ambrosio, M., Cuomo, V., Vanzanella, F. and Pietra, F. 1988. *Dendryphiellin A,* the first fungal *trinor-eremophilane*: Isolation from the marine deuteromycete *Dendryphiella salina* (Sutherland) Puch et Nicot. *Helvetica Chmica Acta 71*, 57-61.

Haefner, B. 2003. Drugs from the Deep. *Drug Discov. Today 8*, 536-544.

Han, F., Yao, W., Yang, X., Liu, X., Gao, X. 2005. Experimental study on anticoagulant and antiplatelet aggregation activity of a chemically sulfated marine polysaccharide YCP. *Int. J. Biol. Macromol. 36*, 201-207.

Hawksworth, D.L. 1991. The fungal dimension of biodiversity: magnitude, significance and conservation. *Mycol. Res. 95*, 641-655.

Huang,Y.F., Li, L.H., Tian, L., Qiao, L., Hua, H.M., Pei, Y.H. 2006. Sg17-1-4, a novel isocoumarin from a marine fungus *Alternaria tenuis* Sg17-1. *J. Antibiot. 59*, 355-357.

Hwang, Y., Rowley, D., Rhodes, D., Gertsch, J., Fenical, W. and Bushman, F. 1999. Mechanism of inhibition of a poxvirus *topoisomerase* by the marine natural product Sansalvamide A. *Mol. Pharmacol. 55*, 1049-1053.

Hyde, K.D. 1992. The effect of incubation on the occurrence of marine fungi on randomly collected lignocellulose samples. *Sydowia 44*, 131-136.

Hyde, K.D., Farrant, C.A. and Jones, E.B.G. 1987. Isolation and culture of marine fungi. *Bot. Mar. 30*, 1-96.

Hyde, K.D., Greenwood, R. and Jones, E.B.G. 1993. Spore attachment in marine fungi. *Mycol. Res. 1*, 7-14.

Isaka, M., Suyarnsestakorn, C., Tanticharoen, M., Kongsaeree, P. and Thebtaranonth, Y. 2002. Aigialomycins A-E, new resorcylic macrolides from the marine mangrove fungus *Aigialus parvus. J. Org. Chem. 67*, 1561-1566.

Jennings, D.H. 1983. Some aspects of the physiology and biochemistry of marine fungi. *Biol. Rev. 58*, 423-459.

Jensen, P.R. and Fenical, W. 2000. Marine microorganisms and drug discovery: Current status and future potential. In: *Drugs from the Sea* (Ed. Fusetani, N.). Karger, Basel, 6-29.

Jensen, P.R. and Fenical, W. 2002. Secondary metabolites from marine fungi. In: *Fungi in Marine Environments* (Ed. Hyde, K.D.). Fungal Diversity Research Series # 7, Hong Kong University Press, Hong Kong, 293-315.

Jiang, Z., Barret, M.C., Boyd, K.G., Adams, D.R., Boyd, A.S.F., Burgess, J.G. 2002. JM 47, a cyclic tetrapeptide HC-toxin analogue from a marine *Fusarium* species. *Phytochem. 60*, 33-38.

Johnson, T.W. and Sparrow, F.K. 1961. *Fungi in Oceans and Estuaries.* J. Cramer, Weinheim. Germany.

Johnson, R.G., Jones, E.B.G. and Moss, S.T. 1991. Histochemical and electron microscope studies of ascospore ontogeny in three marine fungi with Cleistothecia: *Amylocarpus encephaloides* Currey, *Dryosphaera navigans* Koch *et* Jones and *Eiona tunicata* Kohlm. *Bot. Mar. 34*, 229-239.

Jones, E.B.G. 1993. Tropical marine fungi. In: *Aspects of Tropical Mycology* (Ed. Isaac, S., Frankland, J.C., Watling, R. and Whalley. A.J.S.). Cambridge University Press, Cambridge, 73-89.

Jones, E.B.G. 1995. Ultrastructure and taxonomy of the aquatic ascomycetous order Halosphaeriales. *Can. J. Bot. 73*, S790-S801.

Jones, E.B.G. 1988. Do fungi occur in the sea? *Mycologist 2*, 150-157.

Jones, E.B.G. and Byrne, P.J. 1976. Physiology of the higher marine fungi. In: *Recent Advances in Aquatic Mycology* (Ed. Jones, E.B.G.). Elek, London, 135-176.

Jones, E.B.G., Mitchell, J.I. 1996. Biodiversity of marine fungi. In: *Biodiversity, International Biodiversity Seminar* (Ed. Climerman, A. and Gunde-Climerman, N.). National Institute of Chemistry and Slovenia National Commission for UNESCO, Ljublijana, 31-42.

Jones, E.B.G. and Moss, S.T. 1980. Light and Scanning electron microscope observations of the marine ascomycete *Crinigera maritima. Trans. Br. Mycol. Soc. 74*, 625-631.

Kakeya, H., Takahashi, I., Okada, G., Isono, K. and Hiroyuki, O. 1995. Epolactaene, a novel neuritogenic compound human *neuroblastoma* cells produced by a marine fungus. *J. Antibiot. 48*, 733-735.

Kamlangdee, N. and Fan, K.W. 2003. Polyunsaturated fatty acids production by *Schizochytrium* sp. isolated from mangrove Songklanakarin. *J. Sci. Technol. 25*, 643-650.

Kelecom, A. 1999. Chemistry of marine natural products: Yesterday, today and tomorrow. *Ann. Acad. Bras. Cienc. 71*, 249-263.

Kendrick, A.J. and Ratledge, C. 1992. Microbial polyunsaturated fatty acids of potential commercial interest. *Soc. Indust. Microbiol. News 42*, 59-65.

Khudiakova, Iu.V., Pivkin, M.V., Kuznetsova, T.A. and Svetashev, VI. 2000. Marine fungi in sediments of the Sea of Japan (Russian territory) and their biologically active metabolites. *Mikrobiologia 69*, 722-726.

Kobayashi, J. and Ishibashi, M. 1993. Bioactive Metabolites of Symbiotic Marine Microorganisms. *Chem. Rev. 93,* 1753-1769.

Kobayashi, H., Meguro, S., Yoshimoto, T. and Namikoshi, M. 2003. Absolute structure, biosynthesis and anti-microtubule activity of phomopsidin, isolated from a marine-derived fungus *Phomopsis* sp. *Tetrahedron 59*, 455-459.

Kohlmeyer, J. 1977. New genera and species of higher fungi from the deep sea (1615-5315 m). *Rev. Mycol. 41*, 189-206.

Kohlmeyer, J. and Kohlmeyer, E. 1979. *Marine Mycology - The higher fungi.* Academic Press, New York.

Kohlmeyer, J., Spatafora, J.W. and Volkmann-Kohlmeyer, B. 2000. Lulworthiales, a new order of marine Ascomycota. *Mycologia 92*, 453-458.

Kohlmeyer, J. and Volkmann-Kohlmeyer, B. 1991. Illustrated key to the filamentous higher marine fungi. *Bot. Mar. 34*, 1-61.

Konig, G.M. and Wright, A.D. 1996. Marine natural products research: Current directions and future potential. *Planta Medica 62*, 193-211.

Kumaresan, V. and Suryanarayanan, T.S. 2002. Endophyte assemblage in young, mature and senescent leaves of *Rhizophora apiculata*: Evidence for the role of endophytes in mangrove litter degradation. *Fungal Diversity 9*, 81-91.

Landy, E.T. and Jones, G.M. 2006. What is the Fungal Diversity of Marine Ecosystems in Europe? *Mycologist 20*, 15-21.

Lee, C., Howarth, R.W. and Howes, B.L. 1980. Sterols in decomposing *Spartina alterniflora* and the use of ergosterol in estimating the contribution of fungi to detrital nitrogen. *Limnol. Oceanogr. 25*, 290-303.

Li, J.Y., Sidhu, R., Bollon, A. and Strobel, G.A. 1998. The stimulation of taxol production in liquid cultures of *Pestalotiopsis microspora* Ne32. *Mycol. Res. 102*, 461-464.

Li, J.Y. and Strobel, G.A. 2001. Jesterone and hydroxy-jesterone antioomycete cyclohexenone epoxides from the endophytic fungus *Pestalotiopsis jesteri. Phytochem. 57*, 261-265.

Li, Y., Li, X., Kang, J.S., Choi, H.D. and Son, B.W. 2004. New radical scavenging and ultraviolet-A protecting prenylated dioxopiperazine alkaloid related to isoechinulin A from a marine isolate of the fungus *Aspergillus. J. Antibiot.* 57, 337-340.

Liberra, K. and Lindequist, U. 1995. Marine fungi – A prolific resource of biologically active natural products. *Pharmazie 50*, 583-588.

Liberra, K., Jensen, R. and Lindequist, U. 1998. Corollosporine, a new *phthalide* derivative from the marine fungus *Corollospora maritima* Werderm. 1969. *Pharmazie 53*, 578-581.

Lin X, Huang, Y., Fang, M., Wang, J., Zheng, Z., Su, W. 2005. Cytotoxic and antimicrobial metabolites from marine lignicolous fungi, *Diaporthe* sp. *FEMS Microbiol. Lett. 251*, 53-58.

Liu, W., Gu, Q., Zhu, W., Cui, C., Fan, G., Zhu, T., Liu, H. and Fang, Y. 2005. Penicillones A and B, two novel polyketides with tricyclo [5.3.1.03,8] *undecane* skeleton, from a marine-derived fungus *Penicillium terrestre*, *Tetrahedron Lett. 46*, 4993–4996.

Liu, R., Zhu, W., Zhang, Y., Zhu, T., Liu, H., Fang, Y., Gu, Q. 2006. A new diphenyl ether from marine-derived fungus *Aspergillus* sp. B-F-2. *J. Antibiot. 59*, 362-365.

Lorenz, R. and Molitorius, H.P. 1997. Cultivation of fungi under simulated deep-sea conditions, *Mycol. Res. 101*, 1355-1365.

Malmstrom, J., Christophersen, C., Barrero, A.F., Oltra, J.E., Justica, J. and Rosales, A. 2002. Bioactive metabolites from a marine-derived strain of the fungus *Emericella variecolor. J. Nat Prod. 65*, 364-367.

Masuma, R., Yamaguchi, Y., Noumi, M., Omura, S. And Namikoshi, M. 2001. Effect of sea water concentration on hyphal growth and antimicrobial metabolite production in marine fungi. *MycoScience* 42, 455-459.

Masurekar, P.S. 1992. Therapeutic matabolites. In: *Biotechnology of Filamentous Fungi: Technology and Products* (Ed. Finkelstein, D.B. and Ball, C.). Butterworth-Heinemann, Boston, USA, 241-301.

Mayer, A.M.S and Gustafson, K.R. 2003. Marine pharmacology in 2000: Antitumor and cytotoxic compounds. *Int. J. Canc. 105*, 291-299.

Mayer, A.M.S. and Hamann, M.T. 2002. Marine pharmacology in 1999: Compounds with antibacterial, anticoagulant, antifungal, anti-inflammatory, anthelmintic, anti-inflammatory, antiplatelet, antiprotozoal andantiviral activities; affecting the cardiovascular, endocrine, immune, and nervous systems; and other miscellaneous mechanisms of action. *Comp. Biochem. Physiol. C, Comp. Pharmacol. Toxicol. 132*, 315-339.

Mayer, A.M.S. and Hamann, M.T. 2004. Marine pharmacology in 2000: Marine compounds with antibacterial, anticoagulant, antifungal, anti inflammatory, anti malarial, anti platelet, anti tuberculosis, and antiviral activities; affecting the cardiovascular, immune, and nervous systems and other miscellaneous mechanisms of action. *Mar. Biotechnol. 6*, 37-52.

Mayer, A.M.S. and Hamann, M.T. 2005. Marine pharmacology in 2001–2002: Marine compounds with anthelmintic, antibacterial, anticoagulant, antidiabetic, antifungal, anti-inflammatory, antimalarial, antiplatelet, antiprotozoal, antituberculosis, and antiviral activities; affecting the cardiovascular, immune and nervous systems and other miscellaneous mechanisms of action. *Comp. Biochem. Physiol. C 140*, 265-286.

Mayer, A.M.S. and Lehmann, V.K.B., 2000. Marine pharmacology in 1998: Marine compounds with antibacterial, anticoagulant, antifungal, anti-inflammatory, anthelmintic, antiplatelet, antiprotozoal, and antiviral activities; with actions on the cardiovascular, endocrine, immune, and nervous systems and other miscellaneous mechanisms of action. *Pharmacologist 42*, 62-69.

McCarthy, P.J. and Pomponi, S.A. 2004. A search for new pharmaceutical drugs from marine organisms. *Marine Biomed. Res.* 1-2.

Meyers, S.P. 1996. Fifty years of marine mycology: Highlights of the past, projections for the coming century. *Soc. Indust. Microbiol. News* 46, 119-127.

Miller, J.D. and Blackwell, B.A. 1986. Biosynthesis of 3 *acetyledeoxynivalenol* and related compounds by *Fusarium culmorum* HLX 1503 in a stirred jar fermentor. *Can. J. Bot. 64*, 1-5.

Miller, J.D. and Savard, M.E. 1989. Antibiotic activity of the marine fungus *Leptosphaeria oraemaris. Proc. Nova Scotia Institute of Sci. 39*, 51-58.

Miller, J.D., Savard, M.E. and Rapior, S. 1994. Production and purification of *fumonisins* from a stirred jar fermentor. *Natural toxins 2*, 354-359.

Mohamad, M., Keudell, K.C. and Gessner RV. 1989. Use of ELISA to differentiate marine species of *Dendryphiella. Mycol. Res. 93*, 400-402.

Monaghan, R.L., Polishook, J.D., Pecore, V.J., Bills, G.F., Nallin-Omstead, M. and Streicher, S.L. 1995. Discovery of novel metabolites from fungi-is it really a random walk through a random forest? *Can. J. Bot. 73*, S925-S931.

Morris, G.J., Smith, D. and Coulson, G.E. 1988. A comparative study of the changes in the morphology of hyphae during freezing and viability upon thawing for twenty species of fungi. *J. Gen. Microbiol. 134*, 2897-2906.

Moss, S.T. 1990. The relevance of scanning electron microscopy to the taxonomy of marine *ascomycetes,* In: *Scanning Electron Microscopy in Taxonomy and Functional Morphology* (Ed. Claugher, D.). Systematics Association, Special Volume # 41, Clarendon Press, Oxford, 149-170.

Nagumu, Y., Kadeya, H., Yamaguchi, J., Uno, T., Shoji, M., Hayashgi, Y. and Osada, H. 2004. Structure-activity relationships of *epolactaene* derivatives: structural requirements for inhibition of Hsp60 chaperone activity. *Bioorg. Medic. Chem. Lett. 14*, 4425-4429.

Nakagiri, A. and Jones, E.B.G. 2000. Long-term maintenance of cultures. In: *Marine Mycology - A practical Approach* (Ed. Hyde, K.D. and Pointing, S.B.). Fungal Diversity Research Series 1, Fungal Diversity Press, HongKong, 62-68.

Newell, S.Y.1992. Estimating fungal biomass and productivity in decomposing litter, In: *The Fungal Community*, 2nd Edition (Ed. Carroll, G.C. and Wicklow, D.T.). Marcel-Dekker, New York, 521-561.

Newell, S.Y. 1993. Membrane-containing fungal mass and fungal specific growth rate in natural samples. In : *Handbook of methods in Aquatic Microbial Ecology* (Ed. Kemp, P.F., Sherr, B.F., Sherr, E.B. and Cole, J.J.). Lewis publishers, Boca Raton, USA, 579-586.

Newell, S.Y. 1996. The [^{14}C] acetate-to-ergosterol method: L factors for conversion from acetate incorporated to organic fungal biomass synthesised. *Soil Biol. Biochem. 28*, 681-683.

Newman, D.J., Cragg, G.M., Snader, K.M. 2000. The influence of natural products upon drug discovery. *Nat. Prod. Rep. 17*, 215-34.

Nielsen, J., Nielsen, P.H. and Frisvad, J.C. 1999. fungal depside, *guisinol* from a marine derived strain of *Emericella unguis. Phytochem. 50*, 263-265.

Nishii, T. and Nakagiri, A. 1991. Cryopreservation of *oomycetous* fungi in liquid nitrogen. *Inst. Ferment. Osaka Res. Com. 15*, 105-118.

Numata, A., Iritani. M., Yamada, P., Minoura, K., Matsumura, E., Yamori, T. and Tsuruo, T. 1997. Novel antitumor metabolites produced by a fungal strain from a sea hare. *Tetrahedron Lett. 38*, 8215-8218.

Nylund, J.E. and Wallander, H. 1992. Ergosterol analysis as a means of quantifying *mycorrhizal* biomass, In: *Methods in Microbiology*, Volume 24 (Ed. Norris, J.R., Read, D.J. and Varma A.K.). Academic Press, London, 77-88.

Okada, G., Talematsu, A. and Talamura, Y. 1997. Phylogenetic relationships of the *hyphomycete* genera *Chaetopsina* and *Kinochaeta* based on 18S rDNA sequences. *Mycoscience 38*, 409-420.

Osterhage, C., Kaminsky R., Konig G.M. and Wright A.D. 2000. *Ascosalipyrrolidinone A,* an anti microbial alkaloid, from the obligate marine fungus *Ascochyta salicorniae. J. Org. Chem. 65*, 6412-6417.

Kirk, P.W. Jr. 1983. Direct enumeration of marine *arenicolous* fungi. *Mycologia 75*, 670-682.

Panebianco, C., Tam, W.Y. and Jones, E.B.G. 2002. The effect of pre-inoculation of blasa wood by selected marine fungi and their effect on subsequent colonization in the sea. *Fungal Diversity 10*, 77-88.

Paul, E.A. and Johnson, R.L. 1977. Microscopic counting and adenosine 5'-triphosphate measurement in determining microbial growth in soils. *Appl. Environ. Microbiol. 34*, 263-269.

Pisano, M.A., Mihalik, L.A. and Cataland, G.R. 1964. Gelatinase activity by marine fungi. *Appl. Microbiol. 12*, 470-474.

Pointing, S.B., Buswell, J.A., Jones, E.B.G. and Vrijmoed, L.L.P. 1999. Extracellular *cellulolytic* enzyme profiles of lignicolous mangrove fungi. *Mycol. Res. 103*, 696-700.

Pointing, S.B. and Hyde, K.D. 2000. Lignocellose-degrading marine fungi. *Biofouling 15*, 221-229.

Pointing, S.B., Vrijmoed, L.L.P. and Jones, E.B.G. 1998. A qualitative assessment of *lignocellulose* degrading enzyme activity in marine fungi. *Bot. Mar. 41*, 293-298.

Porter, D. 1986. Mycoses of marine organisms: an overview of pathogenic fungi. In: *The Biology of Marine Fungi* (Ed. Moss, S.T.). Cambridge University Press, UK, 141-153.

Prasannarai, K. and Sridhar, K.R. 2003. Abundance and diversity of marine fungi on Intertidal woody litter of the west coast of India on prolonged incubation. *Fungal Diversity 14*, 127-141.

Prasannarai, K., Ananda, K. and Sridhar, K.R. 1999. Intertidal fungi in Mangalore Harbour, Southern India. *Bot. Mar. 42*, 117-122.

Proksch, P., Ebel, R., Edrada, R.A., Schupp, P., Lin, W.H., Sudarsono, V.W. and Steube, K. 2003. Detection of pharmacologically active natural products using ecology. Selected examples from Indopacific marine invertebrates and sponge-derived fungi. *Pure Appl. Chem. 75*, 343-352.

Raghukumar, C. and Raghukumar, S. 1998. Bartolerance of fungi isolated from deep-sea sediments of the Indian Ocean. *Aqua. Microbiol. Ecol. 15*, 153-163.

Raghukumar, C., D'Souza, T.M., Thorn, R.G. and Reddy, C.A. 1999. Lignin-modifying enzymes of *Flavodon flavus*, a basidiomycete isolated from a coastal marine environment. *Appl. Environ. Microbiol. 65*, 2103-2111.

Raghukumar, C., Raghukumar, S., Chinnaraj, A, Chandramohan, D., D's ouza, T.M. and Reddy, C.A. 1994. Laccase and other ligocellulose modifying enzmes of marine fungi isoalted from the coast of India. *Bot. Mar. 37*, 515-523.

Read, S.J., Jones, E.B.G., Moss, S.T. and Johnson, R.G. 1992. Ultrastructural observations of the marine Ascomycotina *Corollospora angusta*, *Corollospora colossa*, *Corollospora lacera* and *Chaetosphaeria chaetosa. Bot. Mar. 35*, 553-560.

Rees, G. and Jones, E.B.G., 1984. Observations on the attachment of spores of Marine fungi, *Bot. Mar. 27*, 145-160.

Rees, G. and Jones, E.B.G. 1985. The fungi of coastal sand dune system. *Bot. Mar.* 28, 213-220.

Rohrmann, S. and Molitoris, H.P. 1992. Screening for wood decay enzymes in marine fungi. *Can. J. Bot. 70*, 2116-2123.

Shearer, C.A. and Von Bodman, S.B. 1983. Patterns of occurrence of *Ascomycetes* associated with decomposing twigs in a mid-western stream. *Mycologia 75*, 518-530.

Shearer, C.A., Descals, E., Kohlmeyer, B., Kohlmeyer, J., Marvanova, L. Padgett, D.E. Porter, D., Raja, H.A., Schmit, J.P., Thorton, H.A. and Voglymayr, H. 2007. Fungal biodiversity in aquatic habitats. *Biodivers. Conserv. 16*, 49-67.

Smith, D. and Onions, A.H.S. 1994. *The Preservation and Maintenance of Living Fungi.* CAB International, Wallingford.

Spatafora, J.W., Volkmann-Kohlmeyer, B. and Kohlmeyer, J. 1998. Independent terrestrial origins of the Halosphaeriales (marine ascomycota). *Am. J. Bot. 85*, 1569-1580.

Sridhar, K.R. and Maria, G.L. 2006. Fungal diversity on woody litter of *Rhizophora mucronata* in a southwest Indian mangrove. *Indian J. Mar. Sci. 35,* 318-325.

Srinivasan, M.C., Rale, M.V. and Ingle, S. 1999. Production and application of enzymes stable to and active under extreme environments: an overview. *Proc. Indian Nat. Acad. 65*, 143-162.

Strobel, G.A., Woapong, F.J., Harper, J.K., Arif, A.M., Grant, D.M., Fung, P.C.W. and Chan, K. 2002. Isopestacin, an isobenzopuranone from *Pestalotiopsis microspora*, possessing antifungal and antioxidant activities. *Phytochem. 60*, 179-183.

Strongman, D.B., Miller, J.D., Calhoun, L., Findlay, J.A. and Whitney, N.J. 1987. The biochemical basis of interference competition among some *lignicolous* marine fungi. *Bot. Mar. 30*, 21-26.

Sugano, M., Sato, A., Iijima, Y., Oshima, T., Furuya, K., Kuwano, H. and Hanzawa, H. 1991. Phomactin A: A novel PAF antagonists from marine fungus, *Phoma* sp. *J. Am. Chem. Soc. 113*, 5463-5464.

Sugano, M., Sato, A., Iijima, Y., Furuya, K., Haruyama, H., Yoda, K. and Hato, T. 1994. Phomactins novel PAF antagonists from marine fungus, *Phoma* sp. *J. Org. Chem. 59*, 564-569.

Sugano, M., Sato, A., Iijima, Y., Furuya, K., Kuwano, H. and Hato, T. 1995. Phomactin E, F, and G: New Phomactin-group PAF antagonists from a marine fungus, *Phoma* sp. *J. Antibiot. 48*, 1188-1190.

Sun, C., Shan, C.Y., Gao, X.D., Tan, R.X. 2005. Protection of PC12 cells from hydrogen peroxide-induced injury by EPS2, an *exopolysaccharide* from a marine filamentous fungus *Keissleriella* sp. YS4108. *J. Biotech. 115*, 137-144

Suryanarayanan, T.S. and Kumaresan, V. 1998. A simple method for storing and transporting fungal cultures. *Mycologist 12*, 173.

Vandamme, E.J. 2003. Bioflavours and fragrances via fungi and their enzymes. *Fungal Diversity 13*, 153-166.

Varoglu, M., Corbett T.H., Valeriote F.A. and Crews, P. 1997. Asperazine, a selective cytotoxic alkaloid from a sponge-derived culture of *Aspergillus niger. J. Org. Chem. 62*, 7078-7079.

Wildman, H.G. 1997. Potential of tropical microfungi within the pharmaceutical industry. In: *Biodiversity of Tropical Microfungi* (Ed. Hyde, K.D.). Hong Kong University press, Hong Kong, 29-46.

Yanagihara, M., Sasaki-Takahashi, N., Sugahara, T., Yamamoto, S., Shinomi, M., Yamashita, I., Hayashida, M., Yamanoha, B., Numata, A., Yamori, T. and Andoh, T. 2005, Leptosins isolated from marine fungus *Leptoshaeria* species inhibit DNA topoisomerases I and/or II and induce apoptosis by inactivation of Akt/protein kinase B. *Cancer Sci. 96*, 816-824.

Yang, X.B., Gao, X.D., Han, F. and Tan, R.X. 2005. Sulfation of a polysaccharide produced by a marine filamentous fungus *Phoma herbarum* YS4108 alters its antioxidant properties *in vitro. Biochim. Biophys. Acta 1725*, 120-127.

Yu, C.M., Curtis, J.M., Walter, J.A., Wright, J.L.C., Ayer, S.W., Kaleta, J., Querengesser, L. and Fathi-Afshar, Z.R. 1996a. Potent inhibitors of cysteine proteases from the marine fungus *Microascus longirostris. J. Antibiot. 49*, 395-397.

Yu, C.M., Curtis, J.M., Wright, J.L.C., Ayer, S.W. and Fathiafshar, Z.R. 1996b. An unusual fatty acid and its glyceride from the marine fungus *Microsphaeropis olivacea. Can. J. Chem. 74*, 730-735.

Yuen, T.K., Hyde, K.D. and Hodgkiss, I.J. 1999. Interspecific interactions among tropical and subtropical freshwater fungi. *Microbial Ecol. 37*, 257-262.

Frontiers in Fungal Ecology, Diversity and Metabolites (edited by K.R. Sridhar), I.K. International Publishing House Pvt. Ltd., New Delhi, 2009.

5

Trichomycetes

J.K. Misra
Mycological Research Unit, Department of Botany, Sri Jai Narain Postgraduate College, Lucknow 226 001, India
(E-mail: jitrachravi@airtelmail.in)

Introduction

Trichomycetes belong to the phylum Zygomycota of the kingdom 'Fungi' and exist as endobionts of adults and larvae of terrestrial, marine, and freshwater arthropods, insects, crustaceans, and millipedes. These fungi remain firmly attached to the gut cuticle of the host through holdfasts (see Figure 2). The preferred region of these fungi is the hindgut in the digestive tract, but some are also found in the foregut or midgut. Trichomycetes possess branched or unbranched thalli (Fig. 1-4,7,8), and asexually reproduce by means of arthrospores (Fig. 7), sporangiospores (Fig. 8), or trichospores (Figs. 1-5). Some members belong to Harpellales, the dominant order of the class, reproduce sexually and produce biconical zygospores (Fig. 6). Although trichomycetes are cosmopolitan in distribution, their hosts have been sought mainly in Europe and the United States and thus most information on trichomycetes is confined to these regions of the world (Lichtwardt, 1986; Lichtwardt *et al.*, 2001; Misra, 1998; Misra and Lichtwardt, 2000). Many parts of the world, particularly the tropics, including the neo-tropics are still awaiting investigations on trichomycetes (Misra, 2001; White *et al.,* 2000). Reports by Rajagopalan (1967) and surveys undertaken by Misra and co-workers in India (Misra and Tiwari, 2002) and Strongman in China (Doug Strongman, Canada, personal comm.), among others are supportive of the view that trichomycetes are cosmopolitan in distribution and can be traced wherever their hosts exist (Lichtwardt, 1986; Misra, 1998, 2001; Misra and Lichtwardt, 2000; White *et al.*, 2000). In view of encouraging mycologists to pursue this fascinating field, the current chapter provides a brief review of research on trichomycetes with a status report with special emphasis on the Indian subcontinent.

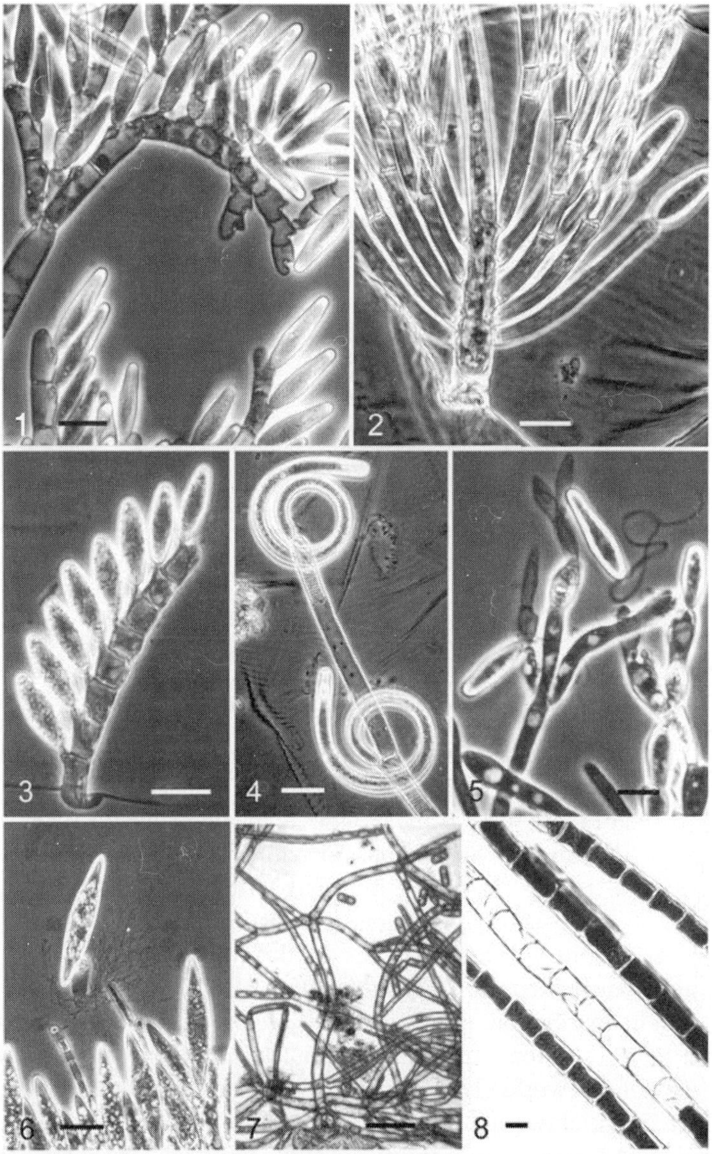

Fig. 1–8. Trichospores produced on sporulating branchlets of Zygopolaris ephemeridarum from an Ephemeroptera (mayfly) nymph (1); Base of a *Genistellospora homothallica* thallus showing holdfast attached to the hindgut cuticle of a Simuliidae (black fly) larva (2); Sporulating thallus of *Stachylina pedifer* as seen through the transparent peritrophic membrane from the midgut of a Chironomidae (midge) larva (3); Coiled trichospores of *Harpella melusinae* produced on an unbranched thallus within the peritrophic membrane of a Simuliidae larva (4); Released trichospore of *Legeriomyces ramosus* showing two appendages from an Ephemeroptera nymph (5); Released zygospore of *Trichozygospora chironomidarum* with a collar and multiple appendages, from a Chironomidae larva (6); Branched thallus of *Asellaria ligiae* attached to the hindgut cuticle of a marine isopod, releasing some cylindrical arthrospores (7); Unbranched thalli of *Arundinula orconectis* from a freshwater crayfish (Crustacea, Astacidae) producing sporangiospores, and some empty sporangia (8) (Scale bars: 1-6, 8 = 20 μm; 7 = 100 μm)

History and Systematics

Joseph Leidy (1849), an American naturalist is credited for initiating the study of trichomycetes by reporting species of the genus *Enterobrus* (now *Enterobryus*) in the hindguts of millipedes and beetles. Later, European biologists, mainly protozoologists, enriched the information about these fungi. Léger and Duboscq (1929) reported the first species of Harpellales (*Harpella melusinae*). During the mid 20th century, a few American investigators began to discover these fungi. Most investigations were carried out by Professor R.W. Lichtwardt, University of Kansas, Lawrence, and his students, trainees and co-investigators in different parts of the world. Lichtwardt is credited with studying these fungi in detail and resolving many taxonomic puzzles that existed before the publication of his monograph (Lichtwardt, 1986).

Currently, 57 genera belonging to 249 species of trichomycetes are known from all over the world. These figures are bound to increase because of the efforts of the investigators who are busy exploring areas that were poorly surveyed. The class, trichomycetes traditionally had three orders (*Asellariales, Eccrinales* and *Harpellales*) and six families (*Asellariaceae, Eccrinaceae, Palavasciaceae, Parataeniellaceae, Harpellaceae* and *Legriomycetaceae*). However, recent molecular data indicate that *Eccrinales*-like amoebidiales formerly classified in trichomycetes are protista rather than fungi (Cafaro, 2005). Now it has been included in *Eccrinales*, which are also gut fungi, while consolidating the information on the genera and species of trichomycetes known up-to-date (Table 1). There are a few ultrastructural studies on the members of the order *Harpellales*, which revealed that the trichospore is a dehiscent monosporous sporangium (Farr and Lichtwardt, 1967; Moss and Lichtwardt, 1976). The early relationships among the orders of trichomycetes (*Asellariales* and *Eccrinales* as derivatives of *Harpellales*) were suggested by Lichtwardt (1986). On the basis of serology and rDNA sequences, *Kickxellales* (*Zygomycetes*) and *Harpellales* share a common ancestry as indicated by Sangar *et al.* (1972) and O'Donnell *et al.* (1998). A recent study by Cafaro (2005) involving sequence analysis of 18S and 25S ribosomal genes clearly suggests that *Eccrinales* like Amoebidiales (Benny and O'Donnell, 2000) are protists belonging to the Mesomycetozoan clade.

Physiology

Studies on the physiology of trichomycetes are limited as a majority of these fungi are unable to grow on artificial media and the difficulty in raising their hosts without other gut microbial consortium. However, *Smittium culisetae* have been used for studies on nutrition and physiology. Studies by Williams (1971), and Williams and Lichtwardt (1972) revealed that tryptone stimulates growth of *Sm. culisetae* and glucose is a better carbon source than 18 other carbohydrate sources. The maximum growth and dry mass was obtained in shake cultures at lower temperatures (Williams and Lichtwardt, 1972). Farr and Lichtwardt (1967) reported that an initial pH of 8.3 of tryptone-glucose-salt medium resulted in the better growth of *Sm. culisetae*

Table 1. *Currently known genera and species of the six families of trichomycetes (Asellariales, Eccrinales, Palavasciaceae, Harpellales, Parataeniellaceae and Legeriomycetaceae)*

Genus	Species	Total species	Species known from India
Asellariaceae			
Asellaria	*A. armadillidii, A. aselli, A. caulleryi, A. gramenei, A. ligiae, A. unguiformis*	6	-
Baltomyces	*B. styrax*	1	-
Orchesellaria	*O. lattesi, O. mauguioi, O. podurae, O. pelta*	4	-
Total genera: 3	Total species: 11	11	-
Eccrinaceae			
Alacrinella	*A. limnoriae, A. sanjuanensis*	2	-
Arundinula	*A. abyssicola, A. capitata, A. galatheae, A. hapalogaster, A. orconectis, A. washingtoniensis*	6	-
Astreptonema	*A. corophii, A. gammari, A. longispora, A. pacificum, A. typica*	5	-
Eccrinidus	*E. flexilis*	1	-
Eccrinoides	*E. broelemanni, E. helleriae, E. henneguyi, E. monticolae*	4	-
Enterobryus	*E. adjanohouni, E. ahlesi, E. apheloriae, E. bifurcates, E. borariae, E. centroboli, E. chaleponci, E. cherikiae, E. cingaloboli, E. cylindroiuli, E. dixidesmi, E. elegans, E. euryuri, E. halophilus, E. hydrophilorum, E. isoporostrepti, E. leptoiuli, E. moniliformis, E. onychostrepti, E. oxidi, E. oxydesmi, E. peridontopygei, E. tuzetae, E. zinophorae*	24	1
Enteromyces	*E. callianassae*	1	-
Enteropogon	*E. sexuale*	1	-
Lajasiella	*L. aphodii*	1	-
Leidyomyces	*L. attenuatus*	1	-
Paramacrinella	*P. microdeutopi*	1	-
Passalomyces	*P. compressus*	1	-
Ramacrinella	*R. raibauti*	1	-
Taeniella	*T. carcini*	1	-
Taeniellopsis	*T. flexilis, T. orchestiae, T. susplugasi*	3	-
Total genera: 15	Total species: 53	53	1
Palavasciaceae			
Palavascia	*P. patagonica , P. philosciae, P. sphaeromae*	3	-
Total genera: 1	Total species: 3	3	-
Parataeniellaceae			
Parataeniella	*P. armadillidii, P. dilatata, P. flavospora, P. latrobi, P. mercieri, P. scotonisci*	6	-
Total genera: 1	Total species: 6	6	-
Harpellaceae			
Harpella	*H. amazonica, H. leptosa, H. melusinae, H. meridianalis, H. tica*	5	1
Stachylina	*St. acutibasilaris, St. ceratopogonidarum, St. chironomidarum, St. dolichospora, St. euthena, St. grandispora, St. jujuyensis, St. lentica, St. litoralis, St. longa, St. lotica, St. macrospora , St. magna, St. manicata, St. minima, St. minuta, St. nana, St. paludosa, St. paucispora, St. pedifer,*	27	3

	St. penetralis, St. platensis, St. prolifica, St. queenslandiae, St. reflexa, St. robusta, St. thaumaleidarum		
Total genera: 2	Total species: 32	32	4
Legeriomycetaceae			
Allantomyces	*A. caenidarum*	1	-
Austrosmittium	*A. aussiorum, A. biforme, A. kiwiorum, A. norinsulare*	4	-
Baetimyces	*B. ancorae*	1	-
Bojamyces	*B. repens, B. transfuga*	2	-
Capniomyces	*C. stellatus*	1	-
Carouxella	*C. coemeteriensis, C. scalaris*	2	-
Caudomyces	*C. japonicus*	1	-
Coleopteromyces	*C. amnicus*	1	-
Ejectosporus	*E. spica, E. trisporus*	2	-
Ephemerellomyces	*E. aquilonius*	1	-
Furculomyces	*F. boomerangus, F. septentrionalis, F. westraliensis*	3	-
Gauthieromyces	*G. indicus, G. microsporus,*	2	1
Genistelloides	*G. amplispora, G. communis, G. helicoids, G. hibernus*	4	-
Genistellospora	*G. guanacastensis, G. homothallica, G. nubile, G. tepidarum, G. tropicalis*	5	-
Glotzia	*G. centroptili, G. coloradense, G. ephemeridarum, G. plecopterorum, G. stenospora, G. tasmaniensis*	6	-
Graminella	*G. bulbosa, G. microspora*	2	-
Graminelloides	*G. biconica*	1	-
Harpellomyces	*H. abruptus, H. eccentricus*	2	-
Lancisporomyces	*L. vernalis*	1	-
Legerioides	*L. tumidus*	1	-
Legeriomyces	*L. aenigmaticus, L. ramosus, L. rarus*	3	1
Legeriosimilis	*L. europaeus, L. tricaudata*	2	-
Orphella	*O. avalonensis , O. catalaunica, O. coronata, O. haysii, O. helicospora, O. hiemalis*	6	-
Pennella	*P. angustispora, P. arctica, P. asymmetrica, P. grassei, P. hovassi, P. montana, P. simulii*	7	-
Plecopteromyces	*P. leptoperlarum, P. patagoniensis, P. trinotoperlarum*	3	-
Pseudoharpella	*P. arcolamylica*	1	-
Pteromaktron	*P. protrudens*	1	-
Simuliomyces	*S. microsporus*	1	-
Smittium	*S. aciculare, S. acutum, S. alpinum, S. angustum, S. annulatum, S. arcticum, S. arvernense, S. biforme, S. bisporum, S. brasiliense, S. brevisporum, S. bulbosporophorus, S. bullatum, S. caudatum, S. cellaspora, S. chironomi, S. coloradense, S. commune, S. compactum, S. culicis, S. culicisoides, S. culisetae, S. cylindrosporum, S. delicatum, S. dimorphum, S. dipterorum, S. elongatum, S. esteparum, S. fasciculatum, S. fastigatum, S. fecundum, S. fruticosum, S. gigasporus, S. gracilis, S. gravimetallum, S. hecatei, S. heterosporum, S. imitatum, S. incrassatum, S. inexpectans, S. kansense, S. longisporum, S. macrosporum, S. magnosporum, S. megazygosporum, S. microsporum, S. morbosum var. rioplatensis, S. morbosum, S. mucronatum, S. orthocladii, S. ouseli, S. paludis, S. parvum, S. peforatum,*	68	-

	S. pennelli, S. phytotelmatum, S. porstratum, S. precipitiorum, S. pseudodimorphum, S. pusillum, S. rarum, S. rupestre, S. simulatum, S. simulii, S. tipulidarum, S. tronadorium, S. typhellum, S. urbanum		
Spartiella	*S. animae, S. barbata*	2	-
Stachylinoides	*S. arctata*	1	-
Stipella	*S. vigilans*	1	-
Tectimyces	*T. leptophlebiidarum, T. robustum*	2	-
Trichozygospora	*T. chironomidarum*	1	-
Zygopolaris	*Z. borealis, Z. ephemeridarum*	2	1
Total genera: 35	Total species: 144	144	3

than the lower initial pH. Growth, sporulation and pH tolerance have been carried out for *Furculomyces boomerangus* and *Austrosmittium biforme* in axenic cultures and it was concluded that greater biomass production in shake cultures was present than in stationary cultures (Misra, 2000). Both species showed different pH optima for biomass and trichospore production. However, viable trichospores were produced in abundance at pH 8.5. Experiments by Horn (1990) on *Sm. culisetae* have also indicated that sporangiospore extrusion from trichospores (a process of new thalli initiation) is influenced by the hydrogen ion concentration of the medium. Horn (1980), and Horn and Lichtwardt (1981) established that in media deficient of vitamins and sterol, the larvae of the mosquito (*Aedes aegypti*) survived better when infested with *Sm. culisetae*. This indicates the mutualistic relationship between the host and gut fungus under nutritionally deficient conditions. Mosquito larvae killing *Smittium morbosum* was reported by Sweeny (1981). *Harpella melusinae* can render female blackflies infertile due to the formation of cysts in their ovaries (Labeyrie *et al.*, 1996; Moss and Descals, 1986), which is one of the mechanisms of fungal dissemination by the adult female. Thus, deleterious relationships of some of these fungi with their hosts are also evident.

Conclusions and Outlook

In India, only a fraction of the total known genera (10.5%) and species (3.2%) of trichomycetes are investigated. There is almost no information on the occurrence of trichomycetes from the neighboring countries (e.g., Nepal, Pakistan, Bangladesh, Bhutan, Sri Lanka). Undoubtedly, many new taxa await discovery in tropical region. There is an urgent need to intensify investigations on trichomycetes to understand the symbiotic/mutualistic/antagonistic relationship with arthropods in view of possible exploitation as potential insect control agents.

Acknowledgements

The author is grateful to Prof. Robert W. Lichtwardt, Department of Ecology and Evolutionary Biology, University of Kansas, Lawrence, USA for suggesting improvements and providing the plate of figures. The financial assistance for inventorying trichomycetes in India by the Department of Science and Technology, Government of India, New Delhi is highly admired (DST No. SP/50/A 75/98).

References

Benny, G.L. and O'Donnell, K. 2000. *Amoebidium parasiticum* is a protozoan, not a Trichomycete. *Mycologia 92*, 1133-1137.

Cafáro, M.J. 2005. Eccrinales (Trichomycetes) are not fungi, but a clade of protists at the early divergence of animals and fungi. *Mol. Phylo. Evol. 35*, 21-34.

Farr, D.F. and Lichtwardt, R.W. 1967. Some cultural and ultrastructural aspects of *Smittium culisetae* (Trichomycetes) from mosquito larvae. *Mycologia 59*, 172-182.

Horn, B.W. and Lichtwardt R.W. 1981. Studies on the nutritional relationship of larval *Aedes aegypti* (Diptera, Culicidae) with *Smittium culisetae* (Trichomycetes). *Mycologia 73*, 724-740.

Horn, B.W. 1990. Physiological changes associated with sporangiospore extrusion from trichospores of *Smittium culisetae. Exp. Mycol. 14*, 113-123.

Horn, B.W. 1980. *Studies on the Nutritional Relationship of Larval Aedes aegypti (Diptera, Culicidae) with Smittium culisetae (Trichomycetes).* Master of Arts Thesis, University of Kansas, USA.

Labeyrie, E.S., Molloy, D.P. and Lichtwardt, R.W. 1996. An investigation of Harpellales (Trichomycetes) in New York State blackflies (Diptera, Simuliidae). *J. Inverte. Pathol. 68*, 293-298.

Léger, L. and Duboscq, O. 1929. Harpella melusinae n. g. n. sp. Entophyte eccriniforme parasite des larves de Simulie. *Comptes Rendus Hebdomadaires des Séances de l'Académie des Sciences Paris 188*, 951-954.

Leidy, J. 1849. *Enterobrus*, a new genus of Confervaceae. *Proc. Acad. Nat. Sci. Philadel. 4*, 225-233.

Lichtwardt, R.W. 1986. *The Trichomycetes, Fungal Associates of Arthropods*. Springer-Verlag, New York.

Lichtwardt, R.W., Cafáro M.J. and White M.M. 2001. *The Trichomycetes, Fungal Associates of Arthropods.* www.nhm.ku.edu/~fungi.

Misra, J.K. and Lichtwardt, R. W. 2000. *Illustrated Genera of Trichomycetes, Fungal Symbionts of Insects and other Arthropods.* Science Publishers, Inc., Enfield, New Hampshire.

Misra, J. K. 1998. Trichomycetes - Fungi associated with arthropods: Review and world literature. *Symbiosis 24*, 179-219.

Misra, J.K. 2000. Growth, sporulation and pH tolerance of *Furculomyces boomerangus* and *Austrosmittium biforme* in axenic culture. *Mycologia 92*, 1051-1056.

Misra, J. K. 2001. Trichomycetes - Fungi associated with arthropods, An introduction and state-of-the-art in the tropics, In, *Trichomycetes and Other Fungal Groups* (Ed. Misra, J. K. and Bruce, W.H.). Science Publishers Inc., Enfield, USA, 1-13.

Misra, J.K. and Tiwari, V.K. 2002. Harpellales (Trichomycetes) do occur in India? First report of two species of *Stachylina. Kavaka* 30, 71-75.

Moss, S.T. and Descals, E. 1986. A previously undescribed stage in the life cycle of Harpellales (Trichomycetes). *Mycologia 78*, 213-222.

Moss, S.T. and Lichtwardt, R.W. 1976. Development of trichospores and their appendages in *Genistellospora homothallica* and other Harpellales and fine-structural evidence for the sporangial nature of trichospores. *Can. J. Bot. 54*, 2346-2364.

O'Donnell, K., Cigelnick, E. and Benny, G. L. 1998. Phylogenetic relationships among the Harpellales and Kickxellales. *Mycologia 90*, 624-639.

Rajagopalan, C. 1967. An *Enterobryus* (Trichomycetes, Eccrinales) in a millipede. *Current Science 36*, 20-22.

Sangar, V.K., Lichtwardt, R.W., Kirsch, J.A.W. and Lester, R.N. 1972. Immunological studies on the fungal genus *Smittium* (Trichomycetes). *Mycologia 64*, 342-358.

Sweeney, A.W. 1981. An undescribed species of *Smittium* (Trichomycetes) pathogenic to mosquito larvae in Australia. *Trans. Br. Mycol. Soc. 77*, 55-60.

White, M.M., Cafáro, M.J. and Lichtwardt, R.W. 2000. Arthropod gut fungi from Puerto Rico and summary of tropical Trichomycetes worldwide. *Caribb. J. Sci. 36*, 210-220.

Williams, M.C. and Lichtwardt, R.W. 1972. Physiological studies on the cultured Trichomycetes, *Smittium culisetae. Mycologia 64*, 806-815.

Williams, M.C. 1971. *Studies on Trichomycetes and their Relationships to Arthropod Hosts.* Ph.D. Dissertation, University of Kansas, USA.

Frontiers in Fungal Ecology, Diversity and Metabolites (edited by K.R. Sridhar), I.K. International Publishing House Pvt. Ltd., New Delhi, 2009.

6

Forest Pathology in India

V. Ravishankar Rai* and A. Bhanumathi
Department of Microbiology, University of Mysore, Manasagangotri, Mysore 570 006, Karnataka, India
*(*E-mail: rrai33@hotmail.com)*

Introduction

Forests are being increasingly subjected to deforestation and degradation with adverse socio-economic and environmental impacts. In India, degraded forests constitute about 41% of the total forest cover. On realizing the importance of conservation and regeneration, several programmes have been implemented at the government and non-government levels for which healthy seeds and seedlings of forest tree species are essential (Joshi and Singh, 2003). Seed is the main prerequisite for tree regeneration in about 90% of the afforestation programme throughout the world (Naik *et al.*, 2001). Huge quantities of forest tree seeds are collected to raise nursery seedlings to meet the large-scale afforestation targets in India. In most instances, collection of seeds will be made from the natural forests, seed stands and seed orchards. Seeds are also exploited by seed industries to extract oils and fats (e.g., *Shorea robusta, Schleichera oleosa*) and some seeds and fruits are also the major source of food and medicine (Jamaluddin, 1999). Seeds of the forest trees are reported to harbor a variety of mycoflora (Mittal *et al.,* 1990). These infected seeds can be divided into three categories: (1) severely infected seeds, (2) partially infected seeds, and (3) surface infected seeds. In severely infected seeds, the fungi attack the embryo and consequently these seeds fail to germinate resulting in the loss of a huge quantity of seeds. On the other hand, partially infected seeds germinate and serve as infection centers to spread disease in the field thereby causing a significant loss at a later stage. Surface infections may cause seedling blight on the emergence of seedlings (Mehrotra and Mehrotra, 2000).

Forest seed pathology in India is still in its infancy. The importance of seed pathology of forest tree species has increased in view of the growing demand for quality seeds produced in seed orchards and established at different parts of the country. In addition, seeds of fast-growing plant species are imported due to the increasing demand. Seed pathological investigations will help in checking the loss of seeds in storage, seed transmissible diseases of seedlings in the nursery at pre- and post-emergence stages and the spread of diseases from the nursery to the plantations. Apart from this, post-entry quarantine will also help prevent inadvertent entry of harmful exotic pathogens in forest nurseries and plantations (Mehrotra and Mehrotra, 2000).

Seed infecting fungi are divided into field fungi and storage fungi. The former, infect seeds during development or after maturity, while the latter grow and damage seeds during storage. Seed-borne fungi of forest tree species belong to diverse taxonomic groups. The current knowledge on fungal taxonomy that is available in India, does not often permit the precise separation of the individual species of fungi occurring on seeds. However, it is clear that many species usually considered as unimportant, also cause considerable damage under certain conditions (Mehrotra and Mehrotra, 2000). Seeds can be damaged on the trees, during collection, transportation and storage by several factors (e.g., biotic, physical) affecting the seed quality. Although voluminous information is available on seed-borne fungi in cultivated crops in India, the role of seed-borne fungi in deteriorating seed quality of forest tree species in the field and during storage is yet to be established (Ghyre, 2005). Very little is known about the various types of pathogens associated with forest tree species (Mittal and Sharma, 1981). Therefore, the current chapter discusses the seed-borne fungi, nursery diseases (associated with seeds, seedlings, leaf, stem and roots) of forest tree species and disease management strategies with special reference to India.

Seed-borne Fungi of Forest Trees

Seed mycoflora of a number of tree species were studied and recorded by Mathur (1974). Table 1 shows over 60 Indian forest tree species affected by a variety of mycoflora, particularly at seed and seedling stages. Different nature of fungal infection of seeds and seedlings has been reported (seed decay, seed deterioration, inhibition of germination, reduced germination, post harvest loss, loss in storage, loss of seed viability, loss of seedlings, pre-emergence damping, post-emergence damping, and damage of young seedlings). Vaartaja (1952) proved the pathogenic nature of some strains of *Trichoderma* on the seeds and seedlings of *Pinus* species. A heavy loss of *Pinus roxburghii* was attributed to *Curvularia pallescens* (Munjal and Sharma, 1976). Seed decay of *Albizia lebbeck* and *Cassia fistula* was reported by Mittal and Sharma (1979) by *Alternaria, Aspergillus, Mamnoniella, Penicillium* and *Rhizopus*. Seeds of *Shorea robusta* in post harvest storage were infected by *Aspergillus niger* up to 18-100% (Singh *et al.,* 1979). Singh *et al.* (1979) reported germination inhibition of seeds of *Quercus* by fungi. Interestingly, it was found that *Aspergillus fumigatus, Penicillium islandicum* and *P. ingricans* stimulated growth of the plumule in *Quercus*. Mittal and Sharma (1983) reported internal infection and seed deterioration of *Shorea robusta* seeds by *Aspergillus fumiculosus, A. niger, A. wentii, Penicillium albicans, P. canadense, P. frequentans* and *Rhizopus oryzae.* They isolated 25 fungal species from the seeds of *Pinus roxburghii,* of which *Aspergillus niger* and *Rhizopus oryzae* caused deep-seated infection and checked seedling development considerably. Out of 44 fungi isolated from the seeds of *Acacia catechu*, *Bauhinia variegata, Cassia siamea* and *Celtis australis,* three fungi (*Alternaria, Aspergillus* and *Curvularia*) caused seed and seedling rot of up to 30-60% (Paul and Bhardwaj, 1987). Vijayan (1988) also observed rapid deterioration and loss of viability of seeds of *Shorea robusta* due to infection by *Alternaria* sp., *Aspergillus niger, Botryodiplodia* sp., *Cylindrocladium* sp., *Fusarium* sp., *Penicillium* sp. and *Schizophyllum commune.* Germination of seeds of *Oroxylum indicum* was reduced as they were infected by *Aspergillus* and *Penicillium*, while they were still inside the partially opened pods on the trees following pre-monsoon rains in April in Assam (Mehrotra, 1990). They also cause seed decay and affected seed germination adversely. *Verticillium* on seeds of *Eucalyptus* hybrid was detected and the damage of young seedlings was noticed by Harsh *et al.* (1992). Seed mycoflora of stored seeds of *Butea monosperma* showed the presence of eleven fungi, *Aspergillus sulphureus* and two unidentified species were predominant (Purohit and Jamaluddin, 1993). Significant seedling mortality due to pre-emergence damping off by *Fusarium avenaceum, F. proliferatum* and *F. sambucinium* was recorded by Motta *et al.* (1996). Mehrotra (1996) studied the seeds of *Michelia champaca* and reported that the seeds were infected in the nursery by *Fusarium*, while still inside the partly opened fruits spread on the ground for seed extraction.

Nursery Diseases

The launching of a large-scale plantation programme in recent years in India (under social forestry, agroforestry and farm forestry), a large number of tree species are now being raised in the nurseries in different parts of the country for propagation and maintenance of healthy seedlings (Mehrotra, 1998). Many nursery diseases have been reported in the forest nurseries throughout India. Following are the description of important nursery diseases of forest tree species:

Acacia

The trees can grow under even the most difficult conditions in tropics. The rapid early growth rate, ability to fix nitrogen, tolerance to soil conditions (infertile, acid, alkaline and saline) makes it an excellent plant in agroforestry systems. Infusion of tender leaves of *Acacia nilotica* is used as an astringent remedy for diarrhea and dysentery (Prajapati *et al.,* 2003).

Seed and Seedling: Seedling blight, defoliation and dieback due to *Glomerella cingulata* (anamorph *Colletotrichum gloeosporioides*) caused serious losses to *Acacia* spp. in India (Mohanan and Sharma, 1988; Sharma and Bhardwaj, 1988) especially under humid conditions (Mohanan and Sharma, 1988). Leaf spot caused by *Cylindrocladium quinqueseptatum*, a common foliar pathogen of trees,

Table 1: *Seed-borne pathogenic mycoflora of forest trees of India*

Host	Pathogen	Nature of infection	Reference
Acacia auriculiformis	*Fusarium semitectum* and *Phoma* sp.	Seeds	Mathur (1974)
A. catechu	*Alternaria alternata, Aspergillus flavus, A. niger, A. ochraceus, Fusarium avenaceum* and *Rhizopus stolonifer*	Seeds	Punam-Singh *et al.* (1999)
A. modesta	*Fusarium semitectum* and *Phoma* sp.	Seeds	Mathur (1974)
A. nilotica	*Alternaria alternata, Aspergillus flavus, A. niger, A. ochraceus, Fusarium avenaceum* and *Rhizopus stolonifer*	Seeds	Punam-Singh *et al.* (1999)
Acer caesium	*Chlamydomyces palmarum, F. semitectum, Melanospora lagenaria* and *Rhizopus stolonifer*	Seeds	Mehrotra and Singh (1999)
Acrocarpus fraxinifolius	*Botrytis cineria, Fusarium equiseti, F. moniliforme, F. semitectum,* and *Phoma* sp.	Seeds	Mathur (1974)
Adenanathera microsperma	*Botryodiplodia* sp., *Fusarium moniliforme, F. semitectum, Pestalotia* sp. and *Phoma* sp.	Seeds	Mathur (1974)
Albizia lebbeck	*Alternaria* sp., *Aspergillus* sp., *Mamnoniella* sp., *Penicillium* sp. and *Rhizopus* sp.	Seed decay	Mittal and Sharma (1979)
A. procera	*Alternaria alternata, Aspergillus ochraceus, Fusarium avenaceum, Penicillium citrinum, Penicillium* sp. and *Trichothecium roseum*	Seeds	Punam-Singh *et al.* (1999)
A. stipulata	*Fusarium moniliforme, F. solani, Macrophomina phaseolina* and *Phoma* sp.	Seeds	Mathur (1974)
Anthocephalus chinensis	*Alternaria alternata, Aspergillus flavus, A. niger, A. ochraceus, Fusarium avenaceum, F. semitectum, Mucor hiemalis, Penicillium citrinum, Rhizopus stolonifer* and *Thielaviopsis bassicola*	Seeds	Mehrotra and Singh (1999)
Azadirachta indica	*Botryodiplodia* sp., *Cephalosporium* sp., *Drechslera sorokiana, Fusarium equiseti, F. moniliforme, F. semitectum, F. solani, Macrophomina phaseolina, Myrothecium roridum, Phoma* sp., *Phomopsis* sp. and *Verticillium* sp.	Seeds	Mathur (1974)

Table 1 cont.

Host	Pathogen	Nature of infection	Reference
Bauhinia purpurea	*Alternaria alternata, Aspergillus flavus, A. niger, A. ochraceus, Fusarium avenaceum, Penicillium citrinum* and *Spicaria* sp.	Seeds	Punam-Singh *et al.* (1999)
	Pestalotiopsis sp.	Seeds	Mathur (1974)
B. variegata	*Pestalotiopsis* sp. and *Phoma* sp.	Seeds	Mathur (1974)
Buchanania lanzan	*Aspergillus flavus, A. ochraceus, Fusarium avenaceum, Penicillium citrinum* and *Rhizopus stolonifer*	Seeds	Mehrotra and Singh (1999)
Butea monosperma	*Aspergillus flavus, A. fumigatus, A. niger, A. nidulans, A. sulphureus, Chaetomium* sp., *Penicillium* sp. and *Rhizopus stolonifer*	Stored seeds	Purohit and Jamaluddin (1993)
Callistemon viminalis	*Pestalotia* sp.	Seeds	Mathur (1974)
Cassia fistula	*Alternaria* sp., *Aspergillus* sp., *Mamnoniella* sp., *Penicillium* sp. and *Rhizopus* sp.	Seed decay	Mittal and Sharma (1979)
Cedrela toona	*Botrytis cineria* and *Fusarium moniliforme*	Seeds	Mathur (1974)
C. serrata	*Phoma* sp.	Seeds	Mathur (1974)
Cedrus deodara	*Fusarium botryocoides, F. monoliforme* var. *minus* and *Oedocephalum lineatum*	Seeds	Munjal and Sharma (1975)
Chukrasia tabularis	*Fusarium moniliforme, Macrophoma* sp. and *Phoma* sp.	Seeds	Mathur (1974)
Chuckrasia velutina	*Alternaria alternata, Aspergillus flavus, A. niger, A. ochraceus, Penicillium citrinum* and *Rhizopus stolonifer*	Seeds	Mehrotra and Singh (1999)
Dalbergia sissoo	*Alternaria alternata, Fusarium moniliforme, F. oxysporum* and *Tricothecium roseum*	Seeds	Mathur (1974)
Eucalyptus camaldulensis	*Fusarium semitectum*	Seeds	Mathur (1974)
E. citriodora	*Aspergillus niger, Curvularia* sp., *Penicillium albicans* and *Rhizopus oryzae*	Seeds	Sahai and Mehrotra (1982)

Table 1 cont.

Host	Pathogen	Nature of infection	Reference
E. grandis	*Alternaria alternata, Aspergillus niger, Curvularia lunata,* Drechslera austrialensis, Fusarium equiseti, F. monoliforme, *F. oxysporum, F. poae, F. semitectum, Helminthosporium tetramera, Macrophomina phaseolina, Myrothecium roridum* and *Trichothecium roseum*	Seeds	Saxena (1985)
E. tereticornis	*Aspergillus niger, Alternaria alternata, Curvularia lunata,* Drechslera austrialensis, Fusarium equiseti, F. monoliforme, F. oxysporum, F. poae, F. semitectum, Helminthosporium tetramera, *Macrophomina phaseolina, Myrothecium roridum* and *Trichothecium roseum*	Seeds	Saxena (1985)
Eucalyptus sp.	*Verticillium* sp.	Seedling damage	Harsh *et al.* (1992)
Eucalyptus sp.	*Aspergillus flavus, A. niger, A. ochraceus, A. sulphureus, Cladosporium cladosporioides, Curvularia lunata,* Fusarium graminearum, F. semitectum, Penicillium brefeldianum, *P. citrinum, Spicaria* sp. and *Xylaria* sp.	Seeds Seeds	Mehrotra and Singh (1999) Mathur (1974)
Ficus benjamina	*Botryodiplodia* sp., *Fusarium moniliforme, F. semitectum* and *Verticillium* sp.		
F. krishnae	*Cephalosporium clavariaeformis, Pestalotiopsis funera* and *Phoma* sp.	Seeds	Mathur (1974)
Gmelina arborea	*Cephalosporium* sp., *Fusarium equiseti, F. moniliforme, F. semitectum, F. solani* and *Phoma* sp.	Seeds	Mathur (1974)
Grewia optiva	*Aspergillus flavus, Cladosporium cladosporioides, Fusarium graminearum, F. semitectum, Penicillium citrinum* and *Rhizopus stolonifer*	Seeds	Mehrotra and Singh (1999)
Haldina cordifolia	*Fusarium moniliforme* and *Phoma* sp.	Seeds	Mathur (1974)
Jacaranda mimosaefolia	*Botrytis cineria, Drechslera* sp., *Fusarium* sp. and *Phoma* sp.	Seeds	Mathur (1974)

Table 1 cont.

Host	Pathogen	Nature of infection	Reference
Acacia catechu, Bauhinia variegata, Cassia siamea and *Celtis australis*	*Alternaria* sp., *Aspergillus* sp. and *Curvularia* sp.	Seed rotting	Paul and Bhardwaj (1987)
Chlorophora excelsa, Ectandrophragma angolensis, Khaya senegalensis and *Triplochiton ecleroxylon*	*Aspergillus* spp., *Curvularia* spp., *Fusarium* spp. and *Rhizoctonia* spp.	Seeds	Adegeya *et al.* (1985)
Acacia auriculiformis, Cassia fistula, C. glauca, C. nodosa, Dalbergia sissoo, Leucaena leucocephala and *Shorea robusta*	*Alternaria* sp., *Aspergillus* sp., *Chaetomium* sp., *Cladosporium* sp., *Curvularia* sp., *Fusarium* sp., *Mucor* sp., *Myrothecium* sp., *Paecilomyces* sp., *Penicillium* sp., *Phoma* sp., *Rhizopus* sp., *Schizophyllum* sp., *Trichoderma* sp., *Trichothecium* sp. and *Verticillium* sp.	Seeds	Vijayan (1991)
Albizia lebbeck, A. lucida, A. procera, Cassia fistula, C. laevigata, C. nodosa, Eucalyptus sp., *Polyalthia longifolia, Schima wallichii, Terminalia chebula* and *T. myriocarpa*	*Fusarium oxysporum*	Seeds	Tiwari and Sharma (1981)

caused defoliation of seedlings and young trees of *Acacia auriculiformis* and *A. mangium* in India (Mohanan and Sharma, 1988). *Fusarium solani, Phytophthora* spp., *Pythium* spp. and *Rhizoctonia solani* cause serious damping off of tree legume seedlings (Sharma and Bhardwaj, 1988). A new vascular wilt disease of *A. nilotica* seedlings caused by *Fusarium oxysporum* that caused high mortality in seedlings raised in polythene bags (33.7%) than those in root trainers (16.9%) was reported by Kapoor *et al.* (2004) and the source of infection was found to be soil-borne rather than seed-borne.

Foliage: Many rusts occur on leaves, twigs and pods of *Acacia* spp. Information on the host range, symptoms, distribution, biology and ecology has been reported by Bakshi and Singh (1967). More than 60 species of rusts are confined to *Acacia* spp. and the most common rust genus in India is *Ravenelia.* Information about rusts is well documented by Bakshi (1976) in India. However, in most cases, their importance has not been realized. Scanty information is available on the importance of the foliage (leaf, twig and branch) diseases caused by at least ten fungal genera, which have been widely recorded on *Acacia* spp. by Mohanan and Sharma (1988). *Exserohilum rostratum* cause dark irregular lesions with pale center on foliage of young *A. auriculiformis* in India (Mohanan and Sharma, 1988). *Rhizoctonia solani* causes web blight of *A. auriculiformis* and *A. nilotica* resulting in premature defoliation as severe as 30-70% (Mehrotra, 1990).

Stem: Sandal spike caused by a *Mycoplasma*-like organism affects *Acacia* spp. in India (Ghosh, 1981; Nayar, 1988; Sen Sarma, 1984). The symptoms include: foliage dwarfing, shortened internodes, excessive branching, pale green or reddish leaves standing out stiffly from twigs, suppressed flowering, and small fruit. Spiked trees usually die within 3-5 years of attack (Nayar, 1988). An extensive review of the disease is provided by Nayar (1988). Although virus-like disorders have been observed on *Acacia* spp., they were not well characterized (Seliskar, 1964). A series of cankers were formed all over the stem on 7-month-old seedlings of *A. nilotica* of Sangrur and Bhatinda (Punjab), causing severe distortion to the stem. *Fusarium lateritium* f. sp. *acaciae* and perithecial stage belonging to the genus *Gibberella* were found together on the same diseased plants. The fungus was observed in the xylem and phloem resulting in wilting and dieback of young seedlings (Uniyal *et al.,* 2004).

Root: Root diseases caused by *Armillaria mellea, Ganoderma* spp. and *Macrophomina phaseolina* have been recorded on *A. auriculiformis* and *A. nilotica* (Bagchee, 1945). Wilt and root rot due to *Fusarium solani* has been reported on *A. auriculiformis* and *A. nilotica* in India (Bagchee, 1945, 1958), while *Botryodiplodia theobromae* is associated with root disease (Gibson, 1975).

Albizia

It is one of the most promising fodder trees of the semi-arid regions. It is an excellent wood for fuel and charcoal and is suitable for construction and furniture as well. The shallow root system helps in soil binding and is recommended for soil conservation and erosion control. The bark is used for tanning and the seeds are

Bauhinia

Bauhinia variegata is a very popular ornamental tree species in tropical and subtropical regions. The roots and the bark possess several pharmaceutical values (e.g., to treat skin diseases, leprosy, wounds, tumors, cough, diabetes, dysentery) (Prajapati *et al.,* 2003).

Foliage: Leaf web blight of *B. variegata* caused by *Rhizoctonia solani* was reported to cause disease incidence up to 80% in seedlings, whereas defoliation varied from 20-100% (Mehrotra, 1998). Leaf spot caused by *Colletotrichum* and leaf blight by *Rhizoctonia* (Shivanna, 2005).

Cassia

Cassia fistula is a deciduous tree commonly called golden shower or Indian laburnum. The sweet blackish pulp of pod is used as a mild laxative. Leaves are useful to treat leprosy, skin diseases, ulcers, and intermittent fevers (Prajapati *et al.,* 2003).

Foliage: Leaf blight disease of *C. fistula* caused by *Rhizoctonia solani* resulted in defoliation of seedlings raised in polypots to an extent of 90-100% (Mehrotra, 1998). Similarly, Shivanna (2005) reported leaf blight of *C. fistula* caused by *R. solani* in forest nurseries in Shimoga.

Dalbergia

It is one of the most useful multipurpose tree species mainly grown for the timber, which is among the finest for cabinets and furniture. Its habit of developing root suckers and runners makes it useful in prevention of soil erosion. The leaves are used to treat gonorrhoea, dyspepsia, colic, diarrhoea, dysentery, vomiting and burning sensation (Prajapati *et al.,* 2003).

Foliage: Incidence of disease was 40% and defoliation ranging from 30-75% was reported by Mehrotra (1998) in *Dalbergia sissoo* by a biotype MS of *Rhizoctonia solani*. This biotype was less damaging to seedlings compared to the biotype TS found attacking plants in the nurseries in Doon valley (Mehrotra, 1991). Tar spot caused by *Phyllachora* and leaf spot by *Colletotrichum* on *D. latifolia* seedlings (Shivanna, 2005).

Dendrocalamus

Bamboo species are one of the most versatile and widely utilized groups of plants. It is of great importance in Asia, where it is considered as the timber of the poor. It is used for making furniture, agricultural implements, baskets, houses and bridges, scaffolding, fencing, fodder, fuel wood and water supply pipes (Das, 1998).

Seed and seedling: The rust fungus, *Tunicospora bagchii* was attacking the leaf sheaths of *Dendrocalamus strictus* (Singh and Pandey, 1971). In a nursery of *D. strictus* raised in Visakhapatnam, the seedlings were found to be attacked by *Dasturella divina* (Mundkur and Kheswala, 1943).

Foliage: Leaf spot and blight observed in 10-month-old seedlings of *D. strictus* caused by *Myrothecium roridum* resulted in defoliation and weakening of seedling (Ravishankar Rai and Mamatha, 2005a). Leaf spot of *D. strictus* caused by *Cercospora apii* was reported from 6-month-old seedlings in forest nurseries of Mysore (Ravishankar Rai and Mamatha, 2005a).

Dipterocarpus

Dipterocarpus is of considerable importance as timber trees. The wood is suitable for plywood production and it is thus widely used in plywood industries in North-eastern India (Rawat, 1998).

Seed and seedling: A 30% germination loss in *Dipterocarpus retusus* due to pre-emergence damping off by species of *Phoma, Fusarium* and *Rhizoctonia* (Singh *et al.,* 2004)

Foliage: Leaf blight caused by *Colletotrichum crassipes* is characterized by a brown lesion or patch initiating at the tip and margin of old leaves (Singh *et al.,* 2004). Severity varies from 50-70% among *D. retusus* seedlings. Similarly, *C. crassipes* also causes leaf blight on *D. macrocarpus* (Singh, 2002). Leaf spot caused by *Cercospora arachidicola* on 2-month-old seedlings of *D. retusus* showed low disease incidence (3- 5%) (Singh *et al.*, 2004).

Stem: Shoot blights caused by *Phoma tropica* affects 15-30-day-old *D. retusus* seedlings resulting in 70-80% mortality (Singh *et al.*, 2004). *Phoma tropica* is basically a soil borne fungus and may also be transmitted through the air (Bilgrami *et al*., 1991; Sutton, 1980). Shoot blights caused by *Phoma eupyrena* affects 1-2-month-old *D. retusus* seedlings resulting in over 90% mortality in nurseries (Singh *et al.,* 2004).

Erythrina

Erythrina is a tropical and subtropical flowering tree species. Powdered leaves of this tree are used to treat wounds and ulcers. The bark of the stem is used against rheumatism in the form of a decoction (Prajapati *et al.,* 2003).

Root: Erythrina jorientatlis and *E. indica* succumb to root rot disease caused by *Botryodiplodia theobromae* and *Fusarium* spp. (Lenné, 1992). Wilt caused by *Fusarium solani* has been recorded on *Erythrina stricta* (Lenné, 1992).

Eucalyptus

Various uses of *Eucalyptus* have made them commercially important tree species. They grow fast and provide many desirable characteristics for use as ornament, timber, firewood, and pulpwood. They are also useful as effective wind breaks.

Seed and seedling: Pre-emergence damping off caused by *Pythium* and *Rhizoctonia* is characterized by the rotting of radicle and subsequently, cotyledons inside the seed coat (Anonymous, 1984). High temperatures (24-30°C) and long showery humid weather (over 4 days) promote leaf blight of *Cyclindrocladium* (Anonymous, 1981). *Corticium salmonocolor* causing 'pink disease' results in severe loss of *Eucalyptus globulus* and *E. tereticornis* in high rainfall and temperature areas of Southern India (Chaturvedi, 1983). Post-emergence damping off and seedling fall of *Eucalyptus* is a complex disease caused by either one or a consortium of pathogens (*Cylindrocladium, Fusarium, Pythium* and *Rhizoctonia*) (Anonymous, 1984). Web blight caused by *Rhizoctonia solani* is characterized by profuse mycelial growth entangling the seedlings and giving an appearance of a cobweb (Anonymous, 1984). Seedling blight caused by three species of *Cylindrocladium* (*C. clavatum, C. ilicicola* and *C. quinqueseptatum*) and *Rhizoctonia solani* results in heavy mortality of seedlings (Anonymous, 1984). Seedling wilt caused by *Sclerotium rolfsii* causes decay of stem and consequently, the wilting of seedlings that is characterized by flaccid leaves and drooping terminal shoot (Anonymous, 1984).

Foliage: Leaf spot disease caused by *Cylindrocladium* spp. is one of the important and widespread diseases in *Eucalyptus* nurseries in Kerala (Anonymous, 1984). Leaf spot caused by *Phaeoseptoria eucalypti* appears initially as purple to brownish-purple spots on mature leaves (Anonymous, 1984). In Punjab, *Cylindrocladium* is one of the common leaf blight and stem canker diseases during periods of high temperature (30°C) and relative humidity (80%). Pathogens associated with die back and cankers of *Eucalyptus* are *Alternaria alternata, Botryodiplodia theobromae, Colletotrichum* sp. and *Phomopsis eucalypti* (Rattan and Dhanda, 2002). *Alternaria* blight disease of *Eucalyptus* hybrid has been identified by symptoms like necrosis of leaves and shoots starting from tip and progressing downwards giving a blighted appearance was reported by Shukla *et al.* (2005).

Root: Root rot, caused by *R. solani,* causes slow wilt of seedlings in scattered patches (Anonymous, 1984). Root rot caused by *Ganoderma lucidum* affects a wide range of Indian *Eucalyptus* plantings and it is likely to isolate this disease by trenching (Chaturvedi, 1983).

Faidherbia

The ability to fix atmospheric nitrogen makes *Faidherbia albida* an excellent tree species in agro-forestry systems. It also has a high value as fodder tree.

Foliage: Rhizoctonia solani causes leaf web blight of *F. albida* (Mehrotra, 1990). *Faidherbia albida* is a good host of root-knot nematodes, *Meloidogyne javanica* and *M. incognita* (Prot, 1986).

Gmelina

Gmelina arborea is a beautiful fast growing deciduous tree that grows naturally throughout India. It is a highly valuable source of timber, pulp and fodder.

Root: Jamaluddin *et al.* (1988) reported root rot disease of *G. arborea* caused by *Fusarium oxysporum* from Madhya Pradesh. Severe root rot of *G. arborea* plants growing in a nursery in Assam was reported to be caused by *Fusarium semitectum* (Singh, 2002).

Hardwickia

Hardwickia binata constitutes the important tree species of forest in the afforestation programme of India.

Seedling: Wilt disease of *Hardwickia binata* causes defoliation and death of seedling due to the infection of *Fusarium oxysporum* (Ravishankar Rai and Mamatha, 2005a).

Leucaena

Leucaena leucocephala is the most widely used forage tree. It is also used as fuel wood and generates an excellent quality charcoal. The bark exudes gum of *Leucaena,* which is an excellent substitute for Arabic gum and has immense uses in Asian continent.

Pod: A high incidence of pod rot caused by *Fusarium* sp. has been recorded in India on *Leucaena collinsii, L. diversifolia, L. macrophylla, L. pulverulenta* and *L. shannoni* (Lenné, 1991).

Foliage: Camptomeris leaf spots (CLS) caused by *Camptomeris leucaenae,* which reduces forage production as well as quality of *L. leucocephala* (Lenné, 1991).

Stem: Gummosis, the exudation of gum from the main trunk, stems and branches of *Leucaena* spp. associated with leaf senescence, dieback, and stunted growth has been described as the most serious disease of *L. leucocephala* in India (NAS, 1984). *Pirex subrinosus*, causal organism of stem canker was first reported as pathogen of *L. leucocephala* in India (Sankaran and Sharma, 1986).

Root: Root rot caused by *Ganoderma lucidum* has been reported on *L. leucocephala* (Raina, 1980, 1983; Pathak, 1986).

Prosopis

Prosopis species form a major component in dry land ecosystem in Asia. The timber is valued for furniture, doors, cobblestones and parquet floors (Felker, 1991).

Foliage: Although a large number of fungi have been recorded on *Prosopis juliflora*, their impact on diseases is not clearly known. Leaf spots caused by *Pleospora scirrhioides* and *Septoria prosopodis* has reported from India, while dieback by *Diplodia dalbergiae, D. prosopidina, Nectria flavistroma* and *N. purtonii* in India and Pakistan (Browne, 1968).

Sesbania

Sesbania provides forage, firewood, pulp and paper, green manure and also facilitates landscape decoration.

Seedling: Colletotrichum capsici, the collar and seedling blight pathogen has been recorded on *Sesbania grandiflora* (Srinivasan, 1952).

Foliage: Severe leaf spot and defoliation by *Pseudocercospora sesbaniae* occurs widely on *S. grandiflora* (Kumar and Joshi, 1983) and on *S. sesban* (Joshi and Kumar, 1986). The rusts *Uromyces poonensis* on *S. grandiflora* (Joshi and Kumar, 1986) and *Uredo sesbaniae* on *S. sesban* in India have been reported (Evans and Rotar, 1987). *Protomycopsis thirumalacharii* causes angular black leaf spot of *S. grandiflora* (Haware and Pavgi, 1976). The powdery mildew, *Erysiphe polygon* has also been recorded on *S. sesban* in India (Evans and Rotar, 1987). *Xanthomonas campestris* pv. *sesbaniae* causes leaf and stem spots and defoliation of *S. sesban* in India (Bradbury, 1986). An unnamed, sap-transmitted virus causes mild mosaic and mottling on leaves and defoliation of *Sesbania* spp. (Singh and Srivastava, 1985).

Root: Sesbania mosaic virus causes decreased nitrate reductase activity and interferes with nodulation of *Sesbania* spp. (Rao and Shukla, 1988; Srivastava, 1986; Sreenivasulu and Nayudu, 1982). Root-knot nematode, *Meloidogyne incognita* is potentially destructive to *Sesbania* spp. (Trivedi *et al.,* 1986).

Tectona

The teak, *Tectona grandis* is widely used as plantation species on sites with a seasonal tropical climate. It is often grown in agro-forestry systems and is one of the most versatile timber species that is used for heavy and light construction work, house building, carpentry and wood carvings.

Foliage: Leaves of *T. grandis* are affected by *Olivae tectonae*, a rust fungus forming orange coloured scattered uredial and telial sori on the lower surface (Shukla, 1995). *Uncinula tectonae*, a powdery mildew is also found in the nurseries and young plants of teak (Shukla, 1995). Collar rot of teak seedlings, a serious nursery disease caused by *Rhizoctonia solani* (anamorph, *Thanatephorus cucumeris*) was recorded for the first time from Tamil Nadu. The disease incidence ranges from 20-100% and is the maximum during July-September (Ramesh, 2000). Leaf spot and leaf blight of *T. grandis* caused by *Myrothecium roridum* was reported by Murthy *et al.* (2004).

Terminalia

Trees of this genus are known especially as a source of secondary metabolites. Leaves of *Terminalia catappa* are used to treat indigestion (Prajapati *et al.,* 2003), *T. chebula* is well known for its valuable medicinal properties (e.g., antidermatophytic) (Dutta *et al.,* 1998; Tanhane *et al.,* 1997).

Foliage: Leaf blight in *Terminalia catappa* was caused due to *Fusarium solani* (Ravishankar Rai and Mamatha, 2005a). *Terminalia tomentosa* suffer from leaf spot disease by *Alternaria* (Shivanna, 2005). A serious leaf rust disease of *Terminalia* is caused by *Uredo terminaliae*. The clonal bank of nursery of *T. chebula* of the State Forest Department Genetics Division, Coimbatore, showed initial infection in the month of November (25%) that increases during January (95%) (Mohan and Manokaran, 2005).

Management Strategies

Seed spoilage during storage and mortality of seedlings in the nursery by seed-borne fungi is a serious drawback that has attracted the attention of silviculturists and forest pathologists. These problems can be effectively tackled through integrated management strategies (e.g. quarantine, seed production site, storage, seed treatment) (Mehrotra and Mehrotra, 2000). In recent years, there has been a spurt in the exchange of seeds of fast-growing species of forest trees (*Acacia mangium, Azadirachta indica, Dalbergia sissoo, Eucalyptus, Paulownia fortunei, Poplars* and *Gmelina arborea*) and tropical pines between India and other Asian, European, American countries besides New Zealand and Australia for the purpose of improving the available genetic material in the country and also to enhance forest productivity by introducing some of these tree species.

Exotic pathogens

Exchange of seeds and cuttings pose new disease problems as the possibilities of introduction of new or exotic pathogens. The exotics, on introduction to a new environment, may either be threatened by the pathogens that are of no consequence in their native place or they may be attacked by the indigenous pathogens against which they have no resistance. There are also fair chances of exotic pathogens posing serious problems to indigenous trees if introduced inadvertently. These exotic pathogens have severely damaged the forests and developmental plans of the country. Imposition of quarantine is, therefore, the first safeguard against inadvertent entry of harmful pathogens into the importing country (Mehrotra and Mehrotra, 2000). For instance, seeds of *Hevea brasiliensis* imported from Malaysia, on examination at Madras Quarantine Station were carrying seed-borne pathogens (*Lasidiplodia theobromae* and *Phomopsis heaveae*) known to cause die-back or collar rot and stem die-back diseases in Malaysia (Srivastava, 1956). During the last three decades, several *Pinus* and *Cupressus* diseases caused by exotic pathogens have been recorded in India: needle blight of *Pinus radiata* (*Dothistroma pini*) in Tamil Nadu (Bakshi and Singh, 1968); twig blight of *P. petula* and *P. pseudostrobus* (*Diplodia pinea*) in Dehradun (Bakshi *et al.,* 1972); needle cast of *P. elliottii* and *P. radiata* (*Lophodepinastri*) (Bakshi *et al.*, 1972); stem canker of *Cupressus arizonica, C. lusitanica, C. macrimiumcarpa* and *C. sempervirens* (*Monochaetia unicornis*) raised in a trial plantation (Bakshi *et al.,* 1970) and needle blight of *Pinus roxburghii, P. radiata* and *P. kesiya* (*Pseudocercospora pini-densiflorae*) (Mehrotra, 1986; Reddy and Pandey, 1973). Import of seeds of five needle pines is banned in India to check or prevent entry of *Cronartium ribicola*, a serious pathogen causing stem blister rust in USA. *Pinus wallichiana*, an indigenous five-needle pine, showed susceptibility to *C. ribicola* and the introduction of a virulent strain of the rust might pose a serious disease problem in the forests. *Phomopsis* sp. colonizing imported cuttings of *Pinus deltoids* has been recorded by Mehrotra and Pandey (1995). This pathogen is highly virulent and attacks the cuttings that result in failure of sprouting and death (up to 60%). *Melampsora laricipopulina*, a rust prevalent on an evergreen clone of *P. deltoids* in Doon Valley and Kalsi, Dehradun is reported to have been inadvertently entered the country through imported cuttings (Pandey *et al.,* 1996). Recently, Mehrotra and Singh (1999) recorded three *Phoma* spp. on seeds of *Paulownia fortunei* imported from China.

Quarantine

Development of post-entry quarantine in India has been emphasized by Bakshi (1976). According to the Plants, Fruits and Seeds Regulation of Import of India order 1984 (issued by the Department of Agriculture and Cooperation, Ministry of Agriculture, Government of India), post-entry quarantine means growing of plants in isolation for any specified period in a glasshouse, any facility area or a nursery approved by the plant protection advisor in order to detect the presence or absence of harmful exotic pathogens associated with the plant materials including the seeds. In view of the growing demand in favor of industrial wood, introduction of fast-growing exotic species has become a necessity. As the diseases of forest plant species are generally different from those of agricultural crops, it is essential to have specially trained staff having adequate knowledge of forest diseases for quarantine purposes and to prevent inadvertent entry of exotic pathogens through seeds and other inert materials (e.g., chaff, needle pieces) accompanying seeds (Mehrotra and Mehrotra, 2000). In India, the Directorate of Plant Protection Quarantine and Storage (Ministry of Agriculture) is the official organization for quarantine task. The Destructive Insects and Pests (DIP) Act in India was passed by the Government in 1914 to restrict or prohibit import of certain plant materials from other countries and also to regulate their movement from one part to another even within the country.

Seed treatments

Seed treatment will be performed for three purposes: firstly, to minimize the loss of seeds due to fungal deterioration; secondly, to raise healthy seedling stock devoid of seed-borne pathogens; thirdly, to prevent or to avoid further spread of the pathogens in plantations. Seed treatment is one of the cheapest and often the safest methods of disease control. Seeds may be treated mechanically, physically, chemically or biologically. Munjal and Sharma (1976) reported that hot water treatment (57°C, 40 min) was effective in elimination of a large number of seed-borne fungi associated with seeds of *Pinus roxburghii* and *P. wallichiana*. Five fungicides were used for seed treatment of four tree species (*Butea monosperma, Dendrocalamus strictus, Lagerstroemia parviflora* and *Pongamia pinnata*). Out of them Bavistin (Carbendazim), Dithane M-45 (Mancozeb; dithiocarbomate) and Thyride (Thiram) at 3 g/kg were highly effective in controlling the seed mycoflora (Mamata-Purohit *et al.,* 1998). Mamatha *et al.* (2000) worked on the quality aspects of germination and vigour of dominant mycoflora isolated from four tree species (*Dalbergia latifolia, Dendrocalamus strictus, Hardwickia binata* and *Phyllanthus emblic*) reported that biological treatment was found to be most effective in reducing the incidence of mycoflora and enhancing the seed germination and vigor.

Dry seed treatment is recommended for seeds likely to be damaged during processing and in such cases slurry or dip methods are not suitable because of absorption of water by the seed. In the seed dip treatment, seeds are dipped in fungicidal solution for a specific duration depending on the thickness of the seed coat. In slurry treatment, the water dispersible fungicide is mixed with water to make the slurry and then applied to seeds. A combination of systemic and contact fungicides is preferred for a broad-spectrum control (Mehrotra and Mehrotra, 2000). Chemotherapy is one of the effective strategies to control serious fungal diseases. Bavistin, Blitox, Bayleton, Dithane M-45 and Topsin-M are commonly used fungicides to treat many diseases (Ravishankar Rai and Mamatha, 2005b). Seed treatment with *Trichoderma harzianum* and *T. pseudokoningii* is an ecofriendly approach superior over other treatments to reduce pathogenic fungi, improve seed germination and for vigor (Mamatha *et al.,* 2000).

Seedling Diseases

The following strategies and techniques may be followed for effective management of seedling diseases of nurseries and plantation. These strategies can be divided into three phases: general strategies, chemical strategies and biological control strategies.

General strategies

- Clear site selection is recommended to avoid soil-borne root rots (Bakshi, 1976). Other measures including trenching to isolate infection centers and the treatment of roots with fungicides were also recommended.
- Solar sterilization of soil during the months of May and June by covering nursery beds with polythene sheet after watering raises the temperature of topsoil to 50-60°C. This kills the pathogenic soil fungi and replaces them with thermophilic species bringing about microclimatic changes in soil. Seeds sown in such soil show better germination and growth than the seeds sown in soils without solar sterilization (Shukla *et al.,* 1990).
- Cultural control by reducing the density of seedlings, restricting watering and modifying the seedling environment less suitable for disease development are the valuable control practices (Sharma and Sankaran, 1987).
- Avoidance and eradication have been used for common stem and root diseases. Application of fungicides directly to cankers successfully reduces gummosis of *Leucaena*. However, the economics of such practice must be compared with the value of wood and forage. When branches are cut for forage, care should be taken to avoid injuries, which are potential sites for fungal invasion (Sharma and Sankaran, 1987).
- Management of the re-growth of tree legumes primarily used for forage is

a widely applicable disease control measure. Natural biocontrol fungi have been recorded on rusts (Khan *et al.,* 1989) and those may help in the management of rust diseases.

Chemical strategies

- Chemicals are commonly, successfully and economically used to control seed and seedling diseases of tree legumes in nurseries (Mohanan and Sharma, 1988).
- Poison food technique is commonly used to screen fungicides. Shukla (1992) used this technique against *Fusarium solani* and reported that Ziram (*zinc dimethyl dithio carbomate*), Shield (coordinate product of zinc, iron and manganese *ethylene bisdithio-carbomate*), Karathane (mixture of 2-4 *dinitro*-6 *octylphenyl crotonate* and 2-6 *dinitro octophenyl crotonate*) and Bavistin (*carbendazim*) were the most effective fungicides for colony growth inhibition.
- About 25-30 g fungicides like Captan (1, 2, 3, 6-tetra hydro-N-(*trichloromethyl thio*) *phthalimide*), Thiram (1-(*dimethylthiocar-bamayldisulfanyl*)-N, *N-dimethyl-methanethioamide*), Dithane Z-78 (*zinc ethylenebidithiocarbonate*), Blitox 50 (copper oxychloride) and Brassicol (1, 2, 3, 4, 5-*pentachloro-6-nitro-benzene*) can be used to disinfect topsoil (15 cm) depending on the texture of the soil for soil-borne pathogens like *Rhizoctonia solani* and *Fusarium* spp. (Shukla, 1995).
- Captan and Thiram at 1-2 g/kg of seeds are commonly used as seed coat disinfectants to coat seeds of forest tree species superficially (Shukla, 1995).
- Integrated control of the disease is recommended, which includes: (i) sanitation to remove the infected seedlings as soon as the disease symptoms appear. This will prevent lateral spread of the disease through contact of the overlapping foliage; (ii) weeding to remove alternate hosts which otherwise may serve as a potential source of infection; (iii) use of Bayleton [1-(4-chlorophenoxy)-3, 3-dimethyl-1-(1,2,4-triazol-1-yl)butan-2-one)] (0.08%), a systemic fungicide, as foliar spray in case the disease is not brought down below the economic injury level by the first two methods (Mehrotra, 1998). Bayleton has also been found highly effective in controlling *Rhizoctonia* leaf web blight in *Albizia lebbeck*.
- *Fusarium* spp. in the soil can be checked by drenching the soil with 0.2% Thiram and Captan. *Rhizoctonia*-sick soils can be amended with 0.2% Brassicol (1,2,3,4,5-pentachloro-6-nitro-benzene) solution. *Pythium* and *Phytophthora* infections can be controlled by 0.5% solution of Blitox 50 (Shukla, 1995).
- Most of the foliar pathogens (rusts, powdery mildew, blight and necrosis in the leaf and tip) can be treated by spraying 0.1-0.2% solution of Dithane M-45 or 0.05% solution of Bavistin (Shukla, 1995).

- Bordeux mixture is effective against various foliar diseases of *Eucalyptus* (Shukla, 1995).
- Rusts and powdery mildew of *Dalbergia sissoo* and *Tectona grandis* can be effectively controlled by three fortnightly sprays of 0.05% solution of Bayleton or sulphur dust on the foliage (Shukla, 1995).
- Ziram and Bavistin were found to be most effective in checking collar rot of *Azadirachta indica* by *Fusarium semitectum* (Uniyal, 1999).
- Seed dressing with fungicides like Bavistin, Topsin-M (thiophanate methyl), Benlate [(methyl 1-(butyl carbamoyl)-2-benzimidazole carbamate)] or Thiram at 0.2% by weight of seeds commonly practiced for forest tree species before sowing mainly to minimize diseases (Rawat and Harsh, 2003).
- Soil drenching with fungicides at 0.05% (Bavistin, Topsin-M, Benlate or Thiram) (10 liter solution should be applied for 10 × 1m bed) with repeated doses after an interval of 7–10 days in diseased areas of *Fusarium* spp., *Macrophomena phaseolina* and *Rhizoctonia solani* (Rawat and Harsh, 2003).
- Root dip in 0.02% solution of Bavistin, Topsin-M, Benlate or Thiram for 3.5 min prior to pricking out in containers provides protection against root diseases like damping off, wilt and root rot that is common among conifers, *Acacia, Albizia, Azadirachta indica, Cassia, Dendrocalamus strictus, Leucaena leucocephala* and *Tectona grandis*. Dipping of cuttings in the same solution for 5–7 min prior to placing them in beds has been found to control the set rot disease (Rawat and Harsh, 2003).
- Application of foliar spray with 0.2% Bavistin, Dithane M-45, Fytolan (copper oxychloride) or Captafol (tetrachloroethyl thio tetrahydrophthalimide) as and when foliar diseases appear (leaf spot, leaf blight, rusts, powdery mildew, blotch, leaf curl). A second spray should be given after a fortnight. To control rusts and powdery mildews of *Dalbergia sissoo* and *Tectona grandis*, foliar spray with 0.2% Sulfex or sulphur powder followed by a second spray after a fortnight is recommended (Rawat and Harsh, 2003).
- The efficacy of seven fungicides against soil-borne *Fusarium oxysporum* causing vascular wilt disease of *Acacia nilotica* has been studied by Kapoor *et al.,* 2004. They reported that Bavistin and Benlate inhibited pathogen growth completely in culture at 0.1% concentration. Soil drenching with Thiram (0.05%) followed by seed dressing with Bavistin (0.05%) gave the best growth performance of seedlings as well as inhibition of disease in root trainers.
- Murthy *et al.* (2004) reported that Bavistin is good in managing leaf spots, leaf blights and seedling decay of teak saplings.
- *In vitro* screening of seven systemic and non-systemic fungicides in controlling mycelial growth of *Fusarium solani* (causes root rot of *Prosopis cineraria*) revealed the superiority of all the fungicides although their efficacy varied. Topsin-M showed 11% inhibition at 50ppm, whereas Emisan and Bavistin were at 100ppm. Bavistin and Topsin-M also retained their supremacy in seed treatment and pre-sowing drenching (Sharma *et al.,* 2004).

- Best control of canker formation was obtained by spraying mixture of 0.2% Devistin and 0.1% Tagcop (copper oxychloride) followed by mixture of 0.1% Devistin and 0.1% Dithane M-45 on *Acacia nilotica* seedlings. Even small patches on the shoot disappeared after the application of fungicides (Uniyal *et al.,* 2004).
- Mohan and Manokaran (2005) reported that trimming of affected portions, application of sulfur-based fungicide by foliar spray followed by soil drenching and use of biocontrol agents (*Trichoderma viride* and *Pseudomonas fluorescence*) effectively controlled leaf rust of *Terminalia chebula* caused by *Uredo terminaliae.*
- Application of Brestan 60 (0.1%) was found to be the most effective fungicide, which decreased the disease index to less than 57% against 86.6% in control against *Alternaria* blight disease of *Eucalyptus* (Shukla *et al.,* 2005).
- The effect of five systemic and two contact fungicides were evaluated against the pathogen *Pestalotiopsis* sp. *in vitro* in case of *Syzygium cumini.* Among the seven fungicides studied *in vitro,* only two systemic fungicides (Bavistin and Roko) were proven to be effective against the pathogen (Bhanumathi and Ravishankar Rai, 2007b).

Biological strategies

- In most instances, the source of disease causing organisms will be in the soil or seed itself. Hence, prophylactic treatment of planting media and material will prove an effective tool in the management of these diseases at an initial stage and the prevention of pathogen buildup for secondary infections (Singh *et al.,* 2004).
- Fungal antagonists by dual culture method showed that *Trichoderma viride* caused maximum inhibition of *Fusarium oxysporum, Pythium aphanidermatum* and *Rhizoctonia solani* followed by *Gliocladium virens, Trichoderma harzianum, T. hamatum* and *T. longibrachiatum*, whereas in pot culture experiments, seed pelleting and soil incorporation of *Trichoderma viride* recorded the highest seed germination (86.6%) and seedling survival (83.4%) in *Acacia nilotica.* Fungi such as *G. virens*, *T. harzianum, T. longibrachiatum* and *T. viride* significantly improved the seed germination and survival of *Albizia lebbek, Dalbergia sissoo, Leucaena leucocephala* and *Prosopis juliflora* (Arya and Kaushik, 2001).
- The antagonistic fungus, *Trichoderma viride* controlled root rot disease of *Prosopis cineraria* caused by *Fusarium solani* up to 56% *in vivo* (Sharma *et al.,* 2004).
- The biological agents *Trichoderma harzianum* and *Pseudomonas fluorescens* proved to diminish the disease up to some extent.

Conclusions and Outlook

Enormous losses are incurred in the country due to forest tree diseases. Seeds and seedlings of forest tree species are attacked by various fungal pathogens. The damages and diseases caused by them are severe and limit the whole afforestation programme. Integrated approaches for management of important forest diseases demanding improved silvicultural practices, avoiding disease-prone areas, use of healthy planting stocks, disease detection, damage assessment, surveillance and need-based application of chemical or biological pesticides. An effort at the international level is therefore required to review the existing knowledge to develop suitable techniques and guidelines for the study of microorganisms on tree seeds, evaluation of their effects on seed health during storage and germination, precise identification of pathogens, risk analysis for quarantine, optimization of seed storage requirements, management of seed pathogens and the formulation and revision of seed health and quarantine regulations for the important tree species (Mathur and Mittal, 2002). Future studies need to address the mutualistic association of microorganisms with seeds and tissues of forest plant species and their role in disease prevention. Several questions need to be answered in future, for instance, whether the microorganism associated with the forest tree species acts as a mutualist or pathogen or saprophyte throughout or under specific environmental or geographical conditions?

References

Adegeye, A.O., Amakiri, M.A., Usiagwu, I.O. and Okojie, J.A. 1995. Seed mycoflora of some indigenous tree species in Nigeria. *Proc. 10 Annual Conference of the Forestry Association of Nigeria*, Yola. 1995, 216 – 224.

Anonymous 1981. Nursery diseases of *Eucalypts* in Kerala and their control. *Kerala Forest Research Institute Information Bulletin. # 6*, Peechi, Kerala, India

Anonymous 1984. Nursery diseases of *Eucalyptus* in Kerala and their control. *Kerala Forest Research Institute Information Bulletin. # 6*, Peechi, Kerala, India, 1-16.

Arya, S. and Kaushik, J.C. 2001. Efficacy of *Trichoderma* species and *Gliocladium virens* as biocontrol agents against damping-off of forest nurseries. *Plant Dis. Res. 16,* 46-51.

Bagchee, K. 1945. Pathological notes # 2: Wilt and dieback on Shisam (*Dalbergia sissoo*), babul (*Acacia nilotica*) and Khair (*A. catechu*) in the artificial regeneration under agriculture-cum-forestry management. *Indian Forester 71*, 20-24.

Bagchee, K. 1958. Diseases of forest trees in India – causes, effects and methods of prevention. *Indian Forester 84*, 407-417.

Bakshi, B.K. 1976. *Forest Pathology, Principles and Practice in Forestry.* Delhi, India.

Bakshi, B.K. and Singh, S. 1967. Rusts on Indian forest trees - Indian forest records (new series). *Forest Pathol. 62*, 139-198.

Bakshi, B.K. and Singh, S. 1968. *Dothistroma* blight. A potential threat to *Pinus radiata* plantations in India. *Indian Forester 94*, 824-825.

Bakshi, B.K., Reddy, M.A.R., Singh, S. and Pandey, P. 1970. Disease situation in Indian forests 1. Stem diseases due to *Coricum salmonicolor* and *Monochaetia unicornis. Indian Forester 98*, 135-144.

Bakshi, B.K., Reddy, M.A.R., Puri, Y.N. and Singh, S. 1972. *Forest Disease Survey - Final Technical Report.* Forest Pathology Branch, Forest Research Institute, Dehra Dun.

Banerjee, S.N. 1947. Fungus flora of Calcutta and Suburbs - I. *Bull. Bot, Soc. Bengal 1*, 37-54.

Bhanumathi, A. and Ravishankar Rai, V. (2007a). Leaf blight of *Azadirachta indica* and its management *in vitro. Afr. J. Agric. Res. 2*, 538-543.

Bhanumathi, A. and Ravishankar Rai, V. (2007b). Leaf blight of *Syzygium cumini* and its management *in vitro. Australian Pl. Dis. Notes 2*, 117-121.

Bhanumathi, A. and Ravishankar Rai, V. (2007c). Testing seeds of *Thespesia papulnea* for mycoflora, germination and pathogenecity. *The Indian Forester 9,* 1241-1246.

Bhat, S.S., Sateesh, M.K. and Devaki, N.S. 1998. A new destructive disease of neem (*Azadirachta indica*) incited by *Phomopsis azadirachtae. Current Science 74*, 17-19.

Bilgrami, K.S., Jamaluddin, S. and Rizwi, M.A. 1991. *Fungi of India, List and References.* Edition 2, Today and Tomorrow's Printers and Publishers, New Delhi.

Bose, S.R. 1919. Descriptions of fungi in Bengal. *Proc. Indian Assoc. Cult. Sci. 4*, 109.

Bradbury, J.F. 1986. *Guide to Plant Pathogenic Bacteria.* CAB International Mycological Institute, Kew, Survey, UK.

Browne, F.G. 1968. *Pests and Diseases of Forest Plantation Trees: An Annotated List of the Principal Species Occurring in the British Commonwealth.* Clarendon Press, Oxford.

Chaturvedi, A.N. 1983. *Eucalypts* for farming. Uttar Pradesh Forest Bulletin # 48, Uttar Pradesh.

Chowdhary, K., Khan, S.N. and Mehrotra, M.D. 1992. Occurrence of virus diseases on *Albizia lebbeck. Ind. J. Forestry 15*, 283-284.

Das, A.N. 1998. Bamboo research in Nepal. *Proceedings of International Bamboo Workshop.* November 14-18, Cochin.

Dutta, B.K., Rahman, I. and Das, T.K. 1998. Hypolipidemic activity of three indigenous drugs in experimentally induced atherosclerosis. *Int. J. Cazrdiol. Decl. 67*, 119-124.

Evans, D.O. and Rotar, P.P. 1987. Productivity of *Sesbania* species. *Trop. Agric. 64*, 193-200.

Felker, P. 1991. *Prosopis alba* and *Prosopis chilensis*: Subtropical semiarid fuel and fodder trees. In: *Agroforestry Species and Technologies - A Compilation of Highlights and Factsheet by NFTA and FACT Net.* Winrock International, USA, 137-138.

Ghosh, S.K. 1981. Association of mycoplasma and allied pathogens with tree diseases in India. In: *Mycoplasma Diseases of Shrubs and Trees* (Ed. Maramorosch, K. and Rayachaudhuri, S.P.). Academic Press, New York, 321-243.

Ghyre, B.P. 2005. Testing of seeds of some forest tree species for germination and seed-borne fungi. *Indian Forester 131*, 121-123.

Gibson, I.A.S. 1975. *Diseases of Forest Trees widely Planted as Exotics in the Tropics and the Southern Hemisphere, Part I: Important members of the Myrtaceae, Leguminosae, Verbenaceae and Meliaceae.* Commonwealth Mycological Institute, Kew, Survey, UK.

Harsh, S.K., Dadwal, V.S. and Jamaluddin. 1992. A new post-emergence damping-off disease of *Eucalyptus* seedlings. *Indian Forester 118*, 279-283.

Harsh, N.S.K., Tiwari, C.K. and Nath, V. 1989. Foliage diseases in forest nurseries and their control. *J. Tropical Forestry 5*, 66-69.

Haware, M.P. and Pavgi, M.S. 1976. Host ranger of two *Protomycopsis* species from India. *Sydowia 29*, 272-274.

Jamaluddin, S. 1980. Problems and prospects of seed pathology of forest trees. *Proceedings of Second Forestry Conference*, Dehra Dun.

Jamaluddin, S. 1999. Status of forest seed pathology in India. *Adv. Forest. Res. India 21*, 227-261.

Jamaluddin, S. Dadwal, V.S., and Soni, K.K. 1988. Some new and noteworthy diseases of *Gmelina arborea* Roxb. from Madhya Pradesh. *J. Trop. Forestry 4*, 297-299

Joshi, H.K. and Kumar, A. 1986. On the occurrence of fungal diseases in some fodder trees. *Nitrogen Fixing Tree Research Reports # 4*, 18-19.

Joshi, M. and Singh, P.P. 2003. *Tropical Deforestation and Forest Degradation: A Case Study from India.* XII World Forestry Congress, Quebec, Canada.

Kapoor, S., Harsh, N. S. K. and Sharma, S. K. 2004. A new wilt disease of *Acacia nilotica* caused by *Fusarium oxysporum. J. Tropical Forest Sci. 16*, 453-462.

Khan, S. N. 2003. Diseases of *Albizia* – An overview. *Ind. J. Forestry 26*, 123-134.

Khan, S. N. and Misra, B. M. 1999. A new *Cladosporium* leaf spot of *Albizia. Indian Forester 125*, 745-746.

Khan, S.N., Misra, B.M., Tiwari, R.K. and Sehgal, H.S. 1989. New records of fungi and new host records from India. *Indian Forester 115*, 40-43.

Khan, S. N., Misra, B. M. and Tivari, R. K. 1993. *Ravenelia sessilis*, a destructive rust of *A. chinensis* in nurseries. *Ind. J. Forestry 16*, 94-95.

Kumar, A. 2006. Pit Canker disease of Siris (*Albizzia procera*) – a new report from Himachal Pradesh. *Indian Phytopathology 59*, 124-125.

Kumar, A. and Joshi, H.K . 1983. Occurrence, incidence and appraisal of grey leaf spots on *Sesbania grandiflora*. *Nitrogen Fixing Tree Research Reports # 1*, 7-8.

Lenné J.M. 1991. Diseases of *Leucaena* species. *Trop. Pest Manag. 37*, 281-289.

Lenné, J.M. 1992. Diseases of multi-purpose woody legumes in the tropics: A review. *Nitrogen Fixing Tree Research Reports # 10*, 13-29.

Mamatha, T., Lokesh, S. and Ravishankar Rai, V. 2000. Impact of seed mycoflora of forest tree seeds on seed quality and their management. *Seed Res. 28*, 5-67.

Mamata-Purohit, Jamaluddin, Mistra, G. P. and Purohit, M. 1998. Studies on germination and seed-borne fungi of some forest tree species and their control. *Indian Forester 124*, 315-320.

Mathur, S.B. 1974. *Fungi recorded in seeds of forest tree species at Danish Government Institute of Seed Pathology*. Copenhagen, Denmark.

Mathur, S. B. and Mittal, R.K. 2002. Future needs in Tree seed health Management Research in Asia. In: *Asian Congress of Mycology and Plant Pathology*. University of Mysore, Mysore, 255-256.

Mehrotra, M.D. 1986. *Pseudocercopsora* needle blight of Khasi pine in Meghalaya. *Annual Research Report*, SFS College and Research Centre, Burnihat, Assam.

Mehrotra, M.D. 1989. *Rhizoctonia* leaf web blight of *Albizia lebbeck*, a destructive disease in forest nurseries in India. *Eur. J. Forest Pathol. 19*, 382-384.

Mehrotra, M.D. 1990. A study on fungal deterioration of seed and damping off in *Oroxylum indicum* and their control. *Indian Forester 116*, 977-979.

Mehrotra, M.D. 1991. *Annual Research Report on Forest diseases*. Indian Council of Forestry Research and Education,Dehradun.

Mehrotra, M.D. 1996. Some destructive nursery diseases and their management. In: *Proceedings of Diseases of Insect Pests in Tropical Forests* (Ed. Nair, K.S.S., Sharma, J.K. and Verma, V.V.). Kerala Forest Research Institute, Peechi, Kerala, 143-152.

Mehrotra, M.D. 1998. *Rhizoctonia* Aerial blight – A destructive nursery disease and its management. *Indian Forester 124*, 637-645.

Mehrotra, M.D. and Mehrotra, A. 2000. Seed-borne fungi of forest trees and their management – An overview. *Ind. J. Forestry 23*, 78-97.

Mehrotra, M.D. and Pandey, P.C. 1995. *Phomopsis set rot of Populus deltoids. Annual Research Report.* Indian Council of Forestry Research and Education, Dehradun, India.

Mehrotra, M.D. and Singh, P. 1999. Study on seed-borne fungi of some forest trees and their management. *Ind. J. Forestry 21*, 345-354.

Mehrotra, M. D., Khan, S. N., Tivari, R. K. and Misra, B. M. 1992. A study on the periodicity of foliage diseases of *Albizia lebbeck* and control of Camptomeris leaf spot in nursery. *Indian Forester 118*, 594-598.

Mittal, R.K. and Sharma, M.R. 1979. Seed mycoflora of *Albizia lebbeck* Benth. *Proceedings of 66th Indian Science Congress,* India.

Mittal, R.K. and Sharma, M.R. 1981. Seed mycoflora of *Cassia fistula* L. *Ind. J. Forestry 4*, 70.

Mittal, R.K. and Sharma, M.R. 1983. Studies on the mycoflora and its control on the seeds of some forest trees V. *Pinus wallichiana. Ind. J. Mycol. Pl. Pathol. 12*, 142-147.

Mittal, R.K., Anderson, R.L. and Mathur, S.B. 1990. Microorganisms associated with tree seeds: World Checklist 1990. *Information Report Patawawa National Forestry Institute*, Chalk River, Ontario, Canada.

Mitter, J.H. and Tandon, R.N. 1932. Fungus flora of Nainital I. *J. Ind. Bot. Soc. 11*, 178-180.

Mohan, V. and Manokaran, P. 2005. Control of leaf Rust of *Terminalia chebula* caused by *Uredo terminaliae. Indian Forester 131*, 115-117.

Mohanan, C. 1988. Diseases of bamboos in Kerala, India. *International Bamboo Seminar*, November 14-17, Cochin, Kerala.

Mohanan, C. and Sharma, J.K. 1988. Diseases of exotic acacias in India. *J. Trop. Forestry 4*, 357-361.

Motta, E., Annesi, T. and Balmas, V. 1996. Seed-borne fungi in Norway spruce: Testing methods and pathogen control by seed dressing. *Eur. J. Forest Pathol. 26*, 307-314.

Mundkur, B.B. and Kheswala, K.F. 1943. *Dasturella* a new genus of *Uredinalla*. *Mycologia 35*, 201-246.

Munjal, R.L. and Sharma, A.D. 1975. Mycoflora of conifer seeds. *Ind. J. Mycol. Pl. Pathol. 5*, 145-148.

Munjal, R.L. and Sharma, A.D. 1976. Control of seed mycoflora of some important conifers. *Ind. J. Mycol. Pl. Pathol. 6*, 135-139.

Murthy, N., Lokesh, S. and Ravishankar Rai, V. 2004. Pathogenicity of *Myrothecium roridum* in Teak saplings and its Management. *Adv. Pl. Sci. 17*, 541-548.

Naik, S.T., Devar, K.V., Suryanarayana, V. and Raichur, R.G. 2001. Seed mycoflora of some important tree species. *Ind. J. Forestry 24*, 8-11.

NAS 1984. *Leucaena, Promising Forage and Tree Crop for the Tropics*. National Academy Press, Washington DC.

Nayar, R. 1988. Damage by mycoplasmas in Indian forests. In: *Advances in Forestry in India*, Volume 2 (Ed. Ram Parkash). International Book Distributors, Dehradun, India, 127-158.

Pandey, P.C., Singh, A., Karnatak, D.C. and Bhartari, B.K. 1996. *Melampsora laricipopulina* on Poplars in India and its control in nursery. *Indian Forester 122*, 1062-1067.

Pathak, P.S. 1986. Mortality in *Leucaena* due to *Ganoderma lucidum. Leucaena Res. Rep. 7*, 65.

Paul, Y.S. and Bhardwaj, L.N. 1987. Studies on seed mycoflora of some agroforestry species and their control. *IUFRO Workshop on Agroforestry for Rural Needs*, February 22-26, New Delhi.

Prajapathi, N.D., Purohit, S.S., Sharma, A.K. and Kumar, T. 2003. *A Handbook of Medicinal Plants – A Complete Source Book*. Agrobios, India.

Prot, J.C. 1986. Sensibilitede sept legumineuses arborescentes aux nematodes: *Meloidogyne javanica, M. incognita, Scutellonema cavenessi* et *Dolichorhynchus elegans. Rev. Nematol. 9*, 416-418.

Punam-Singh, Mehrotra, M.D. and Singh, P. 1999. Seed-borne fungi of some forest trees and their control. *Ind. J. Forestry 22*, 320-324.

Purohit, M. and Jamaluddin. 1993. Seed mycoflora of palas (*Butea monosperma*) during storage. *Seed Res. 21*, 126-127.

Raina, A.K. 1980. Epidemiology of *Ganoderma* root rot - a case study. Woodpower 80. In: *Proceedings of the International Symposium of Forestry and Wood Utilization,* Part 3.

Raina, A.K. 1983. Performance of *Leucaena* in the Indian arid zones. 3. *Fusarium* gummosis and *Ganoderma* root rot. *Leucaena Res. Rep. 4*, 35-36.

Ramesh, K. R. 2000. Collar Rot Disease caused by *Rhizoctonia solani* in Teak (*Tectona grandis* Linn. F.) – A New Record from the Nurseries of Tamil Nadu. *Ind. J. Forestry 23*, 322-323.

Rangaswami, G., Sheshadri, V.S. and Lucy Channamma, K. 1970. *Fungi of South India*. University of Agricultural Science, Bangalore.

Rao, G.P. and Shukla, K. 1988. Influence of Sesbania mosaic virus infection on nodulation and nitrogen fixation in pea (*Pisum sativum* L.). *J. Agric. Sci. 110*, 391–394.

Rattan, G.S. and Dhanda, R.S. 2002. Status of diseases in nursery and agro-forestry plantation species in Punjab. In: *Asian Congress of Mycology and Plant Pathology: Plant Health for Food Security*. Souvenir, Indian Society of Mycology and Plant Pathology, University of Mysore, Mysore.

Ravishankar Rai, V. and Mamatha, T. 2005a. Seedling diseases of some important forest tree species and their management. In: *Diseases and Insects in Forest Nurseries.* Proceedings of the 5th Meeting of IUFRO Working Party, May 6-8, 2003, Kerala Forest Research Institute, Peechi, Kerala, 51-63.

Ravishankar Rai, V. and Mamatha, T. 2005b. Fungal Diseases of Forest Tree Species in India and their Management. In: *Plant Diseases Biocontrol Management* (Ed. Nehra, S.). Aavishkar Publishers and Distributors, Jaipur, 69-83.

Rawat, V.R.S . 1998. *Hollong: Its Past, Present and Future.* IRDFR Publication # 1, Rain Forest Research Institute, Jorhat, Assam.

Rawat, P.S. and Harsh, N.S.K. 2003. Management of diseases in Nurseries. In: *Forest Tree Seeds and Nursery Management: Current Trends* (Ed. Singh, V. and Lavania S.K.). Bishen Singh Mahendra Pal Singh, Dehra Dun, 261-268.

Reddy, M.R.S. and Dayanand, R. 1983. Mycoflora associated with seeds of red sanders (*Pterocarpus sntalinus*), *Ind. J. Forestry 6*, 322.

Reddy, M.A.R. and Pandey, P.C. 1973. *Cercospora* needle blight of radiata pine in India. *Indian Forester 99*, 308-309.

Sahai, A. and Mehrotra, B.S. 1982. Mycoflora associated with the seeds of forest trees and their effect on germination. *Proc. Ind. Nat. Sci. Acad. 48*, 706-713.

Sankaran, K.V. and Sharma, J.K. 1986. *Hydnum subvinosum,* a rare parasite of *Leucaena leucocephala* in India. *Trans. Br. Mycol. Soc. 87*, 401-405.

Sankaran, K.V., Balasundaram, M. and Sharma, J.K. 1986. Seedling diseases of *Azadirachta indica* in Kerala, India. *Eur. J. Forest Pathol. 16*, 324-328.

Sateesh, M.K., Bhat, S.S. and Devaki, N.S. 1997. *Phomopsis azadirachtae* sp. nov. from India. *Mycotaxon 65*, 517-520.

Saxena, R.M. 1985. Seedling mortality of *Eucalyptus* spp. caused by Seed mycoflora. *Ind. Phytopath. 38*, 151-152.

Seliskar, C.E. 1964. Virus and virus like disorders of forest trees. In: *FAO/IUFRO Symposium on Internationally Dangerous Forest Diseases and Insects*. Oxford, July, 1964, FAO/ FORPEST-64.

Sen Sarma, P.K. 1984. Mycoplasma and allied diseases of forest trees in India and vector-host-pathogen interactions. *Proc. Ind. Acad. Sci. (Animal Sci.) 93*, 323-333.

Sharma, J.K. and Ali, M.I.M. 1996. Seed Pathology of Tropical hardwoods. In: *Proceedings Tree Seed Pathology Meeting*. Opoeno, Czech Republic, 74-81.

Sharma, R.C. and Bhardwaj, L.N. 1988. Forest Nursery Diseases and their management. In: *Advances in Forestry in India,* Volume II (Ed. Ram Parkash). International Book distributors, Dehra Dun, 91-118.

Sharma, J.K. and Shankaran, K.V. 1985. *Diseases of Albizia falcataria in Kerala and their possible control measures*. Final Report of the Research Project Pathology KFRI – Research Report # 47, Kerala Forest Research Institute, Peechi.

Sharma, J.K. and Sankaran, K.V. 1987. *Diseases of Albizia falcataria in Kerala and their possible control measures*. Kerala Forest Research Institute Research Report # 47, Peechi.

Sharma, S., Pal, R., Gupta, P. P. and Kaushik, J. C. 2004. Management studies in Root rot of *Prosopis cineraria* caused by *Fusarium solani. J. Tropical Forest Sci. 16*, 71-77.

Shivanna, M.B. 2005. Fungal diseases in forest nurseries in Shimoga district, Karnataka, India. In: *Diseases and Insects in forest Nurseries - Proceedings of the 5th Meeting of IUFRO working Party*. May 6-8, 2003, Kerala Forest Research Institute, Peechi, 27-30.

Shukla, A.N. 1992. Seedling blight and Root rot in Neem (*Azadirachta indica* A. Juss). *Ind. J. Forestry 15*, 266-268.

Shukla, A.N. 1995. Management of diseases in the forest nurseries in India. *Van Vigyan33*, 167-173.

Shukla, A.N., Singh, S. and Sega, H.S. 1988. Diseases and deterioration of bamboos in India. *Indian Forester 114*, 714-719.

Shukla, A.N., Uniyal, K. and Bandar, D.S. 1990. Solar sterilization of nursery soil and its effect on soil inhabiting fungi and germination of *Eucalyptus* hybrid seeds. *Ind. J. Pathol. 13*, 14-20.

Shukla, P., Trivedi, S., Srivastava, R. J. and Kumar, A. 2005. Alternaria blight of *Eucalyptus* hybrid and its Management. *Ind. J. Forestry 28*, 59-62.

Singh, Y. 2002. Nursery diseases of *Dipterocarpus macrocarpus* and *Gmelina arborea* in Assam – New records. *Indian Forester 128*, 965-970.

Singh, S. and Pandey, P.C. 1971. *Tunicospora,* a new rust genus on Bamboo. *Trans. Br. Mycol. Soc. 56*, 301-303.

Singh, R. and Srivastava, A.K. 1985. A new sap transmissible virus of Haunch (*Sesbania sesban* (L.) Mer). *Current Science 55*, 249-250.

Singh, S., Verma, V.P.S. and Sure, R.K. 1979. Protection of Sal seeds in storage against moulds. *Indian Forester 105*, 811-815.

Singh, S., Khan, S.N. and Mishra, B.M. 1982. Gummosis brown spot and seedling mortality in Subaru I – Disease incidence and pathology of the host. *Indian forester 109*, 185-192.

Singh, S. Khan, S.N. and Mishra, B.M. 1983. Gummosis brown spot and seedling mortality in Subabul II – Epidemiology and control of disease. *Indian Forester109*, 810-821.

Singh, A.N., Borax, T.R., Sharma, G.S. and Borax, N.J. 2004. Nursery diseases of *Dipterocarpus retusus. Indian Forester 130*, 742-748.

Sreenivasulu, P. and Nayudu, M.V. 1982. Purification and partial characterization of *Sesbania* Mosaic Virus. *Current Science 51*, 6-87.

Srinivasan, K.V. 1952. Seedling blight of *Sesbania grandiflora* Peers. *Current Science 21*, 318.

Srivastava, S.N.S. 1956. Two fungal infections of *Hevea* seeds. *Current Science 25*, 157–158.

Srivastava, A.K. 1986. Effect of *Sesbania* mosaic virus in the nitrate reductase activity in leaf tissues of haunch *Sesbania sesban* L. Mer. *Current Science 55*, 249-250.

Srivastava, K. K. and Soni, K. K. 1993. Seedling blight of *Albizia falcataria* and its control. *Annals of forestry 1*, 82-84.

Stoney, W. 1997. *Azadirachta indica*: Neem, A versatile Tree for the Tropic and Subtropics. In: *Agroforestry Species and Technologies A compilation of highlights and factsheet by NFTA and FACT Net.* Winrock International, USA. 49-50.

Sutton, B.C. 1980. *The Coelomycetes – Fungi Imperfecti with pycnidia acervuli and stromata.* Commonwealth Mycological Institute, Kew, Survey, UK, 383-384.

Tanhane, M.D., Thorat, S.P., Rege, N.N. and Dhanukar, S.A. 1997. Effect of oral administration of *Terminalia chebula* on gastric emptying: an experimental study. *J. Post grad. Med. 43*, 12-13.

Tiwari, D.N. 1994. *A Monograph on Bamboo.* International Book Distributors, Dehra Dun, India.

Tiwari, B.K. and Sharma, G.D. 1981. Seed mycoflora of eleven tree species of North Eastern India. *Ind. Phytopathol. 34*, 83.

Trivedi, P.C., Sharma, C. and Datta, S. 1986. New records of the root-knot nematode *Meloidogyne incognita* Kofoid and White. *Ind. J. Nematol. 16*, 279.

Uniyal, K. 1999. Collar rot in *Azadirachta indica* and its control. *Indian Forester 125*, 513-516.

Uniyal, K., Harsh, N.S.K., Uniyal, D.P. and Verma, S. 2004. Canker and die back of *Acacia nilotica* seedlings in nursery. *Indian Forester 130*, 95-100.

Vaartaja, O. 1952. Forest humus quality and light conditions as factors influencing damping off. *Phytopathol. 42*, 501-506.

Vijayan, A.K. 1988. *Studies on Seed Mycoflora of some Important Tree Species of Northern India.* Ph.D. Thesis, University of Garhwal, Srinagar.

Vijayan, A.K. 1991. *Recent Advances in Forest Seed Pathology.* Bishen Singh Mahendra Pal Singh, Dehra Dun.

Frontiers in Fungal Ecology, Diversity and Metabolites (edited by K.R. Sridhar), I.K. International Publishing House Pvt. Ltd., New Delhi, 2009.

7

Diversity and Adaptive Response of Lichens in Antarctica

S. Nayaka*, D.K. Upreti and R. Bajpai
[1]*Lichenology Laboratory, National Botanical Research Institute, Rana Pratap Marg, Lucknow 226 001, Uttar Pradesh, India*
*(*Email: sanjeeva_n@yahoo.com)*

Introduction

Two hundred million years ago Antarctica was united to Africa, Australia, India, New Zealand, and South America forming the super continent called Gondwana. Forces within the earth caused these continents to separate and drift apart. About thirty million years ago, the continents reached their approximate present positions.

Antarctica is like a different planet, a region of the world where the driving forces are associated with conservation and adventure. The discovery of the ozone hole above the Antarctic in 1985 led the world leaders to consider pollution control measures on a global scale for the first time. Antarctica is now recognized as an important region for research on the state of global environment and ocean systems, which control the sustenance of humans on earth. It is a land of several unusual phenomena such as the Aurora, whiteout, mirage, halo, Sun pillar, blizzard, catabatic winds, the midwinter moon and the midnight sun. These phenomena provide amazing facts for research and hence, the Antartica can be called a 'Scientist's Paradise'.

Antarctica possesses a unique biodiversity where a variety of biota has adaptated and survived in the harsh environment. Penguins, skuas, albatross, snow petrels, seals, whales, dolphins and krill are the major components of wildlife of Antarctica. Only two species of native angiosperm are found in Antarctica, which have their distribution restricted to the Peninsula region that has a warmer climate. A few species of mites, insects and invertebrates are also known in Antarctica, while the number of bacteria for the whole continent is unknown. More than 300 taxa of non-marine algae are known from Antarctica, which sometimes form spectacular red, yellow or green patches on areas of permanent snow. About 70 taxa of microfungi (Vishniac, 1996), 104 taxa of mosses and 25 taxa of liverworts (Ochyra *et al.,* 1998) have been reported from the continent. Lichens with 439 taxa (Øvstedal and Smith, 2001, 2004) are the most diverse group of visible terrestrial organisms numerically and the most conspicuous in the landscapes of ice-free regions of Antarctica. Lichens individually or together with bryophytes form the largest amount of standing biomass in Antarctic landscapes with up to 950 gm^{-2} in continental and up to 1300 gm^{-2} in maritime Antarctic habitats (Kappen, 2000). The fruticose thalli of the genus *Usnea* form shrublets of up to 20 cm length and the foliose plate-shaped *Umbilicaria* reach a diameter of 20 cm. Thus lichens are the largest primary producers of the Antarctic continent (Kappen *et al.,* 1996).

Reports on the transport and deposition of pollen grains from other continents (Marshall, 1996) indicate that more organisms would have potentially establish themselves in the Antarctic regions than actually exist. Hence, the unequal presence of plants and microorganisms in Antarctica cannot be explained because of their limited immigration into the continent due to the large distance, weaker air streams or other meteorological conditions. In the absence of a fossil record for lichen from the polar regions, any remark on the past distribution and transformation into a modern kind will be speculative. According to Lamb (1970) and Rudolph (1967), evolution in polar lichens is extremely slow and genetic recombination being a rare event, a little evolution may be possible in the Antarctic lichen flora as a whole. The establishment of lichen thalli depends on the simultaneous presence of the suitable photobiont and the mycobiont, either free-living or as vegetative diaspores such as isidia and soredia. It is however, interesting to note that a major part of the Antarctic lichen vegetation is recruited from crustose species that usually form spores instead of vegetative diaspores. It is still poorly understood how these species find their suitable photobiont, particularly under the hostile conditions of Antarctica (Kappen, 2000). The major factor that affects the present composition of Antarctic lichen flora has been the Late Cenozoic glaciations. The continental drift had already moved Antarctica to its isolated position near the pole and the rich subtropical and temperate floras that existed in many localities declined and perished (Barton, 1963). It can be assumed that the glaciations further drastically influenced the lichens and in some localities, they became extinct. The lichens that survived the glaciations and harsh conditions took refuge in nunataks and transformed from temperate to continental forms. As a result, the plant life of the Antarctic regions has been profoundly modified. The Antarctic Peninsula and adjacent islands being closest to a neighboring continent (South America, 2000 km) appear to be a pathway for species immigration (Kappen and Straka, 1988; Marshall, 1996). Among all the theories, the 'adaptation theory' best explains the dominance of lichens in Antarctica. The current chapter deals with diversity, stress tolerance, adaptive response, and endemism of lichens in Antarctica. This study is mainly focused on the lichens collected during different expeditions and housed in the herbarium of the National

Botanical Research Institute, Lucknow (LWG), India and other published reports. An updated checklist for lichens occurring in Schirmacher Oasis is furnished with comments.

Extremes of Antarctica

Antarctica is a land of extremes and 98% of the continent is permanently covered with ice, while the remaining 2% of ice-free land is restricted mostly to the periphery of the continent. These ice-free habitats are very young on the geological time scale. The Antarctic glaciation is believed to have commenced in the earliest Oligocene period after long geological periods with warm climates (Barker *et al.*, 1998). Antarctica is the coldest continent with heat negative balances everywhere (Engelskjön, 1986). The lowest outdoor temperature ever recorded on the earth is –89.6ºC at the Russian Vostok station on the inland ice cap of Antarctica in 1983. It is the windiest of the continents with the highest wind speed measured at d'Urville (327 km/h). The wind acts in three ways: by desiccation, wind-chill and ice-blast action. The absolute humidity in the continent is 0.03% and it is lower than the Sahara Desert. Low average humidity combined with the extreme cold makes the South Pole region the world's driest desert. It rains very rarely in maritime Antarctica. Snowfall, which averages less than one inch annually, is the major precipitation in the continent. Solar radiation becomes zero as the Sun dips below the horizon on March 22 and not seen again until September 22. Owing to the tilt (23.3º) in the earth's axis, relative to its orbit around the sun, the Sun does not shine at the South Pole for about six months of the year. When the Sun shines, much less solar energy reaches the ground at the pole because the Sun's rays pass through a thicker layer of atmosphere than at the Equator. In addition, due to the predominance of ice and the snow cover in Antarctica, most of the sun's rays that do reach the ground are reflected back into space leaving the continent cold with an unmolten ice mass. The freshwater lakes in Antarctica are mostly confined to certain ice-free areas such as Larsemann Hills, Schirmacher Oasis, Bunger Hills, and Vestfold Hills. These lakes remain frozen for the most part of the year and are fed by glacier and snowmelted streams during the short austral summer months. Despite the restricted availability of liquid water, extreme air and water temperatures, and near-perennial ice cover, these lakes provide a unique habitat for the persistence of microbial life throughout the year.

Lichens and Stress

In an environment with predominant stress, the existence and evolution of an organism can be directly attributed to their adaptation to the stress (Kappen, 2000). The adaptive stimulus is clearly expressed in Antarctic lichens as they are the most successful elements of the terrestrial biota in Antarctica and have obvious dominance in ice-free areas. They are superior to bryophytes and perhaps to free-living algae and fungi, particularly in harsh regions (Kappen, 2004). Apparently, lichens did not develop 'unique' adaptive mechanisms in the extreme Antarctic environment. The cold tolerance and morphological features that are present in Antarctic lichens were also been observed in lichens growing in harsh climate of other regions (Kappen, 1988). The adaptive responses of lichens to various stress conditions are interesting.

Habitat and substrate

Lichens are able to grow on all existing substrata such as solid rock, moraines, single stones, concrete, timber of old building material, rusted iron, bones, boots, dead moss and glass. Only a few lichens such as *Cladonia* and *Peltigera* exclusively grow on soil. For the establishment of a lichen thallus, the structure of the substrate is more important than its chemistry. Nutrient availability from the substratum plays a minor role, as many lichens from inland habitats have a low nutrient demand and mainly derive it from airborne dust. However, the availability of nutrients in substratum may facilitate luxuriant growth of lichens. For example, some ornithocoprophilous lichens grow profusely near the bird colony. The rarity of calcareous substrate (except concrete, bones and marble in a few areas) in the Antarctic region and the absence of trees, shrubs and other phorophytes (except drift wood and timber of station) maybe one of the reasons for the lower species diversity in the Antarctic lichens and the possible increase in the endemism. Due to the lack of competition with bryophytes and other plants, lichens are increasingly predominant polewards, up to 86°20'S (Kappen, 2000), unlike in maritime Antarctica with milder climatic conditions.

Lichens mostly prefer to grow in protected or insolated sites that provide heat and facilitate snowmelt (Kappen, 1985). In windy regions, the leeward sites of rocks and sheltered niches are colonized by lichens mostly and are strongly fixed to the substratum. In the maritime Antarctic, the cryptogamic vegetation is more widespread but ceases near glaciers, in places with long-lasting snow and in animal colonies. Most of the macrolichens prefer open sites with shallow snow cover and early snow melt, whereas some crustose lichens are resistant to long-term snow cover in depressions. The occurrence of endolithic lichens on granite and sand stone is common in Antractica, which may be an adaptive mechanism to avoid a hostile environment outside as well as wind erosion. Sometimes, lichens are confined to gaps between boulders or crevices in the rock. Such crevices may facilitate trapping lichen spores and dust particles, retaining moisture for long durations, providing shade and preventing wind blowing at fast speeds.

Temperature and moisture

Lichens are extremely tolerant to freezing stress, exposure to low temperature (Kappen *et al.,* 1996) and can withstand temperature as low as –196°C. The photosynthesis of Antarctic lichens was found to continue even at –20°C. The hydrated thalli of Antarctic lichens withstand gradual or rapid freezing up to 12h at –196°C without any damage. Lichens survived 96 weeks of freezing at –15°C and or ten years at –20°C, while *Alectoria ochroleuca* (Hoffm.) A. Massal. recovered fully from 3.5 years of storage at –60°C. Thus, lichens in the anabiotic state may be able to persist in the most extreme environment (Kappen *et al.,* 1996). The high amount of carbohydrates and polyols in lichens is thought to be responsiblefor their tolerance to freezing stress (Hale, 1967). Frost injury of lichens in their natural habitats has rarely been reported (Barkman, 1958).

Lichens cannot control their water relation as vascular plants. Thus, their metabolic activity depends strictly on an ambient moisture condition (Kappen *et al.,* 1996). Due to the low humidity in Antarctica, the availability of water for photosynthesis in lichens is one of the major constraints. As poikilohydrous system, lichens can be hydrated and activated quickly by the snow, dew, mist or glacier melt water and resume photosynthesis. The snow can melt by the sun's radiation even under an overcast sky. Some lichens (e.g. *Umbilicaria*) possess dark coloured thalli, absorb sunlight, become much warmer and melt snow crystals deposited over them. Further, the moistened thalli can reach an over-air temperature of nearly 23K, which means they become as warm as 20°C maximum that can induce maximal productivity as in the case of *Buellia frigida*. Some lichens are capable of gaining enough moisture from the humidity at the edge of a temporary snow patch on the rock without any visible melting process. In continental Antarctica, the rehydration of epilithic lichens can be seen on inclined or vertical rock walls, where melted water trickles down quite often.

Lichen can photosynthesize at low water potentials (–20 MPa at –20°C). Experiments in hot deserts have shown that lichens are able to photosynthesize even at low water potentials (–38 MPa) (Nash *et al.,* 1990). As soon as the availability of moisture stops, the lichen thallus quickly desiccates, loses water up to 97% and falls into an anabiotic state. Under a low water potential, the rate of photosynthesis is also low and can be considered as a means of keeping the photosynthetic apparatus intact and producing frost-protective carbohydrates than the actual dry matter. However, sometimes in maritime Antarctica, frequent rain, sleet or snow may lead to over-saturation of lichen thalli, which is a major problem for lichens than drought or freezing (Kappen, 1988). In such instances, a marked depression in the net photosynthesis can be seen.

Light and radiation

Strong irradiation may not affect the lichen of Antarctica, but chlorophyll contents and net photosynthetic rate are higher in the lichens growing in the shade than in Sun-exposed dark forms. The occurrence of yellowish or greenish forms of the same species or thallus in the shade, cracks and fissures demonstrates that pigmentation is a response to light. Further, the concentration mycosporine-like amino acid (MAA) derivatives that are capable of absorbing UVB radiation in lichens with cyanobionts were high in exposed lichens (Kappen, 2000). Many lichens growing in open habitats possess strong colored pigments like yellow, red, brown or black. Such pigmentation has been interpreted as a protection against irradiation (Kappen, 1973; Hertel, 1988). The pigmentation also protects the phycobionts from UV radiation (Lindsay, 1977). The pigments, in addition to absorbing light of high intensity by quenching the incident radiation, also increase the use of light for photosynthesis at low intensity by fluorescence (Rao and LeBlanc, 1965). Those species that are devoid of pigmentation have a cortex and/or a necrotic layer (Dodge, 1973), whose thickness varies according to their habitat. Lichens growing in exposed areas have a thicker cortex than their counterparts growing in the shade. The presence of a thick cortex and an epinecral layer, apart from protecting the lichen thallus from strong irradiation may also act as a barrier to water absorption as well as its loss. In maritime Antarctica with far less harsh environment, dark coloured species are less prominent and are less heavily pigmented (Lindsay, 1977). Some lichens develop more conspicuous black or dark colored prothallus, which again helps in the absorption of more solar radiation. The covering of apothecia and thallus with pruina, i.e., a layer of crystals or dust, has been interpreted as a protective device against insolation (Kappen, 1973).

Lichens can even exist in completely dark places (Kappen, 1973) and hence their occurrence in continuous darkness in Antarctic during winter is not surprising. They can photosynthesize in low light intensity and become physiologically inactive when no light is available. A deep snow layer may be protective against temperature oscillations in winter but may be too thick for light penetration. However, a snow layer of less than 40cm can provide photosynthetic activity in early spring. In eastern Antarctica, in the lichens buried at a depth of 15cm, the photosynthetic active quantum flux still reached values that allow lichen photosynthetic activity near saturation at about 0°C, although the quantum flux density of up to 250μM photos $m^{-2}S^{-1}$ was around 10% of the ambient irradiance. The photosynthetic light compensation point is by 30μM photons $m^{-2}S^{-1}$ (Kappen, 1988).

Morphology and anatomy

The phenotypic variations including stunted growth, deformation of the thallus and unusual alterations in gross morphological characters are common in Antarctic lichens. They are generally smaller compared to their counterparts growing in other regions (Dodge, 1973). Isidiate and sorediate taxa are rare, fruit body producing taxa are frequent and the production of a great number of apothecia is common in Antarctic lichens (Kappen, 1988). However, their development is usually arrested and they do not produce mature ascospores. The lichens growing in Peninsula and maritime Antarctica with mild climate sometimes produce fertile apothecia (Lindsay, 1977). Some crustose lichens [e.g., *Bacidia stipata* I.M. Lamb, *Buellia cladocarpiza* I.M. Lamb, *Catilaria corymbosa* (Hue) I.M. Lamb, *Caloplaca regalis* (Vain.) Zahlbr., *Lecania brialmontii* (Vain.) Zahlbr.] are either stipitate or pulvinate (and appear fruticose), which may be an adaptation to the frequently occurring fog and high air humidity in maritime Antarctica (Kappen, 1988) or a result of freeze-thaw condition in continental Antarctica (Filson, 1982). Stipitate thalli can quickly protrude through the snow melting on the tips and a stipitate growth raises the apothecia and thereby, ensures ascospore dispersal

(Kappen, 1988). In fruticose lichens, the apothecia tend to be borne towards the upper parts of the thalli and in many species towards the tips of the branches. Those lichens growing luxuriantly in nitrogen-enriched sites (e.g., orinithocoprophilous) usually develop thick and at times almost squamulose thalli (Weber, 1962). Such luxuriant modifications are sometimes taxonomically disadvantageous as these modifications are important characters for identification.

The deformation of the thallus due to animal feeding is rare in Antarctica, while fungal parasitism is frequent. The parasitic fungus form gall-like structures, cause discoloration and in massive infections may suppress the development of ascocarps or ascospores of the host lichen. Ascocarps of the parasitic fungus sometimes grow on the sterile thallus of the host and the entire system is then misunderstood for normal lichen (Hertel, 1988). The ability to regenerate damaged structures is well developed in crustose lichens. However, the resulting regenerated thalli or ascocarps may differ considerably in morphology than in the undamaged species. Carbonized structures may develop around the damaged tissues of the thallus, while regeneration of hymenia may lead to a gyrodisc and umbonate ascocarp. The development and regeneration are very slow events in Antarctic lichens. Thus, a new wave of damage often occurs before an earlier damage is regenerated and this leads to the development of very strange thallus and ascocarp structures (Hertel, 1988). Due to the low temperature in continental Antarctica, the relative hardness of ice and snow particles increases. When blown by strong winds (ice-blast) these particles become extremely forceful agents of erosion (Fristrup, 1951; Juckes, 1969). In maritime Antarctica, foliose and fruticose lichens are dominant growth forms, but they decrease proportionately in diversity with change in the continental climate. In such situations, they become dwarfed and increasingly adpressed to the substratum to avoid the effects of the wind (Lindsay, 1977).

Besides the sunlight, mineral-rich substratum also influences the coloration of the lichen thallus. The lichens growing on iron-rich rocks often show a rusty coloration (Hertel, 1988) and those growing on copper-rich substratum show green coloration, while the thallus of the same species growing on other substratum show different coloration. Sometimes, the mineral-rich substratum may also modify the chemistry of the lichen, which may be confused entirely as a different species than as a chemotype. Due to such extreme phenotypic variations, the identification and monographing of the lichens of Antarctica, especially the crustose ones is a challenging task. The lack of floras and monographs, the unavailability of important type specimens and the lack of representative collections that are adequate to study the variability and distribution, further complicate the taxonomy of Antarctic lichens. It demands prior experience on lichenology in the alpine, polar or temperate regions. Thus, such difficulties leading to an erroneous identification can be illustrated with the example of Dodge (1973) who reported a total of 429 lichens from Antarctica. Subsequently, Castello and Nimis (1995, 1997) revised these lichens and reduced the figure dramatically to 241. In the same manner, 74 taxa of

Acarospora (subgenus *Xanthothallia*) described from Antarctica turned out to be the environmental modification of only two (Webber, 1968).

Growth, biomass and physiology

Lichens are generally slow growing and their growth rate is further influenced by the harsh climate of Antarctica. Comparatively, the fruiticose and foliose forms grow much faster than the crustose lichens (range, 0.01 – 1.3 mm/year). Lichen growth is not a direct function of photosynthetic productivity. The rates of growth vary greatly with the species and location in Antarctica. The availability of nutrients in ornithocoprophilous lichens growing near a bird colony certainly induces fast growth as well as a large biomass (Kappen, 1988, 2000). The success of lichens in the Antarctica can be explained mainly by their physiological capacity, in particular, their high tolerance of low temperature stress and high resistance to photo stress. Their capacity to be active photosynthetically at subzero temperatures is remarkable. Species from the maritime Antarctica generally have a greater photosynthetic capacity and wider temperature range than those from the continent. Lichens with green algae as photobionts with heteromerous thallus are able to reactivate and to maintain photosynthesis by means of water vapour uptake from the ice and snow, which is an important property of many desert lichens too. Interestingly, the cyanolichens, usually with homeomerous thallus are not able to reactivate photosynthesis by means of water vapour and therefore, require liquid water. Further, they are unable to carry out photosynthesis at temperatures below freezing. Thus, they are restricted mostly to maritime Antarctica, which has warmer climate with frequent availability of rainfall or melted water (Kappen, 1988). Lichen can be metabolically active during winter with a carbon balance negative or positive, but it should not be zero because the endogenous anabiotic rest phase does not exist in lichens (Kappen *et al.,* 1996). However, physiological capacitance of the lichens is not a result of adaptive evolution in the Antarctic region, as it is ubiquitous in high-mountain and desert lichens. It is rather a pre-condition that occur within the lichens.

Lichen Diversity and Endemism

Antarctic lichens have been studied since 1823 (Torrey, 1823). The Antarctic Peninsula and islands of maritime Antarctica are explored in the past hundred years and such information is limited to continental Antarctica. Out of the 20,000 lichen species known around the world, Antarctica represents about 439 (Øvstedal and Smith, 2001, 2004). In general, the dominance of lichens increases with the increase in latitude in Antarctica although, in absolute terms, the species number and standing biomass decreases (Kappen, 2000). The lichen flora is dominated by microlichens with 330 species (75%), including endolithic, crustose, leprose, squamulose, placodioid, effigurate, and subeffigurate forms. The microlichens are capable of colonizing the extreme habitats better than any other macroscopic plants. The foliose and fruticose lichens (macrolichens) are represented by 52 and 57 species respectively. Most of the macrolichens are known from the Antarctic Peninsula and maritime Antarctica. Macrolichen species having a wider distribution include: *Physcia caesia, P. dubia, Pseudephebe minuscula, Umbilicaria aprina, U. deccussata, Usnea antarctica*, and *Xanthoria elegans.*

A total of 177 (40%) lichen species are endemic to Antarctica with a majority of them confined to mild maritime Antarctica. However, the 'continental' Antarctica possesses the maximum proportion (50%) of the endemic taxa when the lichen flora of this region alone is considered. This demonstrates the extent of the endemism in the continent. There is a trend of increasing the endemic element with the increase in latitude or the climatic severity of the site. Their dominance in such regions indicates that they are better adapted to extreme habitats than many other species that have wider geographical amplitude. However, the degree of taxonomic separation of the endemics is low and occurs only at the species level (Kappen, 2004). Apart from the endemism, Antarctic lichens show restricted distribution in the southern hemisphere (24 species) including the alpine, temperate South America, New Zealand and Australia or bipolar (160 species). Only 72 species are cosmopolitan in distribution and most of them grow in a temperate habitat.

At present, it is not correctly known where the endemic lichens originated. It can be noted that for endemic species, the time span for an adaptive radiation and speciation is limited to the Pleistocene and the Holocene era. Antarctica is isolated from other southern hemisphere continents except for South America. The new establishment of immigrant lichens is more limited and scattered. Many tiny islands off shore and small, highly isolated ice-free areas called 'oases' (Smith, 1997) with harsher climate might have acted as a precondition favouring speciation of lichens in Antarctica.

Schirmacher Oasis

Schirmacher Oasis (SO) is a small ice-free landmass of 35 km^2, located in Queen (Dronning) Maud Land Area of continental east Antarctica (70.8° S and 11.8° E) and is surrounded by 10 nunataks (Fig. 1). Lichens from SO have been collected since the 1970s and Golubkova and Simonov (1972) published the first comprehensive list of 21 lichen taxa from the area. Ritcher (1990) opined that the specimens were not statistically represented and the collectors did not consider the variability due to the ecological conditions in different habitats. He provided a list of 25 species of lichens along with distribution maps. Ritcher's (1990) report was based on specimens collected by Russian researchers between 1979 and 1984. Subsequently, Ritcher (1995) published a revised list of lichens comprising 26 species including the specimens collected during 1988-89.

The National Botanical Research Institute (NBRI), Lucknow was identified by Ministry of Earth Science, New Delhi (India) as the centre for carrying out lichenological investigation in SO in 1991. The NBRI participated for the first time in the 11th Indian Antarctic Expedition (IAE) during the year 1991-92 and reported a total of 26 lichen species (Upreti and Pant, 1995; Upreti 1996; 1997). During the

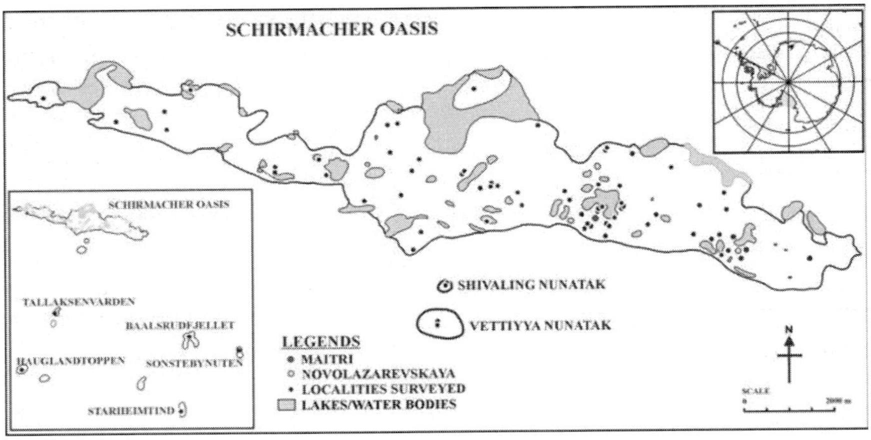

Fig. 1 : Map of Schirmacher Oasis, East Antarctica showing the localities surveyed by various Indian lichenologists.

17th IAE, Pandey and Upreti (2000) reported a total of 19 lichen species collected from the SO and the Vettiya nunatak. Meanwhile, Gupta *et al.* (1999) of the Botanical Survey of India also collected some lichens from SO during the 18th IAE and reported five more species. Recently, the NBRI participated in the 22nd IAE and explored 31 sites in the whole stretch of SO and seven nunataks and identified a total of 35 lichen species (Nayaka and Upreti, 2005). All the published reports on SO lichens accounts to a total of 48 species and out of them 40 species confirm their occurrence in SO and the rest have a high probability of growing in this region of Antarctica (Table 1). The field guide for lichens of Antarctica and South Georgia by Øvstedal and Smith (2001) and their subsequent publication (Øvstedal and Smith, 2004) are used as the main literature source in the present study.

Although some drawbacks exist in the Golubkova and Simonov's (1972) report, the occurrence of a few species such as *Aacarospora macrocyclos*, *Lecanora epibryon, Lecidea auriculata* and *L. rupicola* in SO cannot be ruled out as these species have a wide range of distribution, while *Caloplaca saxicola* and *Buellia pycnogonoides* were recorded from other places within Dronning Maud Land, thus their presence in SO is possible. The report of Ritcher (1995) can be considered authentic, as R. Filson has identified in their materials. *Umbilicaria vellea*, reported by Gupta *et al.* (1999) is temporarily included in the list as it is a temperate species occurring in the Arctic and there is a possibility that it may occur in SO too. Among the various lichens reported by Golubkova and Simonov (1972), *Acarospora petalina* Golubk and Savicz, *Buellia muscicola* Dodge and Baker, *Pyrenodesmia mawsonii* C.W. Dodge, and *Sarcogyne angulosa* C.W. Dodge and G.E. Baker were temporarily excluded from SO as their specimens have to be studied for confirmation, while *Rinodina turfacea* (Wnbg.) Th. Fr. does not occur in SO and may be an erroneous identification. Further, *Acarospora molybdina* Darb. and *Lecanora molybdina* Hook are confused with *Acarospora macrocyclos* as if they are the synonyms of the later species and hence they are

Table 1. *List of lichens reported from Schirmacher Oasis and nearby nunataks by various authors. The original names mentioned by these authors are included in the list as synonyms whenever relevant for better understanding (Growth: Cr, crustose; Fo, foliose; Fr, fruticose; Distribution: En, endemic; Co, cosmopolitan; Bi, bipolar; Source: 1, Golubkova and Simonov, 1972; 2, Richter, 1990; 3, Richter, 1995; 4, Upreti and Pant, 1995; 5, Upreti, 1996; 6, Upreti, 1997; 7, Gupta et al., 1999; 8, Pandey and Upreti, 2000; 9, Nayaka and Upreti, 2005; +, reported; Remarks: Y, occurrence confirmed; P, occurrence possible)*

	Lichen taxa	Growth	Distribution	1	2	3	4	5	6	7	8	9	Remarks
1.	*Acarospora gwynnii* C.W. Dodge and E.D. Rudolph	Cr	En	-	+	+	+	-	+	-	+	+	Y
2.	*A. macrocyclos* Vain.	Cr	En	+	-	-	-	-	-	-	-	-	P
3.	*A. williamsii* Filson	Cr	En	-	+	+	+	-	-	-	-	+	Y
4.	*Amandinea petermannii* (Hue) Matzer, H. Mayerhofer and Scheid. – *Rinodina petermanii* (Hue) Darb.	Cr	En	-	-	-	+	-	-	-	-	+	Y
5.	*Buellia frigida* Darb.	Cr	En	-	+	+	-	-	-	-	+	-	Y
6.	*B. grimmiae* Filson	Cr	En	-	-	-	+	+	-	-	+	+	Y
7.	*B. illaetabilis* I.M. Lamb.	Cr	En	-	-	-	-	-	-	-	-	+	Y
8.	*B. lignoides* Filson	Cr	En	-	+	+	-	-	-	-	-	-	Y
9.	*B. pallida* C.W. Dodge and G.E. Baker	Cr	En	-	-	-	+	+	-	-	+	-	Y
10.	*B. pycnogonoides* Darb.	Cr	En	+	-	-	-	-	-	-	-	-	P
11.	*Caloplaca citrina* (Hoffm.) Th.Fr.	Cr	Co	-	+	+	-	-	-	-	-	+	Y
12.	*C. saxicola* (Hoffm.) Nordin – *Gasparrinia murorum* Tornab.	Cr	Bi	+	-	-	-	-	-	-	-	-	P
13.	*Candelariella flava* (C.W. Dodge and G.E. Baker) Castello and Nimis – *C. antarctica* Filson – *C. hallettensis* (J.S. Murray) Øvstedal – *Protoblastenia citrina* C.W. Dodge	Cr	En	+	-	+	-	-	+	-	+	+	Y
14.	*Carbonea assentiens* (Nyl.) Hertel	Cr	En	-	-	-	-	-	-	-	-	+	Y
15.	*C. vorticosa* (Flörke) Hertel – *C. capsulata* (C.W. Dodge and G.E. Baker) Hale – *Lecidea capsulata* C.W. Dodge and G.E. Baker	Cr	Bi	-	+	+	+	+	-	-	+	+	Y
16.	*Lecanora* cf. *geophila* (Th. Fr.) Poelt	Cr	Bi	-	-	-	-	-	-	-	-	+	Y
17.	*L. epibryon* (Ach.) Ach. – *L. broccha* Nyl.	Cr	Bi	+	-	-	-	-	-	-	-	-	P
18.	*L. expectans* Darb.	Cr	En	+	+	+	-	-	+	-	+	+	Y
19.	*L. fuscobrunnea* C.W. Dodge and G.E. Baker	Cr	En	-	-	-	-	-	-	-	-	+	Y
20.	*L. orosthea* (Ach.) Ach.	Cr	Bi	-	-	-	-	-	-	-	-	+	Y
21.	*L. polytropa* (Hoffm.) Rabenh.	Cr	Bi	+	-	-	-	-	-	-	-	+	Y
22.	*Lecidea andersonii* Filson	Cr	En	-	+	+	-	-	-	-	-	+	Y
23.	*L. auriculata* Th. Fr.	Cr	Co	+	-	-	-	-	-	-	-	-	P
24.	*L. cancriformis* C.W. Doge and G.E. Baker – *L. phillipsiana* Filson	Cr	En	-	+	+	+	+	-	+	+	+	Y
25.	*L. lapicida* (Ach.) Ach. - *L. rupicola* Vain.	Cr	Co	+	-	-	-	-	-	-	-	-	P
26.	*Lecidella siplei* (C.W. Dodge and G.E. Baker) May. Inoue												

	– *Lecidea siplei* C.W. Dodge and G.E. Baker	Cr	En	-	-	-	+	+	-	-	+	+	Y
27.	*L. stigmatea* (Ach.) Hertel and Leuckert	Cr	Bi	-	-	-	-	-	-	-	-	+	Y
28.	*Lepraria neglecta* (Nyl.) Lettau	Cr	Bi	+	-	+	-	-	-	-	-	-	Y
29.	*Leproloma cacuminum* (A. Massal) J.R. Laundon	Cr	Co	-	-	-	-	-	-	-	-	+	Y
30.	*Physcia caesia* (Hoffm.) Fürner. – *P. wainioi* Räsänen	Fo	Co	-	+	+	+	-	-	-	+	+	Y
31.	*P. dubia* (Hoffm.) Lettau	Fo	Co	-	+	+	-	-	-	-	-	-	Y
32.	*Pleopsidium chlorophanum* (Wahlenb.) Zopf – *Acarospora chlorophana* (Wahlenb.) A. Massal – *Biatorella antarctica* J.S. Murray – *B. cerebriformis* (C.W. Dodge) Filson	Cr	Bi	+	-	+	-	-	-	-	-	-	Y
33.	*Pseudephebe minuscula* (Nyl. ex Arnold) Brodo and Hawks. – *Alectoria minuscula* (Nyl. ex Arnold) Degel.	Fr	Bi	+	+	+	+	-	-	+	+	+	Y
34.	*Rhizocarpon geograhicum* (L.) DC – *R. flavum* C.W. Doge and G.E. Baker	Cr	Co	+	+	+	-	+	-	-	+	+	Y
35.	*R. nidificum* (Hue) Darb.	Cr	En	-	-	-	-	-	-	-	-	+	Y
36.	*Rhizoplaca melanophthalma* (Ram.) Leuckert and Poelt – *L. melanophthalma* (Ram.) Ram.	Cr	Bi	-	+	+	+	-	+	-	+	-	Y
37.	*Rinodina endophragmia* I.M. Lamb.	Cr	Bi	-	-	-	-	-	-	-	-	+	Y
38.	*R. olivaceobrunnea* C.W. Dodge and G.E. Baker	Cr	Bi	-	+	+	+	-	+	-	+	+	Y
39.	*Sarcogyne privigna* (Ach.) A. Massal	Cr	Co	-	-	-	-	-	-	-	-	+	Y
40.	*Umbilicaria africana* (Jatta) Krog and Swinscow	Fo	Co	-	-	-	-	-	-	-	-	+	Y
41.	*U. antarctica* Frey and I.M. Lamb	Fo	En	-	-	-	-	-	-	+	-	+	Y
42.	*U. aprina* Nyl. – *Omphalodiscus spongiosus* var. *subvirginis* (I.M. Lamb and Frey) Golubk.	Fo	Co	+	+	+	+	-	-	+	+	+	Y
43.	*U. decussata* (Vill.) Zahlbr. – *Omphalodiscus decussatus* var. *discolor* Lynge	Fo	Co	+	+	+	+	-	-	-	+	+	Y
44.	*U. vellea* (L.) Ach.	Fo	Co	-	-	-	-	-	-	+	-	-	P
45.	*Usnea antarctica* Du Rietz	Fr	Co	-	+	-	-	-	-	-	+	+	Y
46.	*U. sphacelata* R. Br. – *U. sulphurea* Th. Fr.	Fr	Bi	-	-	+	-	-	-	-	-	-	Y
47.	*Xanthoria elegans* (Link) Th. Fr. – *Gasparrinia elegans* Stein apud Cohn	Fo	Co	+	+	+	+	-	+	-	-	+	Y
48.	*X. mawsonii* C.W. Dodge	Fo	En	-	+	+	-	-	-	-	-	+	P

excluded. From Ritcher's (1995) list, *Buellia pergrina* Darb. has been excluded as the current status of this taxon is untraceable. Further, *Xanthoria candelaria* (L.) Th. Fr. f. *antarctica* (Vain.) Hillm. is replaced with *Xanthoria mawsonii* as there is more chance of the later taxon to occur in SO than the former. *Caloplaca isidioclada* Zahlbr., *Pertusaria* sp., *Polycauliona murrayi* (C.W. Dodge) Poelt,

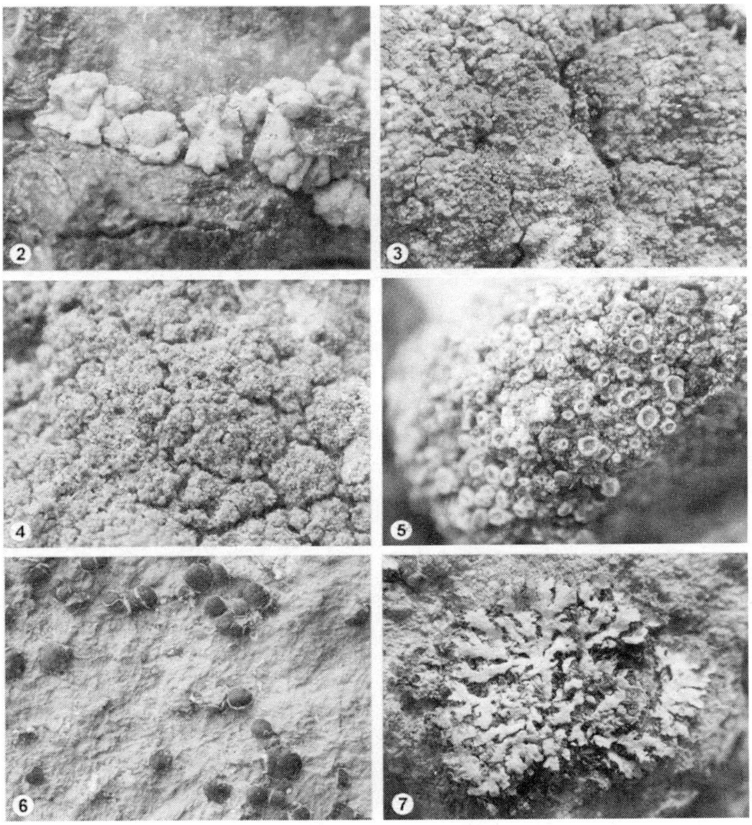

Fig. 2-7 : Some common lichens occurring in Schirmacher Oasis, 2. *Acarospora gwynnii,* 3. *Caloplaca citrina,* 4. *Candelariella flava,* 5. *Lecanora expectans,* 6. *Lecidea cancriformis,* and 7. *Physcia caesia.*

P. regalis Murray and *Porpidia* sp. reported by Upreti and Pant (1995) and Pandey and Upreti (2000) are excluded from SO as they were erroneously identified, while *Lepraria membranacea* (Dicks.) Lett. (also reported by Ritcher, 1995) is found to be *Leproloma cacuminum* as the correct identity. The *L. membranacea* is an Arctic species and its distribution in Antarctica is doubtful (Øvstedal, personal communication). *Verrucaria holizoa* Leight, a semi-aquatic lichen living in the fresh water habitat of SO as described by Nayaka and Upreti (2005), found to be erroneous as the specimens were rarely fertile. *Xanthoria candelaria* (L.) Th. Fr. and *X. mawsonii* are very similar species, which differ mainly in the pycnidial characters and mostly the later species occur in continental Antarctica (Øvstedal, personal communication). The recent collection of this specimen from SO (Nayaka and Upreti, 2005) was not available for critical study due to insect infestation and hence, temporarily excluded from the checklist.

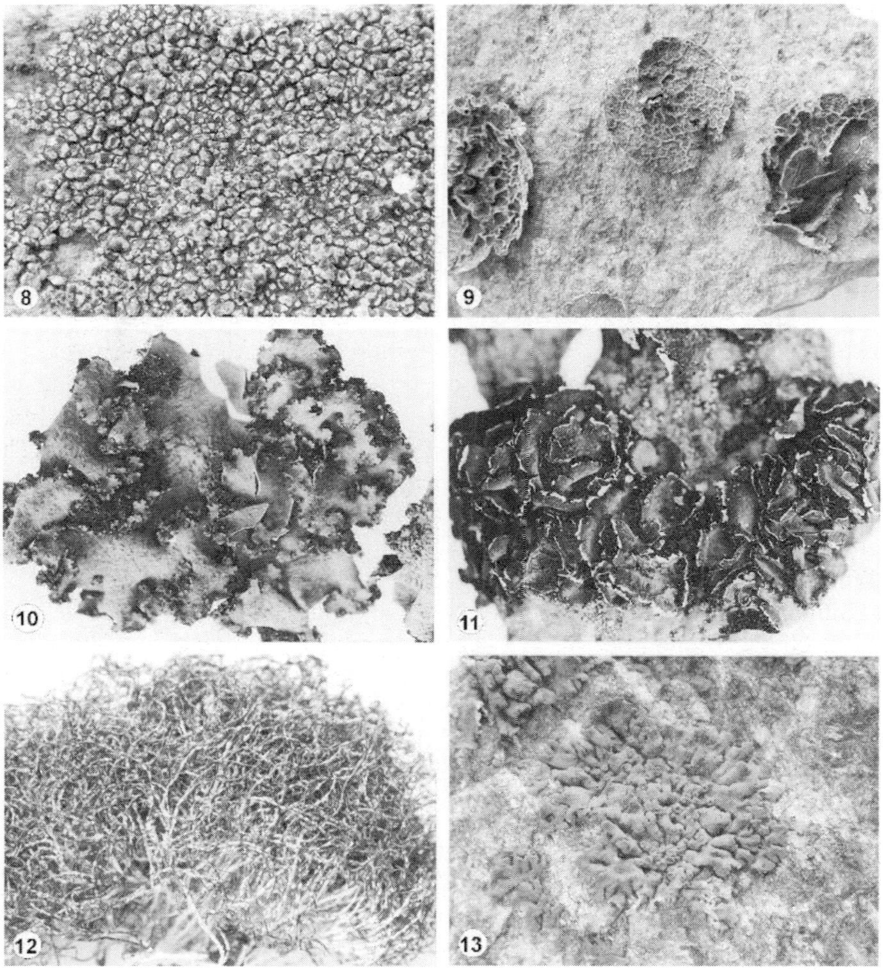

Fig. 7-13 : Some common lichens occurring in Schirmacher Oasis (continued), 8. *Rhizocarpon geographicum*, 9. *Umbilicaria decussata*, 10. *U. africana*, 11. *U. aprina*, 12. *Usnea antarctica*, and 13. *Xanthoria elegans*.

Thus, a total of 16 taxa either erroneously identified or doubtful in occurrence in SO are also excluded.

Out of the 48 lichen species that occur in SO, 19 are endemic to Antarctica, 15 are bipolar and 14 are cosmopolitan in distribution. Thirty-six crustose lichens dominate in SO followed by nine foliose species and three fruticose lichens (Table 1). The most common lichen found in SO include: *Acarospora gwynnii*, *Caloplaca citrina*, *Candelariella flava*, *Carbonea vorticosa*, *Lecanora expectans*, *Lecidea andersonii*, *L. cancriformis*, *Physcia caesia* (Fig. 2-7), *Pseudephebe minuscula*,

Rhizocarpon geographicum, *Rinodina olivaceobrunnea*, *Umbilicaria decussata*, *U. africana*, *U. aprina*, *Usnea antarctica* and *Xanthoria elegans* (Fig. 7-13).

Conclusions and Outlook

Due to its tolerance of harsh climatic conditions (long-term desiccation and storage in high vacuum at temperatures near absolute zero), lichens have been proposed earlier as the living system most likely to exist beyond earth along with tardigrades, unicellular algae and bacteria (Kappen *et al.,* 1996). The wide distribution and ubiquity of the propagules suggest that lichens would potentially colonize all ice-free areas in Antarctica although the actual colonization is not homogeneous. Lichens can colonize fresh substrata after a lag phase, which is required for changing the surface structure of the substratum by weathering and becoming suitable for colonization. Lichen assemblages buried under snow may be stable for centuries if there is none or the least disturbance (Kappen, 2000). The checklist of Schirmacher Oasis represents the diversity and distribution of lichens of Antarctica. However, the number of taxa will certainly increase as there are many unidentified specimens of complex groups that still exist in herbaria of NBRI (LWG) and Russia. It also remains to be reexamined as specimens of Golubkova and Simonov (1972) and Richter (1990, 1995) for further confirmation. Under the present climatic conditions, vegetation structure remains stable, but any drastic change of climatic conditions (e.g., air pollution or mechanical stress by animal and human activities), being highly vulnerable, would most likely alter or destroy the lichen vegetation. .

References

Barker, P.F., Barrett, J.P., Camerlenghi, A., Cooper, A.K., Davey, F.J., Domack, E.W., Escuria, C., Kristoffersen, Y. and O'brien, P.E. 1998. Ice sheet history from Antarctic continental margin sediments: the ANTOSTRAT approach. *Terra Antarctica 5*, 737-760.

Barkman, J.J. 1958. *Phytosociology and Ecology of Cryptogamic Epiphytes.* VaGorcum, Assen, Netherlands.

Barton, C.M. 1963. The significance of two separate Tertiary plant assemblages from King George Island, South Shetland Islands. *Polar Res. 11*, 784-785.

Castello, M. and Nimis, P.L. 1995. A critical revision of Antarctic lichens described by C.w. Dodge. *Biblioth. Lichenol. 57*, 71-92.

Castello, M. and Nimis, P.L. 1997. Diversity of lichen in Antarctica. In: *Antarctic Communities: Species, Structure and Survival* (Ed. Battaglia, B., Valencia, J. and Walton, D.W.H.). Cambridge University Press, Cambridge. 15-21.

Dodge, C.W. 1973. *Lichen flora of Antarctic Continent and Adjacent Islands.* Pheonix Publishing, Canaan, New Hampshire.

Engelskjön, T. 1986. Zonality of climate and plant distributions in some Arctic and Antarctic regions. *Rapportserie Norsk Polarinstitutt* 30, 1-49.

Filson, R.B. 1982. Lichens of continental Antarctica. *J. Hattori Bot. Lab.* 53, 357-360.

Fristrup, B. 1951. Wind erosion in the Arctic deserts. *Geogr. Tidsskr. 52*, 51-65.

Golubkova, N.S. and Simonov, I.M. 1972. Lishayniki Oazisa Shirmakhera. *Trudy Sovet Skoy East Antarkti Cheskoy Eksfeditsii Leningrad 60*, 317-327.

Gupta, B.K., Sinha, G.P. and Singh, D.K. 1999. A note on lichens of Schirmacher Oasis, East Antarctica. *Indian J. Forestry 22*, 292-294.

Hale, M.E.Jr. 1967. *Biology of Lichens.* Arnold, London.

Hertel, H. 1988. Problems in monographing Antarctic crustose lichens. *Polarforschung 58*, 65-76.

Juckes, L.M. 1969. Weathering hollows in charnockite at Mannefaulknausane, Dronning Maud Land. *Bull. Brit. Antarct. Survey 22*, 97-98.

Kappen, L. 1973. Response to extreme environments. In: *The lichens* (Ed. Ahmadjian, V. and Hale, M.E.). Academic Press, New York, 311-380.

Kappen, L. 1985. Lichen-habitats as micro-oases in the Antarctic - the role of temperature. *Polarforschung 55*, 49-54.

Kappen, L. 1988. Ecophysiological relationships in different climatic regions. In: *CRC Handbook of Lichenology,* Volume II (Ed. Galun, M.). CRC Press, Boca Raton, 37-100.

Kappen, L. 2000. Some aspects of great success of lichens in Antarctica. *Antarctic Science 12*, 314-324.

Kappen, L. 2004. The diversity of lichens in Antarctica, a review and comments. *Biblioth. Lichenol. 88*, 331-343.

Kappen, L. and Straka, H. 1988. Pollen and spores transport into the Antarctic. *Polar Biology 8*, 173-180.

Kappen, L., Schroeter, B., Scheidegger, C., Sommerkorn, M. and Hestmark, G. 1996. Cold resistance and metabolic activity of lichens below 0°C. *Adv. Space Res. 18*, 119-128.

Lamb, I.M. 1970. Antarctic terrestrial plants and their ecology. In: *Antarctic ecology* (Ed. Holdgate, M.W.). Academic Press, London and New York, 733-751.

Lindsay, D.C. 1977. Lichens of cold deserts. In: *Lichen Ecology* (Ed. Seaward, M.R.D.). Academic Press, London, 183-209.

Marshall, W.A. 1996. Biological particles over Antarctica. *Nature 383*, 680.

Nash III, T.H., Reiner, A,, Demmig-Adams, B., Kilian, E, Kaiser, W.M. and Lance, O.L. 1990. The effect of atmospheric desiccation and osmotic water stress on photosynthesis and dark respiration of lichens. *New Phytologist 116*, 269-276.

Nayaka, S. and Upreti, D.K. 2005. Schirmacher Oasis, East Antarctic, a lichenologically interesting region. *Current Science 89*, 1059-1060.

Ochyra, R., Bednarek-Ochyra, H. and Smith, R.I.L. 1998. 170 years of research of the Antarctic moss flora. In: *Polish Polar Studies 25th International Polar Symposium.* Institute of Geophysics of the Polish Academy of Sciences, Warszawa, 159-177.

Øvstedal, D.O. and Smith, R.I.L. 2001. *Lichens of Antarctica and South Georgia. A Guide to their Identification and Ecology.* Cambridge University Press, UK.

Øvstedal, D.O. and Smith, R.I.L. 2004. Addition and corrections to the lichens of Antarctica and South Georgia. *Cryptogamie Mycol. 25*, 323-331.

Pandey, V. and Upreti, D.K. 2000. Lichen flora of Schirmacher Oasis and Vettiyya Nunatak. In: *Scientific Report: Seventeenth Indian Expedition to Antarctica.* Department of Ocean development, New Delhi, India, Technical Publication # 15, 185-201.

Rao, D.N. and LeBlanc, F. 1965. A possible role of atranorin in the lichen thallus. *Bryologist l68*, 284-289.

Ritcher, W. 1990. The lichens of the Schirmacher Oasis (East Antarctica). *Geodätische und Geophysikalische Veröffentlichungen 16*, 471-488.

Ritcher, W. 1995. Biology. In: *The Schirmacher Oasis, Queen Maud Land, East Antarctica and its Surroundings* (Ed. Bormann, P. and Fritzsche, D.). Gotha, Jastus Perthes Verlag Gotha, Germany, 321-347.

Rudolph, E.D. 1967. Lichen distribution. In: *Terrestrial life in Antarctica* (Ed. V. Bushnell). *Antarct. Map Fol. Ser. 5*, 9-11.

Smith, R.I.L. 1997. Oases as centres of high plant diversity and dispersal in Antarctica. In: *Ecosystem processes in Arctic Ice-free Landscapes* (Ed. Lyons, W.B., Howard-Williams, C. and Hawes, I.). Rotterdam, Balkema, 119-128.

Torrey, J. 1823. Description of a new species of *Usnea,* from New South Shetland. *The Americal Journal of Science and Arts 6*, 104-106.

Upreti, D.K. 1996. Lecideoid lichens from the Schirmacher Oasis, East Antarctica. *Willdenowia 25*, 681-686.

Upreti, D.K. 1997. Notes on some crustose lichens from Schirmacher Oasis, East Antarctica. *Feddes Repertorium 108*, 281-286.

Upreti, D.K. and Pant, G. 1995. Lichen flora in and around Maitri region, Schirmacher Oasis, East Antarctica. In: *Scientific Report: Eleventh Indian Expedition to Antarctica.* Department of Ocean development, New Delhi, India, Technical Publication # 9, 229-241.

Vishniac, H.S. 1996. Biodiversity of yeasts and filamentous microfungi in terrestrial Antarctic ecosystem. *Biodiversity and Conservation 5*, 1365-1378.

Webber, W.A. 1962. Environmental modification and the taxonomy of the crustose lichens. *Svensk Botan. Tidskr. 56*, 293-333.

Webber, W.A. 1968. A taxonomic revision of *Acarospora,* subgenus *Xanthothallia. Lichenologist 4*, 16-31.

Frontiers in Fungal Ecology, Diversity and Metabolites (edited by K.R. Sridhar),
I.K. International Publishing House Pvt. Ltd., New Delhi, 2009.

8

Diversity of Arbuscular Mycorrhizal Fungi of Endemic Myristicaceae

P. Rama Bhat
Department of Biotechnology, Alva's College,
Moodbidri 574 227, Karnataka, India
(E-mail: ramabhatp@yahoo.com)

Introduction

The Western Ghats of India (8°20'-20°40' N and 73°-77° E) are endowed with a variety of flora, fauna and microorganisms. The vegetation consists of scrub jungle, grasslands, moist and dry deciduous forests, tropical evergreen forests, and shoals. Approximately 5,000 out of 17,000 flowering plants species of India exist in Western Ghats. A large proportion of plant species of the Western Ghats are endemic (1,720 taxa of 54 genera). Nearly one third of endemic taxa are threatened and some believed to be extinct or at serious risk of near extinction (Nayar, 1996; Shetty and Kaveriappa, 1991). Myristicaceae is one of the important medicinally-valued families of flowering plants endemic to the Western Ghat forests in Karnataka and consists of five species belonging to three genera (*Gymnacranthera farquhariana*, *Knema attenuate, Myristica dactyloides, M. fatua var. magnifica* and *M. malabarica*). The association between plants and zygomycetous fungi of the order Glomales is considered as one the most ancient and widespread mycorrhizal symbioses (Allen, 1996; Simon *et al*., 1993). Arbuscular mycorrhizal (AM) fungi are characterized by intraradicle mycelia, arbuscules, vesicles, coils and appressorium. The colonization of roots of the terrestrial plants by AM fungi is very common (Smith and Gianinazzi-Pearson, 1988), and this enhances plant growth (Xavier and Germida, 1997), reproduction (Koide *et al*., 1988) and prevents pathogenesis (Dugassa *et al*., 1996). In view of meager information on the AM fungal association of forest tree species of the Western Ghats, the current chapter presents the results of an investigation on AM fungal association with five endemic plant species belonging to the family Myristicaceae.

Study Area, Sampling and Analysis

The study was carried out at the forest, Kathalkan, situated at an elevation of 500-530 m above mean sea level in Gersoppa of the state of Karnataka (14°15'40"-14°15′50" N; 74°45′25"-74°45′35" E). This area is endowed with two hectares of evergreen forest and consists of a *Myristica* swamp. Maximum rainfall occurs between May and August (range: 68-373 cm) and the monthly minimum and maximum temperature varies between 20.4-26.2 and 28.2-33.5°C. Temperature increases steadily by the end of February and reaches a peak during April-May and then starts declining by the end of May due to pre-monsoon showers. Temperature again rises during August-September and attains a peak in December. To study their association with AM fungi, five tree species of the family Myristicaceae were chosen: *Gymnacranthera farquhariana* (Hook. f. & Thomson) Warb., *Knema attenuata* (Hook. f. & Thomson) Warb., *Myristica dactyloides* Gaertner, *M. fatua* Houtt. var. *magnifica* (Beddome) Sinclair and *M. malabarica* Lam.

Soil and root samples were collected during summer (March), monsoon (July) and post-monsoon (November) during the year 1994-1995. After removing the debris on the surface, about 500g rhizosphere soil (RS) was collected in sterile polythene bags at 15-30cm depth and at about 1m distance from the main stem. Samples collected from ten plants of each species were mixed to form a composite sample of 5kg, air dried and stored at 4°C for further analysis of soil quality and associated AM flora. Similarly, non-rhizosphere soil (NRS) was also sampled and stored for further analysis. The feeder roots of each plant species were collected by retaining soil particles adhering to them and then fixed in formalin-acetic-alcohol (FAA) to assess the AM fungal colonization. Based on the methods outlined in Jackson (1973), soil edaphic features were evaluated. Soil temperature was measured on the spot using a soil thermometer. Soil moisture was estimated gravimetrically on drying at 100°C until constant weight was attained. The pH and conductivity of soil suspension (1:10 w/v) were determined using at the sampling site. Organic matter was estimated by Walkley and Black's rapid titration method, while nitrogen by macro-Kjeldahl method. The Na and K were estimated using flame photometry, while P by spectrophotometry. Soil minerals (Mg, Mn, Fe, Zn, Ca and Cu) were determined after digesting soil samples with a mixture of concentrated HNO_3 and $HClO_4$ (3:1) using atomic absorption spectrophotometry.

The FAA-fixed feeder roots were gently washed and cut into 10 mm segments. Fifty segments per plant were mixed to form a composite sample and a randomly chosen 200 segments were assessed for AM colonization using trypan blue (Koske and Gemma, 1989; Phillips and Hayman, 1970). The percentage of hyphal, vesicular and arbuscular colonization were determined by the gridline intersect method outlined by Giovannetti and Mosse (1980):

Colonization (%) = (Total number of roots intersecting) ÷ (Total number of roots intersecting gridline) × 100

The isolation and enumeration of AM fungal spores of the RS was performed following wet-sieving and decanting method (Gerdemann and Nicolson, 1963). Spores and sporocarps were mounted in a drop of polyvinyl lactic glycerol

(PVLG) on microscopic slides and then identified based on morphological characters (color, size, number and size of the wall layers, nature of inner content and hyphal attachment) (Hall, 1984; Morton and Benny, 1990; Schenck and Perez, 1990; Trappe, 1982). Percent abundance (PA) of spores of each species and percent frequency (PF) of each genus were calculated:

PA = (Total number of spores of each species in all the replicates) ÷ (Total number of spores of all the species) × 100

PF = (Total number of fungal species in each genus/sample) ÷ (Total number of fungal species isolated in each sample) × 100

Sorensen's similarity index (SI) between AM fungi of each plant species was compared (Sorensen, 1948): SI = (2C) ÷ (S1 + S2) × 100

(where, C, Number of common fungal species in plant species 1 and 2; S1 and S2, Total number of fungal species in plant species 1 and 2).

Pearson's correlation was employed to determine the degree of association between the AM indices (root colonization and spore density) *vs*. soil edaphic properties of each tree species in different seasons.

Observations and Discussion

Colonization and spore density

Physicochemical properties of soils (RS and NRS) are presented in Table 1. The moisture and organic matter of NRS were lower than RS, while the temperature was higher. The pH, EC and elements (N, P, K, Na, Ca, Cu, Fe, Mg, Mn and Zn) of the RS of some plant species were higher than NRS. Soil moisture has been positively correlated with root colonization (RC) and spore density (SD) by several investigators (Kaur *et al*., 1997; Klironomos *et al*., 1993; Mohankumar and Mahadevan, 1987; Muthukumar *et al*., 1994b). Udaiyan *et al*. (1996) and Visalakshi (1997), however negatively correlated soil moisture with RC of two species of *Acacia* in the forests of Pondicherry. In Sri Lanka, Hafeel and Gunatilleke (1988) also negatively correlated soil moisture with SD of forest tree species. Low moisture and high temperature of soil are reported to influence RC (Bareen, 1990; Visalakshi, 1997; Furlan and Fortin, 1973; Pugh *et al*., 1981), while SD is influenced by moderate soil moisture and temperature (Mohankumar and Mahadevan, 1987; Raman and Gopinathan, 1992). Visalakshi (1997) showed positive correlation between RC and pH, while Khaliel (1988) showed correlation of RC with SD and pH. Udaiyan (1996) observed negative influence of pH on RC and positive influence on SD. However, in many studies, no significant correlation was seen between soil pH and SD (Bergen and Koske, 1984; Klironomos *et al*., 1993; Mohankumar and Mahadevan, 1987).

Seasonal variations in SD, RC, and species richness of AM fungi of five tree species are given in Figure 1. Spore density was least in the monsoon season for all tree species except for *M. dactyloides*. Root colonization showed an increasing trend from summer to the post monsoon season. The richness of species did not vary too much in different seasons. Table 2 indicates the range of hyphal,

Table 1. *Edaphic features of rhizosphere soil (RS) of five endemic Myristicaceae tree species at Kathalkan in Gersoppa, Western Ghats (Gf, Gymnacranthera farquhariana; Ka, Knema attenuata; Md, Myristica dactyloides; Mf, M. fatua; Mm, M. malabarica) and Non-rhizosphere soil (NRS) during study period (range, n = 30)*

Parameters	Gf	Ka	Md	Mf	Mm	NRS
Moisture (%)	24.62-41.56	12.79-36.67	13.38-36.19	24.01-42.27	14.38-33.66	12.07-23.90
Temperature (°C)	23.40-27.10	22.40-27.90	23.40-27.90	21.50-25.40	23.50-26.90	24.20-28.90
pH	5.80-6.15	5.42-6.37	5.33-6.08	5.73-6.01	5.67-6.18	5.62-5.84
Conductivity (mmohs)	2.00-5.00	2.00-7.00	2.00-5.00	2.00-6.00	3.00-5.00	3.00-5.00
Organic matter (%)	3.90-4.10	3.93-4.28	4.00-4.48	3.79-4.07	3.79-4.07	1.47-2.24
N (%)	0.40-1.20	0.40-1.10	0.40-1.10	0.40-0.70	0.60-0.80	0.50-0.60
P (%)	0.56-1.37	0.56-0.88	0.53-0.94	0.52-1.87	0.52-0.83	0.34-0.60
K (%)	0.028-0.061	0.102-0.139	0.295-0.393	0.054-0.072	0.095-0.148	0.156-0.190
Na (%)	0.232-0.300	0.069-0.195	0.207-0.300	0.365-0.426	0.364-0.531	0.227-0.397
Ca (%)	0.073-0.240	0.013-0.161	0.059-0.115	0.054-0.068	0.0026-0.0033	1.26-1.63
Cu (%)	0.0020-0.0032	0.0014-0.0033	0.0011-0.0013	0.106-0.201	0.0040-0.0072	1.06-3.83
Fe (%)	0.72-1.44	1.04-3.63	0.88-0.94	0.054-0.081	0.0022-0.0045	0.68-1.25
Mg (%)	0.035-0.171	0.137-0.187	0.135-0.189	0.128-0.198	0.071-0.153	0.052-0.128
Mn (%)	0.0030-0.0350	0.0110-0.0290	0.0230-0.0270	0.0110-0.0210	0.0180-0.0510	0.0100-0.0370
Zn (%)	0.0032-0.0035	0.0044-0.0063	0.0035-0.0053	0.0054-0.0059	0.0035-0.0069	0.0026-0.0064

Table 2. *Range of arbuscular mycorrhizal root colonization of five endemic Myristicaceae tree species at Kathalkan in Gersoppa, Western Ghats (Gf, Gymnacranthera farquhariana; Ka, Knema attenuata; Md, Myristica dactyloides; Mf, M. fatua; Mm, M. malabarica) during study period (range, n = 30)*

AM fungi	Gf	Ka	Md	Mf	Mm
Hyphal colonization	44.00-50.30	39.86-47.89	41.03-42.89	46.27-48.09	40.68-53.79
Arbuscular colonization	22.70-40.07	32.05-35.23	31.07-35.58	32.56-37.77	30.38-36.44
Vesicular colonization	15.26-26.99	16.88-27.72	22.40-26.05	17.14-19.77	15.83-22.88
Total colonization	72.34-87.50	81.99-88.12	69.86-84.05	75.35-94.48	73.65-93.10
AM Spores	630-460	450-620	525-860	485-750	480-635

arbuscular, vesicular and total colonization and SD of five tree species. A significant correlation between RC and SD was seen with soil temperature and seasons (summer, monsoon and post-monsoon) (Table 3). The SD was significantly correlated with pH of plant species and seasons. According to Raman and Gopinathan (1992), high organic matter favours maximum sporulation of AM

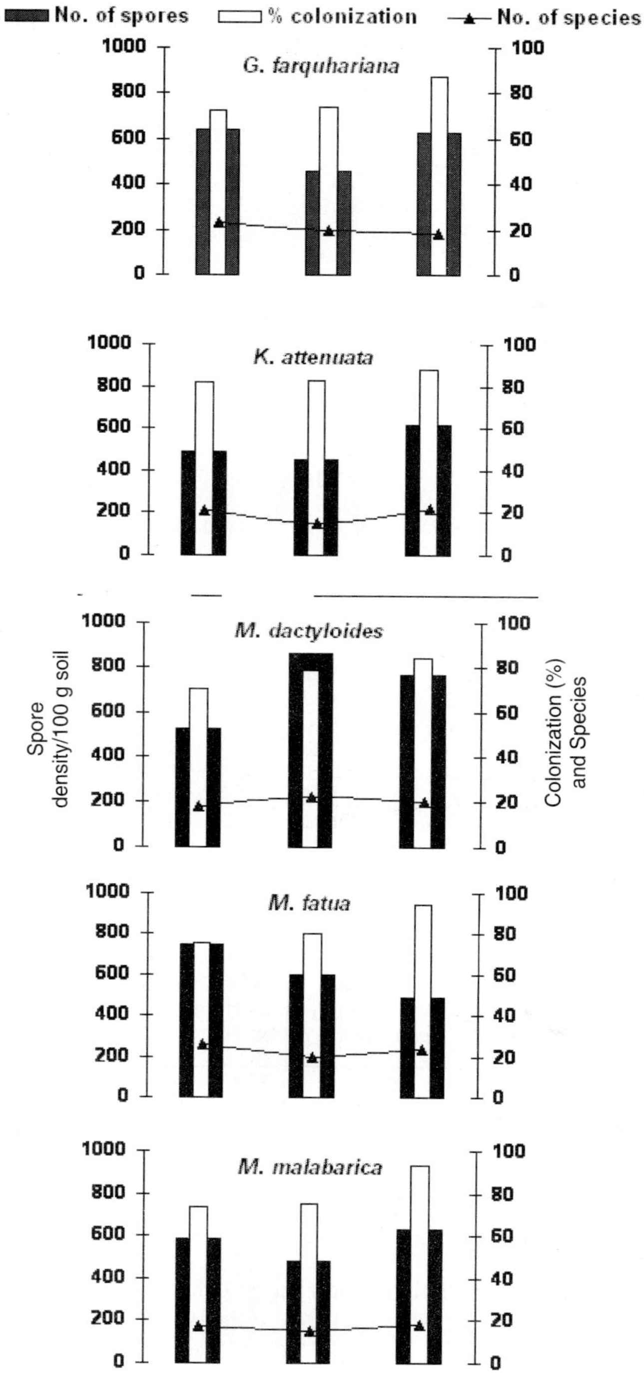

Fig. 1. *Seasonal variations in SD, RC and species richness of AM fungi of five tree species of Myristicaceae.*

Table 3. *Pearson's correlation coefficient (r) between AM spore density (SD) and root colonization (RC) against AMF indices (HC, hyphal colonization; AC, arbuscular colonization; VC, vesicular colonization) soil edaphic features (T, temperature; EC, electrical conductivity; OM, organic matter and minerals) (*Significant at P < 0.001)*

Parameter	March	July	November
Gymnacranthera farquhariana			
SD vs. RC	0.113	0.774*	0.638
SD vs. HC	-0.793*	-0.101	-0.217
SD vs. AC	0.143	0.444	0.789*
SD vs. T	0.359	-0.566	0.786*
SD vs. EC	-0.459	0.511	0.777*
SD vs. Fe	-0.811*	0.243	0.317
SD vs. Mg	0.078	-0.303	0.731*
RC vs. AC	-0.092	0.576	0.810*
RC vs. VC	-0.794*	0.348	0.425
RC vs. OM	0.758*	-0.330	-0.156
RC vs. N	-0.822*	0.300	-0.404
RC vs. Na	0.785*	0.697	-0.305
RC vs. Mg	0.756*	-0.502	-0.731
RC vs. Zn	0.832*	0.612	-0.232
Knema attenuata			
SD vs. RC	0.7 85*	0.508	0.402
SD vs. AC	0.388	0.062	0.805*
SD vs. MC	-0.738	0.654	0.775*
SD vs. Na	0.111	-0.809*	-0.240
SD vs. Mg	-0.408	0.468	-0.768*
RC vs. HC	0.206	0.783*	0.126
RC vs. MC	-0.888*	0.463	-0.294
Myristica dactyloides			
SD vs. RC	0.888*	0.240	0.848*
SD vs. HC	0.797*	0.544	0.380
SD vs. AC	-0.860*	-0.496	-0.483
SD vs. VC	-0.792*	0.407	0.183
SD vs. EC	0.801*	0.586	0.360
SD vs. OM	0.308	-0.827*	0.387
SD vs. P	0.742*	0.767*	-0.705
SD vs. Na	-0.390	-0.893*	0.264
SD vs. Cu	0.279	-0.626	0.863*
SD vs. Fe	0.786*	-0.408	-0.543
SD vs. Mn	-0.217	0.809*	0.695
SD vs. Zn	0.282	-0.915*	0.333
RC vs. HC	-0.838*	0.137	0.888*
RC vs. AC	0.766*	0.586	0.496
RC vs. MC	0.177	0.648	0.799*
RC vs. P	0.854*	-0.198	0.483
Myristica fatua			
SD vs. RC	0.751	-0.782*	0.490
SD vs. VC	0.432	0.866*	-0.758*
SD vs. T	0.298	0.143	0.795*
SD vs. Na	0.080	0.822*	0.428
SD vs. Fe	0.797*	-0.493	-0.278
RC vs. HC	-0.444	0.610	0.844*
RC vs. AC	-0.140	0.149	0.797*
RC vs. VC	0.059	-0.863*	-0.603
RC vs. T	0.177	0.548	0.833*
RC vs. EC	-0.852*	-0.705	-0.933*

RC vs. N	-0.436	-0.619	0.748*
RC vs. P	0.423	-0.129	0.804*
RC vs. Na	0.376	-0.729*	-0.279
RC vs. Ca	0.176	-0.123	0.748*
RC vs. Mn	0.876*	0.349	0.496
RC vs. Zn	0.299	0.667	0.888*
Myristica malabarica			
SD vs. HC	-0.515	0.888*	0.297
SD vs. T	0.755*	0.409	0.600
SD vs. K	0.502	0.746*	-0.722*
SD vs. Cu	0.245	-0.851*	0.177
SD vs. Fe	0.509	0.796*	0.654
SD vs. Mg	0.488	-0.675	0.881*
SD vs. Zn	0.789*	0.088	-0.391
RC vs. HC	0.848*	-0.131	0.313
RC vs. AC	0.178	-0.614	0.606*
RC vs. VC	-0.827*	0.709	0.697
RC vs. MC	0.102	-0.747*	-0.325
RC vs. T	-0.185	0.821*	0.610
RC vs. pH	0.576	-0.797*	0.404
RC vs. OM	0.826*	0.761*	0.400
RC vs. N	-0.756*	0.588	-0.556
RC vs. K	-0.760*	-0.354	-0.171
RC vs. Ca	-0.609	-0.846*	0.659
RC vs. Cu	0.865*	0.238	0.529
RC vs. Fe	-0.770*	0.362	0.271
RC vs. Mn	0.192	0.544	0.759*

fungi. Soil organic matter has been shown to have positive correlation with RC as well as RD (Hepper and Warner, 1981; Klioromos *et al.*, 1993; Mohankumar and Mahadevan, 1987). However, Hafeel and Gunatilleke (1988) showed negative correlation between organic carbon and SD. In the present study, organic matter was significantly correlated with RC and SD in different plant species and seasons. The impact of nitrogen on RC and spore production is believed to be due to modification of soil pH (Sylvia and Neal, 1990; Thompson, 1986). Raman and Gopinathan (1992) indicated that moderate temperature and moisture, acidic pH, high organic matter and low NPK of soil favors maximum sporulation. Soil P negatively influenced RC as well as SD in *Alysicarpus monilifer*, while it was positive in *Indigofera linnaei* (Muthukumar *et al.*, 1994b). The response of P to RC is attributed to changes in the P or cell membrane permeability due to varying cellular P concentration (Ratnayake *et al.*, 1978; Smith, 1982). The influence of K on SD has been reported either positive or negative (Udaiyan, 1996). Raman and Gopinathan (1992) reported lower concentration of NPK that favors maximum sporulation, but Muthukumar *et al.* (1994b) did not find such correlation for NPK against RC and SD. In the present study, Zn, Mn, Cu and Fe in soil showed positive correlation with RC and SD in different tree species and seasons.

This result supports moderate soil N, high temperature and low moisture favor, while high N and P do not favor RC (Ietswaart *et al.*, 1992; Raman and Gopinathan, 1992). The correlation of NPK *vs.* RC and SD was negatively significant in different plant species and seasons in the present study and it has been supported by the observations by Aziz and Habte (1989) and Udaiyan (1996). According to Sanders and Fitter (1992), hyphal, arbuscular and vesicular colonization vary between plant species and seasons. Surveys of the forest plants for AM fungi showed seasonal variations in RC and SD (Klironomos *et al.*, 1993; Louis and Lim, 1987). Some reports showed variations in RC as well as SD between seasons within the same region. For instance, in the evergreen forests of Kalakad (Tamil Nadu), the RC was highest during June-November and lowest during February-April, while the SD was higher during January-April and lowest during May-December (Mohankumar and Mahadevan 1987). In the forests of Servarayan hills (Tamil Nadu), the RC was highest during summer and SD in late winter (January-February), while SD was lowest during the rainy season (July-August) (Raman and Gopinathan, 1992). In the Marakkanam reserve forest and Puthupet sacred grove (Pondicherry), the RC and SD were highest during October (Visalakshi, 1997). Variations in RC and SD in the same plant species between seasons have been reported. For example, Dubey and Negi (1995) found maximum RC during winter and the minimum during summer in *Cupressus torulosa*. Udaiyan *et al.* (1996) showed highest RC during August-November and SD during January-March in *Acacia farnesiana*, while in *A. planifrons* during April-August and September-November. In the present study, RC in all five tree species attained the maximum during the post monsoon (November) and the minimum in summer (March). The SD was highest either in November or July, while the lowest varied among the tree species. According to Udaiyan (1996), cooler months (September-November) followed by rain, favor particularly arbuscule formation which indicates the vigorous vegetative growth of the host as well as the AM fungi. Maximum RC in the month of October (rainy season) was due to the formation of new roots, which resulted in high surface area. Summer months (March-May) favored vesicle formation as well as sporulation thereby suggesting decreased host dependence on AM fungi. Giovannetti (1985) also opined that RC and SD would be low during summer due to poor and slow root growth in temperate regions. The RC and SD in the present study showed positive as well as negative correlation with soil moisture in different seasons. In *G. farquhariana* and *M. fatua*, when the SD was high the RC was low or vice-versa. In other tree species, such inverse relationships were seen only in some seasons. Similar observations were made by many investigators (Giovannetti, 1985; Hayman, 1970; Louis and Lim, 1987; Mohankumar and Mahadevan, 1987; Read *et al.*, 1976; Udaiyan, 1996). However, there are some reports indicating no significant correlation between RC and SD (Giovannetti and Nicolson, 1983; Hayman and Stovold, 1979; Muthukumar *et al.*, 1994b; Sanchez and Honrubia, 1992; Udaiyan *et al.*, 1996).

Colonization and species richness

Gymnacranthera farquhariana: The maximum RC was recorded in November and the minimum in March. The hyphal and vesicular colonization were the maximum in July, while arbuscular colonization in March. The SD was the highest in March and the lowest in July, while species richness in March and November. Among the 38 species of five genera, the order of the first three species based on percent abundance in three seasons (**S**, summer: March; **M**, monsoon: July; **PM**, post-monsoon: November) was: **S**: *Glomus microaggregatum*>*Sclerocystis coremioides*>*Gigaspora gigantea*; **M**: *Glomus multicaule*>*G. macrocarpum*>*G. geosporum*; **PM**: *G. macrocarpum*>*G. fasciculatum*>*S. indica.*

Knema attenuata: Maximum RC (mainly hyphal and arbuscular), SD, and species richness were found in November, while minimum colonization in March (vesicles in July), SD and species richness in July. Out of 31 species belonging to four genera, the order of the first three species based on percent abundance in three seasons was: **S**: *G. fasciculatum*>*G. macrocarpum*>*G. aggregatum*; **M**: *G. macrocarpum*>*G. etunicatum*>*A. scrobiculata*; **PM**: *G. fasciculatum*>*G. macrocarpum*>*G. australe.*

Myristica dactyloides: Maximum RC was found in November (vesicles in June, arbuscules in March) and the minimum in March. Maximum SD as well as species richness was found in July and the minimum in March. In 33 species belonging to five genera, the order of the first three species based on percent abundance in the three seasons was: **S**: *G. macrocarpum*>*G. fasciculatum*>*G. versiforme*; **M**: *G. macrocarpum*>*G. monosporum*>*G. fasciculatum*; **PM**: *G. macrocarpum*>*S. indica*>*G. tenebrosum.*

Myristica fatua: The RC was highest in November (hyphal in March, arbuscules in November and vesicles in July) and the minimum in March. Maximum SD and species richness was recorded in March and the minimum SD in November and species richness in July. Out of 38 species belonging to six genera, it showed the first three species based on percent abundance in three seasons was: **S**: *G. fasciculatum*>*G. microaggregatum*>*G. heterosporum*; **M**: *G. monosporum*>*G. macrocarpum*>*G. fasciculatum*; **PM**: *G. macrocarpum*>*G. fasciculatum*>*G. hoi.*

Myristica malabarica: Maximum RC was recorded in October (hyphal colonization in March, arbusucles and vesicles in July) and minimum in March. The SD and species richness was maximum in November and minimum in July. Out of 28 species belonging to six genera, the order of the first three species based on percent abundance in three seasons was: **S**: *Sclerocystis rubiformis*>*G. fasciculatum* and *G. versiforme*; **M**: *G. geosporum*>*G. macrocarpum*>*S. indica*; **PM**: *G. fasciculatum*>*G. microcarpum* and *Scutellospora calospora.*

Among 57 species of AM fungi belonging to six genera isolated from five tree species, 13 (22.81%) were common to all tree species (Table 4), while 15 (26.32%) restricted to one of the tree species either during summer or post-monsoon. Six species (*Acaulospora laevis*, *Gigaspora gigantea*, *Glomus fasciculatum*, *G. geosporum*, *G. macrocarpum* and *G. monosporum*) were common in each month indicating their activity throughout the year. Species of *Glomus* were dominant in all tree species and months of sampling except for the month of March in *M. malabarica* (*Sclerocystis rubiformis*). Sorensen's similarity between 57 species of five tree species ranged between 58 (*K. attenuata* and *M. malabarica*)

Table 4. *Range of arbuscular mycorrhizal fungi (spores/100 g rhizosphere soil) associated five endemic Myristicaceae tree species at Kathalkan in Gersoppa, Western Ghats (Gf, Gymnacranthera farquhariana; Ka, Knema attenuata; Md, Myristica dactyloides; Mf, M. fatua; Mm, M. malabarica) during study period (range, n = 30)*

AM fungi	Gf	Ka	Md	Mf	Mm
Acaulospora appendicula Spain, Sieverding & Schenck	0-12				
A. bireticulata Rothwell & Trappe	0-15	0-27	20-42	13-39	0-37
A. elegans Trappe & Gerdemann			0-25		
A. foveata Trappe & Janos	0-33			0-10	0-10
A. laevis Gerdeman & Trappe	19-25	15-32	22-30	14-23	18-32
A. scrobiculata Trappe	0-28	17-41	0-40	0-32	0-20
A. thomii Blazkowski					0-24
Entrophospora infrequens (Hall) Ames & Schneider				0-18	0-19
Gigaspora albida Schenck & Smith	0-10		0-26	0-33	0-16
G. decipiens Hall & Abbott		0-13		0-18	
G. gigantea (Nicolson & Gerdemann) Gerdemann & Trappe	11-51	21-37	41-47	10-26	17-36
G. margarita Becker & Hall	0-12	0-12	11-23	13-33	
G. ramisporophora Spain, Sieverding & Schenck	0-39		0-29		
G. roseus Nicolson & Schenck	0-12				0-47
Glomus aggregatum Schenck & Smith	0-33	0-38			0-31
G. ambisporum Smith & Schenck	0-38	11-22	22-41	0-15	
G. australe (Berck.) Berch	0-20	0-49	0-48	0-20	0-30
G. botryoides Rothwell & Victor	0-11		0-33		
G. caledonium (Nicolson & Gerdemann) Trappe & Gerdemann	0-16	0-22		14-27	0-10
G. citricola Tang & Zang		0-28	0-34		
G. claroides Schenck & Smith	0-25	0-20			
G. clarum Nicolson & Schenck		0-12	0-13		
G. constrictum Trappe		0-20			
G. deserticola Trappe, Bloss & Menge	0-21	0-11	0-23	0-20	0-23
G. etunicatum Becker & Gerdemann		0-47			
G. fasciculatum (Thaxter) Gerdemann & Trappe	32-65	37-66	34-73	43-88	37-144
G. flavisporum (M. Lange & Lund) Trappe & Gerdemann					0-21
G. fragilistratum Skou & Jakobsen				0-21	
G. fuegeanum (Sepg.) Trappe & Gerdemann	0-18		0-24	0-17	
G. fulvum (Berkeley & Broome) Trappe & Gerdemann		0-18			
G. geosporum Nicolson & Gerdemann	31-43	25-32	29-53	24-37	33-77
G. gerdemanii Rose, Daniels & Trappe				0-10	
G. heterosporum Smith & Schenck	0-22	0-13	0-20	0-49	
G. hoi Berch & Trappe				0-34	
G. intraradices Schenck & Smith				0-29	
G. macrosporum Tulasne & Tulasne	45-92	42-69	62-122	46-72	49-73
G. maculosum Miller & Walker	0-30			0-16	
G. melanosporum Gerdemann & Trappe	0-22				
G. microaggregatum Koske, Gemma & Olexia	0-77	0-32	0-33	12-53	0-15
G. microcarpum Tulasne & Tulasne	0-22	10-30	0-36	0-31	0-58
G. monosporum Gerdemann & Trappe	19-51	17-33	28-74	15-133	24-31

G. mosseae (Nicolson & Gerdemann) Gerdemann & Trappe	17-20	0-24	0-55	0-27	
G. multicaule Gerdemann & Bakshi	0-53	0-31	0-51	20-22	
G. tenebrosum (Thaxter) Berch			0-55		0-30
G. tenue (Greenall) Hall			0-14		
G. versiforme (Karsen) Berch	0-40	0-31	0-47	0-18	0-52
G. warcupii McGee		0-13			
Sclerocystis coremioides Berkeley & Broome	0-61		0-29		0-14
S. dussii (Patouillard) von Hohnel	0-32		0-41	0-40	0-31
S. indica Bhattacharjee & Mukerji	0-56		13-70		0-49
S. pachycaulis Wu & Chen	0-18	20-36		0-21	12-36
S. rubiformis Gerdemann & Trappe	0-38	0-32		0-27	0-68
S. sinuosa Gerdemann & Bakshi	0-13	11-22	0-22	0-10	
S. taiwanensis Wu & Chen	0-27			0-21	
Scutellospora calospora (Nicolson & Gerdemann) Walker & Sanders			0-26	0-19	0-58
S. nigra (Redhead) Walker & Sanders	0-23				
S. scutata Walker & Diederichs			0-44		

Table 5. *Sorenson's similarity index (%) between AM fungi of the five tree species (Gf, Gymnacranthera farquhariana; Ka, Knema attenuata; Md, Myristica dactyloides; Mf, M. fatua; Mm, M. malabarica)*

	Gf	Ka	Md	Mf	Mm
Gf	0	69.57	73.24	76.32	66.67
	Ka	0	65.63	69.57	57.63
		Md	0	70.42	68.85
			Mf	0	69.7
				Mm	0

and 76% (*G. farquhariana* and *M. fatua*) (Table 5). The generic distribution of AM fungi in five tree species of *Myristicaceae* is given in Figure 2. *Glomus* was most dominant in terms of species maxima in all tree species. The order of dominance of AM fungal genera in five tree species was: *Gymnacranthera farquhariana*: *Glomus*>*Sclerocystis*>*Acaulospora* and *Gigaspora*>*Scutellospora*; *Knema attenuata*: *Glomus*>*Sclerocystis*, *Acaulospora* and *Gigaspora*; *Myristica dactyloides*: *Glomus*>*Sclerocystis*, *Acaulospora* and *Gigaspora*>*Scutellospora*; *Myristica fatua*: *Glomus*>*Sclerocystis*>*Acaulospora* and *Gigaspora*>*Scutellospora* and *Entrophospora*; *Myristica malabarica*: *Glomus*>*Sclerocystis*>*Acaulospora* and *Gigaspora*>*Scutellospora* and *Entrophospora.* The order of plant species in accordance with the number of AM fungi recorded were *G. farquhariana* and *M. fatua*>*M. dactyloides*>*K. attenuata*>*M. malabarica.*

The number of species of AM fungi recorded in the rhizosphere of each plant species investigated was high and ranged from 28-38 per tree species in the present study. In some of the earlier studies on forest trees, it varied between 1 and 19 suggesting that the forest trees harbor diverse AM fungal flora (Bhat, 1993; Manoharachary and Reddy, 1995; Rama Bhat and Kaveriappa, 2005; Santhaguru *et al.*, 1995; Udaiyan, 1996; Udaiyan *et al.*, 1996). According to Mohankumar and Sivaswamy (1992), *Glomus* is the most dominant genus in many forest tree species of India (Bhat, 1993; Hariram and Santhaguru, 1995; Muthukumar *et al.*,

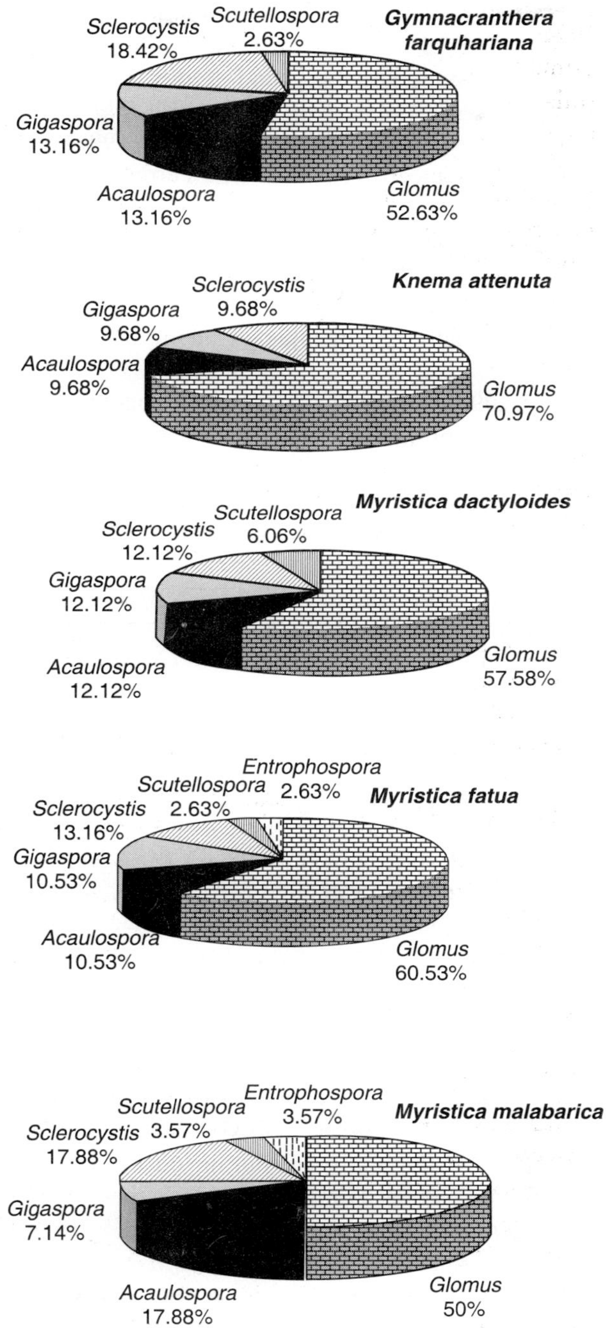

Fig. 2. The generic distribution of AM fungi in five tree species of Myristicaceae

1994a, 1994b; Santhaguru *et al.*, 1995; Sharma *et al.*, 1986; Visalakshi, 1997). Trimurtulu and Johri (1998) also reported maximum *Glomus* species in the Tarai soils of Uttar Pradesh. Manoharachary and Reddy (1995) documented the dominance of *Glomus* in *Dalbergia sissoo, Terminalia alata* and *Tectona grandis*. In *Acacia, Eucalyptus, Grevillea* and *Tectona*, Udaiyan (1996) and Udaiyan *et al.* (1996) observed the dominance of *Glomus*. The current study supports earlier observations as the maximum number of species and spores of *Glomus* in Myrsticaceae members. Among the species recorded, *Glomus fasciculatum* and *G. macrocarpum* were dominant in terms of spore abundance. Rama Bhat and Kaveriappa (2005) also found dominance of *Glomus* in the rhizospheres of *Myristicaceae* members that were maintained at the arboretum of Mangalore University Campus.

Santhaguru *et al.* (1995) and Udaiyan *et al.* (1996) opined that host specificity exists among AM fungi of forest tree species. There are reports on specific association or selective multiplication of AM fungi in plant species (Kruckelmann, 1975; Schenck and Kinloch, 1980; Udaiyan *et al.*, 1996). In the present study (even though cross inoculation experiments were not performed), some AM fungi were rarely encountered. For instance, 15 species of AM fungi belonging to genera *Acaulospora, Gigaspora, Glomus* and *Scutellospora* were recorded in one of the five tree species in only one sample. Such rare occurrence can be attributed to their narrow adaptability in contrast to species of *Glomus* (Abbott and Robson, 1991; Bhat, 1993; Schenck and Kinloch, 1974; Sharma *et al.*, 1986; Sparling and Tinker 1978). According to McGonigle and Fitter (1990), the majority of AM fungi are distributed in diverse habitats, but certain species show a slight degree of habitat or ecological specificity (e.g., soil edaphic features). Among the 16 species considered as uncommon by Udaiyan (1996), *G. geosporum* and *G. macrocarpum* were common to five tree species studied, while *G. constrictum* and *G. intraradices* were confined to single tree species in the present study. According to Udaiyan (1996), some AM fungal species have a broad host range, while some have a narrow one. It can be predicted that each plant species supports the growth of a set of AM fungi including some of the uncommon ones.

Conclusions and Outlook

Rhizosphere of five endemic tree species belonging to the family *Myrticaceae* revealed the occurrence of 57 species of AM fungi belonging to six genera. Overall root colonization was maximum during post-monsoon and minimum during summer in all tree species. A maximum of 38 species each was recorded from *Gymnacranthera farquhariana* and *Myristica fatua*, followed by 33, 31 and 28 species in *M. dactyloides*, *Knema attenuata* and *M. malabarica,* respectively. The maximum number of AM fungi as well as spore density was found during summer in *G. farquhariana* and *M. fatua*, during monsoon in *M. dactyloides* and during post-monsoon in *K. attenuata* and *M. malabarica. Glomus* was the dominant genus in all the tree species in all the seasons. Six AM fungal species were common to all five tree species in all the seasons (*Acaulospora laevis, Gigaspora gigantea, Glomus fasciculatum, G. geosporum, G. macrocarpum* and *G. monosporum*). Fifteen AM fungi were isolated from the rhizosphere of one of the five tree species indicating their dependence on specific species or soil edaphic factors.

It is increasingly realized that the diversity of glomalean fungal communities decreases in natural ecosystems that are disturbed by human interference (Johnston and Pfleger, 1992). The conversion of natural ecosystems into agro-ecosystems, cultural practices and reduction in plant species probably have strong selective pressures on AM fungi resulting in survival of species and strains that are stress tolerant. Notable reductions have been seen in the composition and richness of AM fungal populations in cultivated habitats compared to natural ones (Giovannetti and Gianinazzi-Pearson, 1994). The principal way of safeguarding the diversity of mycorrhizal fungi would be through the protection of the ecosystem or the habitat. It may be achieved perhaps by *in situ* conservation of specific biomes (e.g., nature reserves) with the least or minimum human interference (Hawksworth, 1991). The extent of human interference on the ecosystem forced to practice *ex situ* conservation of fungal diversity through the maintenance of germplasm (Giovannetti and Gianinazzi-Pearson, 1994). Efficient utilization of biodiversity is of crucial importance for sustainable production. In plant conservation program, the role of microbial diversity is not emphasized. The diversity of microorganisms associated particularly with endemic and endangered plant species assumes great significance as their survival mainly depends on the existence and perpetuation of such plant species. Thus, it is very important to emphasize the need for conservation of microbial diversity in a plant conservation program, as they are habitat or host specific.

References

Abbott, L.K. and Robson, A.D. 1991. Factors influencing the occurrence of vesicular-arbuscular mycorrhizas. *Agric. Ecosyst. Environ. 35*, 121-150.

Allen, M.F. 1996. The ecology of arbuscular mycorrhizas: A look back into the 20th century and a peak into the 21st. *Mycol. Res. 100*, 769-782.

Aziz, T. and Habte, M. 1989. Influence of inorganic N on mycorrhizal activity, nodulation, growth of *Leucaena leucocephala* in an oxisol subjected to simulated erosion. *Commun. Soil Sci. Plant Anal. 20*, 239-251.

Bareen, F.E. 1990. Vesicular arbuscular mycorrhiza in aquatics. In: *Current Trends in Mycorrhizal Research* (Eds. Jalali, B.L. and Chand, H.). Haryana Agriculture University, Hisar, India, 1-3.

Bergen, M. and Koske, R.E. 1984. Vesicular-arbuscular mycorrhizal fungi from sand dunes of Cape Cod, Massachusetts. *Trans. Br. Mycol. Soc. 83*, 157-158.

Bhat, M.N. 1993. Vesicular arbuscular mycorrhizae in submountain region. *Acta Bot. Indica 25*, 52-56.

Dubey, R.C. and Negi, C.M.S. 1995. Seasonal occurrence of VA *mycorrhizal* fungi on roots of Cupress trees in relation to edaphic factors. *Acta Bot. Indica 23*, 173-175.

Dugassa, G.D., Von Alten, H. and Schonbeck, F. 1996. Effects of arbuscular mycorrhiza (AM) on the health of *Linum usitatissimum* L. infected by fungal pathogens. *Plant Soil 185*, 173-182.

Furlon, V. and Fortin, A. 1973. Formation of endomycorrhizae by *Endogone calospora* on *Allium cepa* under three temperature regimes. *Nat. Can. 100*, 467-477.

Gerdemann, J.W. and Nicolson, T.H. 1963. Spores of mycorrhizal *Endogone* species extracted by wet sieving and decanting. *Trans. Br. Mycol. Soc. 46*, 235 - 244.

Giovannetti, M. 1985. Seasonal variations of vesicular-*arbuscular mycorrhizas* and endogonaceous spores in a Marine sand dune. *Trans. Br. Mycol. Soc. 84*, 679-684.

Giovannetti, M. and Gianinazzi-Pearson, V. 1994. Biodiversity in arbuscular mycorrhizal fungi. *Mycol. Res. 98*, 705-715.

Giovannetti, M. and Mosse, B. 1980. An evaluation of techniques for measuring vesicular arbuscular mycorrhizal infection in roots. *New Phytol. 84*, 498-500.

Giovannetti, M. and Nicolson, T.H. 1983. *Vesicular-arbuscular mycorrhizas* in Italian sand dunes. *Tans. Br. Mycol. Soc. 80*, 552-557.

Hafeel, K.M. and Gunatilleke, I.A.U.N. 1988. Distribution of endomycorrhizal spores in disturbed sites of a lowland rainforest in Sri Lanka. In: Mycorrhiza For Green Asia (Eds. Mahadevan, A., Raman, N. and Natarajan, K.), Centre for Advanced Studies in Botany, Madras, India, 37-45.

Hall, I.R. 1984. Taxonomy of VA mycorrhizal fungi. In: *VA Mycorrhiza* (Ed. Powell, C.L. and Bhagyaraj, D.J.). CRC Press, Boca Raton, Florida, 57-94.

Hariram, N. and Santhaguru, K. 1995. Vesicular-arbuscular mycorrhiza in rhizosphere soils of some plant species from Madurai. *J. Ecobiol. 7*, 243-248.

Hawksworth, D.L. 1991. The fungal dimension of biodiversity: magnitude, significance and conservation. *Mycol. Res. 95*, 641-655.

Hayman, D.S. 1970. Endogone spore number in soil and vesicular-arbuscular mycorrhiza in wheat as influenced by season and soil treatment. *Trans. Br. Mycol. Soc. 54*, 53-63.

Hayman, D.S. and Stovold, G.E. 1979. Spore populations and infectivity of vesicular-arbuscular mycorrhizal fungi in New South Wales. *Aust. J. Bot. 27*, 227-234.

Hepper, C.M. and Warner, A. 1981. Role of organic matter in growth of a vesicular arbuscular mycorrhizal fungus in soil. *New Phytol. 81*, 155-156.

Ietswaart, J.H., Griffioen, W.A.J. and Ernst, W.H.O. 1992. Seasonality of VAM infection in three populations of *Agrositis capillaris* (Gramineae) on soil with or without heavy metal enrichment. *Plant Soil 139*, 67-73.

Jackson, M.L. 1973. *Soil Chemical Analysis*. Prentice-Hall International, USA.

Johnston, N.C. and Pfleger, F.L. 1992. Vesicular-arbuscular mycorrhizae and cultural stresses. In: *Mycorrhizae in Sustainable Agriculture* (Eds. Bethlenfalvay, G.J. and Linderman, R.G.). ASA Special Publication # 54, Madison, WI, USA, 71-79.

Kaur, M.J., Sagar, A. and Lakhanpal, T.N. 1997. Observation on vesicular arbuscular mycorrhizal association of *Grewia optiva. J. Mycol. Pl. Pathol. 27*, 323-324.

Khaliel, A.S. 1988. Incidence of VAM in some desert plants and correlation with edaphic factors. In: *Mycorrhiza For Green Asia* (Eds. Mahadevan, A., Raman, N. and Natarajan, K.). Centre for Advanced Studies in Botany, Madras, India, 56-59.

Klironomos, J.N., Moutoglis, P., Kendrick, B. and Widden, P. 1993. A comparison of spatial heterogeneity of vesicular-arbuscular mycorrhizal fungi in two maple forest soils. *Can. J. Bot. 71*, 1472-1480.

Koide, R., Li, M., Lewis, J. and Irby, C. 1988. Role of mycorrhizal infection in the growth and reproduction of wild versus cultivated plants. I. Wild vs. cultivated oats. *Oecologia 77*, 537-543.

Koske, R.E. and Gemma, J.N. 1989. A modified procedure for staining roots to detect VA mycorrhizas. *Mycol. Res. 92*, 486-488.

Kruckelmann, H.W. 1975. Effects of fertilizers, soils, soil tillage and plant species on the frequency of *Endogone* chlamydospores and mycorrhizal infection in arable soils. In: *Endomycorrhizae* (Eds. Sanders, F.E., Mosse, B. and Tinker, P.B.), Academic Press, London, 511-525.

Louis, I. and Lim, G. 1987. Spore density and root colonization of vesicular-arbuscular mycorrhizas in tropical soil. *Trans. Br. Mycol. Soc. 88*, 207-212.

Manoharachary, C. and Reddy, P.J.M. 1995. Role of vesicular arbuscular mycorrhizal fungi in forestry. In: *Biofertilizers for the future: Mycorrhiza* (Eds. Adholeya, A. and Singh, S.). Tata Energy Research Institute, New Delhi, India, 297-302.

McGonigle, T.P. and Fitter, A.H. 1990. Ecological specificity of vesicular-arbuscular mycorrhizal associations. *Mycol. Res. 94*, 120-122.

Mohankumar, V. and Mahadevan, A. 1987. Vesicular-arbuscular mycorrhizal association in plants of Kalakad reserve forest, India. *Angew. Botanik 61*, 255-274.

Mohankumar, V. and Sivaswamy, S.N. 1992. Vesicular-arbuscular mycorrhizae. *Adv. Pl. Sci. 50*, 17-28.

Morton, J.B. and Benny, G.L. 1990. Revised classification of arbuscular mycorrhizal fungi (Zygomycetes): A new order, Glomales, two new suborders, *Glomineae* and *Gigasporineae* and two new families, *Aculosporaceae* and *Gigasporaceae*, with an amendation of Glomaceae. *Mycotaxon 37*, 471-491.

Muthukumar, T., Udaiyan, K. and Manian, S. 1994a. Vesicular arbuscular mycorrhizae in certain tropical wild legumes. *Ann. For. 20*, 33-43.

Muthukumar, T., Udaiyan, K. and Manian, S. 1994b. Role of edaphic factors on VAM fungal colonization and spore populations in certain tropical wild legumes. *Pertanika J. Trop. Agric. Sci. 17*, 33-42.

Nayar, M.P. 1996. *Hot spots of Endemic plants of India, Nepal and Bhutan.* Tropical Botanical Garden and Research Institute, Palode, Kerala.

Phillips, J.M. and Hayman, D.S. 1970. Improved procedures for clearing roots and staining parasitic and vesicular arbuscular mycorrhizal fungi for rapid assessment of infection. *Trans. Br. Mycol. Soc. 55*, 158-161.

Pugh, I.M., Roncadon, R.W. and Hussey, R.S. 1981. Factors affecting vesicular arbuscular mycorrhizal development and growth of cotton. *Mycologia 73*, 869-879.

Rama Bhat, P. and Kaveriappa, K.M. 2005. Studies *on* arbuscular mycorrhizal fungi associated with tree species planted at Mangalore University, Mangalore, Karnataka. *Mycorrhiza News 17*, 10-11.

Raman, N. and Gopinathan, S., 1992. Association and activity of vesicular arbuscular mycorrhizae of trees in a tropical forest of Southern India. *J. Trop. For. 8*, 311-322.

Ratnayake, M., Leonard, R.T. and Menge, J.A. 1978. Root exudation in relation to supply of phosphorus and its possible relevance to mycorrhizal formation. *New Phytol. 81*, 543-552.

Read, D.J., Koucheki, H.K. and Hodgson, J. 1976. Vesicular-arbuscular mycorrhiza in natural vegetation systems. *New Phytol. 77*, 641-653.

Sanchez, M.E.L. and Honrubia, M. 1992. Seasonal variation of vesicular-arbuscular mycorrhizae in eroded soils from southern Spain. *Mycorrhiza 2*, 33-39.

Sanders, I.R. and Fitter, A.H. 1992. The ecology and functioning of vesicular-arbuscular mycorrhizas in co-existing grassland species I. Seasonal patterns of *mycorrhizal* occurrence and morphology. *New Phytol. 120*, 517-524.

Santhaguru, K., Ponmalar, S.B.G. and Karunakaran, R. 1995. Vesicular-arbuscular mycorrhizae in tree-legumes and its rhizospheric soils in Alagar hills. *Indian Forester 212*, 817-823.

Schenck, N.C. and Kinloch, R.A. 1974. Pathogenic fungi, parasitic nematodes and endomycorrhizal fungi associated with soybean root in Florida. *Plant Dis. Reptr. 58*, 169-173.

Schenck, N.C. and Kinloch, R.A. 1980. Incidence of mycorrhizal fungi on six field crops in mono-culture on a newly cleared woodland site. *Mycologia 72*, 445-455.

Schenck, N.C. and Perez, Y. 1990. *Manual for the Identification of VA Mycorrhizal Fungi.* Synergistic Publications, Gainseville, Florida.

Sharma, S.K. Sharma, G.D. and Mishra, R.R. 1986. Status of mycorrhizae in sub-tropical forest ecosystem of Meghalaya. *Acta Bot. Indica 14*, 87-92.

Shetty, B.V. and Kaveriappa, K.M. 1991. The Western Ghats - Need for preservation. In: *Perspectives on Dakshina Kannada and Kodagu.* Mangalore University, Karnataka, India, 258-272.

Simon, L., Bousquet, J, Levesque, R.C. and Lalonde, M. 1993. Origin and diversification of endomycorrhizal fungi and coincidence with vascular land plants. *Nature 363*, 67-69.

Smith, E. 1982. Inflow of phosphate into mycorrhizal and non-mycorrhizal plants of *Trifolium subterraneum* at different levels of phosphate. *New Phytol. 90*, 292-303.

Smith, S.E. and Gianinazzi-Pearson, V. 1988. Physiological interactions between symbionts in vesicular-arbuscular mycorrhizal plants. *Ann. Rev. P. Physiol. Mol. Biol. 39*, 211-244.

Sorensen, T. 1948. A method of establishing groups of equal amplitudes in plant sociology based on the similarity of species content and its application to analyse the vegetation of Darnish commons. *Biol. Skr. 5*, 1-34.

Sparling, G.P. and Tinker, P.B. 1978. Mycorrhizas in Pennie grassland. I. Level of infection in the field. *J. Appl. Ecol. 15*, 943-950.

Sylvia, D.M. and Neal, L.H. 1990. Nitrogen affects the phosphorus responses of VA *mycorrhiza. New Phytol. 115*, 303-310.

Thompson, J.P. 1986. Soilless culture of vesicular arbuscular mycorrhizas of cereals: effect of nutrient concentration and nitrogen source. *Can. J. Bot. 64*, 2282-2294.

Trappe, J.M. 1982. Synoptic keys to the genera and species of Zygomycetous mycorrhizal fungi. *Phytopathology 72*, 1102-1107.

Trimurtulu, N. and Johri, B.N. 1998. Prevalence and distribution of vesicular arbuscular mycorrhizal spore population in different Tarai soils of Uttar Pradesh. *J. Mycol. Pl. Pathol. 28*, 236-239.

Udaiyan, K. 1996. Impact of edapho-climatic factors on the dynamics of VAM root colonization and spore density in three forest tree species of Western Ghats, India. *Pertanika J. Trop. Agric. Sci. 19*, 143-162.

Udaiyan, K, Karthikeyan, A. and Muthukumar, T. 1996. Influence of edaphic and climatic factors on the dynamics of root colonization and spore density of vesicular-arbuscular mycorrhizal fungi in *Acacia farnesiana* Willd. and *A. planifrons* Wight & Arn. *Trees Structure Function 11*, 65-71.

Visalakshi, N. 1997. Dynamics of vesicular-arbuscular mycorrhizae in two tropical dry evergreen forests, South India. *Int. J. Ecol. Environ. 23*, 25-36.

Xavier, L.J.C. and Germida, J.J. 1987. Growth response of lentil and wheat to *Glomus clarum* NT4 over a range of P levels in a Saskatchewan soil containing indigenous AM fungi. *Mycorrhiza 7*, 3-8.

Frontiers in Fungal Ecology, Diversity and Metabolites (edited by K.R. Sridhar), I.K. International Publishing House Pvt. Ltd., New Delhi, 2009.

9

Arbuscular Mycorrhizal Fungi of the 'Khazan Land' Agro-Ecosystem

B.F. Rodrigues[1]* and Anuradha N. Naik[2]
[1]*Department of Botany, Goa University, Goa 403 206, India*
[2]*National Institute of Oceanography, Dona Paula, Goa 403 004, India*
*(*E-mail: felinov2001@yahoo.co.uk)*

Introduction

'Khazan Lands' are partially modified mangrove and coastal wetland ecosystems of the west coast of India. Water with variable salinity is the dynamic element of this managed agro-ecosystem. As Khazan Lands are the sites of extreme conditions (e.g., salinity, flooding, physiological drought, siltation), it is interesting to study the interactions between plants and microbes, particularly arbuscular mycorrhizal (AM) fungi associated with pioneering vegetation. The presence of AM fungi and their restricted association with plant species indicates their tolerance and survival under stress conditions of Khazan Lands. The present chapter deals with the occurrence and diversity of AM fungi in association 28 plant taxa belonging to 16 genera that grow in Khazan Lands with a view to understand their ecology, role in rehabilitation and the reclamation of Khazan Lands and similar biomes of coastal region.

Khazans

The term 'Khazan' is a vernacular name in Konkani language of Goa for coastal saline soils. In Maharashtra, similar soil habitats are called 'Khar' or 'Kharvat', in Karnataka as 'Gazzaani', and in Kerala as 'Pokkal' and 'Chemeen Kattu'. Khazans are the integrated agro-aqua ecosystems that are managed in Goa and its territory. These lands have been reclaimed over centuries from marshy mangrove swamps and developed by the local people using their traditional knowledge of climate, tidal cycles, geomorphology, precipitation, runoff, sediment dynamics, soil properties and drainage characteristics of estuarine lands (Sonak, 2005). These

Colonization

The extent of root colonization of AM fungi of 28 plant taxa is presented in Table 1. Roots of sixteen taxa were colonized by AM fungi with high vesicle colonization. Root colonization ranged between 23% (*Paspalum scrobiculatum*) and 81% (*Sphaeranthus africanus*). Some reports are available on the absence of AM fungal colonization in members belonging to the family *Nymphaeaceae* (Bagyaraj *et al*., 1979; Dharmarajan *et al*., 1993; Khan, 1974). Our study also revealed no AM fungal colonization in the roots of *Nymphaea nouchali*. Similarly, no AM fungal colonization was seen in *Acanthus ilicifolius, Acrostichum aureum, Avicennia officinalis* and *Derris uliginosa* as reported earlier (Gupta *et al.,* 2002; Mohankumar and Mahadevan, 1986). The presence of AM fungal colonization in roots of *Sesuvium portulacastrum* confirms the earlier report by Bhaskaran and Selvaraj (1985). Roots of *Ludwigia perennis* were colonized by AM fungi, while *Fimbristylis aestivalis* were devoid of it. However, Dharmarajan *et al.* (1993) reported the absence of AM colonization in *L. perrenis* and colonization of *Fimbrystylis* sp. In our study, four out of six Cyperaceae members showed colonization. The family, *Cyperaceae* has been previously categorized as non-mycorrhizal because roots of a few species were evaluated for AM fungi (Brundrett, 1991; Newman and Reddell, 1987; Peat and Filter, 1993; Powell, 1975; Smith and Read, 1997), but later there have been several reports of mycorrhizal colonization in certain *Cyperaceae* taxa (Lovera and Cuenca, 1996; Wetzel and Van der Valk, 1995). A variation in the mycorrhizal condition in Cyperaceae may be due to the environmental and edaphic factors rather than its phylogenetic constraint per se (Read, 1984), because AM fungal colonization seems to be negatively correlated with soil moisture (Anderson *et al*., 1984). It has been speculated that non-mycorrhizal state of some Cyperaceae members is due to their occurrence in marshy and anaerobic soils rather than their taxonomic position (Tester *et al*., 1987).

Spores

Rhizsphere spore density of AM fungi of 28 plant taxa is given in Table 1. The spore density varied between 53 (*Salvinia molesta*) and 370 (*Ceratopteris thalictroides*) per 100 g rhizosphere soil. Muthukumar *et al.* (2001) suggested that the spore production of AM fungi was strongly influenced by environmental factors with host species and soil types. Khan (1974) also reported that the soil type influences the dissemination of AM fungi. Twelve plant taxa (*Avicennia officinalis, Phyllanthus niruri, Sphenoclea zeylanica, Nymphaea nouchali, Echinochloa colonum, Salvinia molesta, Eclipta alba, Eleocharis geniculata, Acanthus ilicifolius, Fimbristylis aestivalis, Derris uliginosa* and *Acrosticum aureum*) although possess AM fungal spores in the rhizosphere (23-126 spores/ 100 g), were devoid of root colonization. This observation indicates that the mere presence of AM spores in the rhizosphere will not prove root colonization and the possibilities of association with other microbes to overcome the stress conditions prevail in Khazan habitats.

AM fungi

A total of 17 AM fungi belonging to three genera (*Acaulospora, Glomus* and *Scutellospora*) were recovered from the pot cultures (Table 2). Among them, *Acaulospora* and *Glomus* with a total of eight species each dominated the Khazan Lands, *Glomus claroideum* was most abundant followed by *Glomus fasciculatum*

Table 2. *Arbuscular mycorrhizal fungi of rhizosphere soils of selected plants growing in Khazan Lands (see Fig. 1)*

Plant species	AM fungal species
Acanthus ilicifolius L.	*Glomus claroideum, G. geosporum, G. multicaule*
Acrosticum aureum L.	*Acaulospora delicata, A. spinosa, Glomus claroideum*
Ammannia multiflora Roxb.	*Glomus claroideum G. fasciculatum, G. formosanum,*
Avicennia officinalis L.	*Acaulospora dilatata, A. nicolsonii*
Ceratopteris thalictroides (L.) Brongn.	*Glomus aggregatum, G. claroideum, G. geosporum, G. intraradices, Scutellospora weresubiae*
Cyprus cyperoides (L.) O. Kuntze.	*Acaulospora dilatata, A. Delicata, A. mellea, A. myriocarpa, A. spinosa, Glomus aggregatum, G. claroideum, G. fasciculatum*
Cyperus rotundus L.	*Glomus claroideum, G. fasciculatum*
Derris uliginosa Benth.	*Acaulospora delicata, A. spinosa, Glomus fasciculatum*
Echinochloa colonum (L.) Link.	*Glomus claroideum, G. geosporum, G. multicaule*
Eclipta alba (L.) Hassk.	*Glomus claroideum, Glomus formosanum*
Eleocharis geniculata (L.) Roem & Schult.	*Acaulospora nicolsonii, Glomus claroideum, G. multicaule*
Excoecaria agallocha L.	*Acaulospora delicata, A. rehmii, A. spinosa,Glomus claroideum*
Fimbristylis aestivalis (Retz.) Vahl.	*Glomus claroideum, G. formosanum G. geosporum, Glomus mosseae* *Fuirena ciliaris* (L.) Roxb. *Glomus claroideum, G. geosporum*
Ipomoea muricata Jacq.	*Acaulopsora dilatata, Glomus geosporum*
Isachne globosa (Thunb.) O. Kuntze.	*Glomus fasciculatum, Acaulospora nicolsonii*
Ludwigia perennis L.	*Acaulospora dilatata, Glomus aggregatum, G. claroideum*
Nymphaea nouchali Burm.	*Acaulospora dilatata, Glomus formosanum, G. mosseae*
Oryza sativa L. var. Asgo	*Acaulospora delicata, A. mellea, Glomus fasciculatum*
Oryza sativa L. var. Jyoti	*Acaulospora mellea, Glomus formosanum*
Oryza sativa L.var. Korgut	*Glomus claroideum*
Paspalum scrobiculatum L.	*Acaulospora nicolsonii, Glomus fasciculatum*
Phyllanthus niruri L.	*Acaulospora laevis, Glomus claroideum, G. fasciculatum*
Salvinia molesta Mitch.	*Acaulospora mellea, A. spinosa,*
Scirpus lateriflorus Gmel	*Glomus claroideum, G. fasciculatum, G. multicaule*
Sesuvium portulacastrum L.	*Acaulospora delicata, A. dilatata, A. nicolsonii, Glomus aggregatum, Scutellospora weresubiae*
Sphaeranthus africanus L.	*Acaulospora delicata, A. laevis, A. myriocarpa, Glomus claroideum, G. mosseae*
Sphenoclea zeylanica Goert. Frud.	*Glomus formosanum, G. geosporum*

and *Acaulospora delicata*. The genus *Scutellospora* was represented by a single species, *Scutellospora weresubiae*, which was recovered from the pot cultures of *Certopteris thalictroides* and *Sesuvium portulacastrum*. The differences observed in the absence or presence of AM fungal genera might be attributed to the edaphic factors of soil and the host plant interactions at the study sites.

Conclusions and Outlook

Although Khazan Land agro-ecosystems exhibit variations in physicochemical properties, the vegetation consists of a variety of AM fungi. Seventeen AM fungi belonging to three genera in Khazan Land indicate their adaptability to stress conditions (flooding, salinity, physiological drought). It also reveals that these fungi have an important role to play in plant growth promotion and survival under adverse conditions. Abundance of *Acaulospora delicata, Glomus claroideum* and *G. fasciculatum* in Khazan Lands indicates their importance as biofertilizers. Further investigations and developments in Khazan Land may realize the importance of AM fungi and avoid disrupting this valuable plant-microbe link. This study forms a baseline for future investigations on the ecology and application of AM fungal species for rehabilitation, reclamation and to improve productivity of Khazan Land and associated biomes.

Acknowledgements

The authors gratefully acknowledge the financial assistance received from the University Grants Commission (UGC), New Delhi under Special Assistance Programme (SAP) as a sponsored research project.

References

Abbott, L.K. and Robson, A.D. 1982. The role of vesicular arbuscular mycorrhizal fungi in agriculture and the selection of fungi for inoculation. *Aust. J. Agric. Res. 33*, 389-408.

Almeida, R.T. and Schenck, N.C. 1990. A revision of the genus *Sclerocystis* (Glomaceae, Glomales). *Mycologia 82*, 703-714.

Anderson, R.C., Liberta, A.E. and Dickman, L.A. 1984. Interaction of vascular plants and vesicular-arbuscular mycorrhizal fungi across a soil moisture gradient. *Oecologia 64,* 111-117.

Bagyaraj, D. J., Munjunath, A. and Patil, R. B. 1979. Occurrence of vesicular arbuscular mycorrhizae in tropical aquatic plants. *Trans. Br. Mycol. Soc. 72,* 164-167.

Bentivenga, S.P. and Morton, J.B. 1995. A monograph of genus *Gigaspora*, incorporating developmental pattern of morphological characters. *Mycologia 87*, 720-732.

Bhaskaran, C. and Selvaraj, T. 1985. Reports on the colonization in the roots of salt marsh plants by different native VA-mycorrhizal fungi. In: *Proceedings of a National Seminar on Micro Organisms in Sustainable Agriculture*. Thiagarajar College, Madurai, 151-153.

Bower, C.A. and Wilcox, L.V. 1965. Soluble salts. In: *Methods of Soil Analysis* (Ed. Black, C.A.). American Society of Agronomy, Madison, Agronomy # 9, 933-940.

Brundrett, M. 1991. Mycorrhizas in natural ecosystems. *Adv. Ecol. Res. 21,* 171-313.

Dharmarajan, S., Kannan, K. and Lakshminarasimhan, C. 1993. Vesicular-arbuscular (VA) mycorrhizal status of some aquatic and marshy plants. *Acta Bot. Indica 21*, 167-171.

Gaur, A. and Adholeya, A. 1994. Estimation of VAM spores in soil - A modified method. *Mycorrhiza News 6*, 10-11.

Gerdemann, J.W. and Nicolson, T.H. 1963. Spores of mycorrhizal *Endogone* species extracted from soil by wet sieving and decanting. *Trans. Br. Mycol. Soc. 46*, 235-244.

Gilmore, A.E. 1968. Phycomycetous mycorrhizal organisms collected by open pot cultures. *Hilgardia 39*, 87-105.

Godeas, A. Fracchia, S. Mujica, M.T. and Ocampo, J.A. 1999. Influence of soil impoverishment on the interaction between *Glomus mosseae* and saprobe. *Mycorrhiza 9*, 185-189.

Gupta, N., Basak, U.C. and Das, P. 2002. Arbuscular mycorrhizal association of mangroves in saline and non-saline soils. *Mycorrhiza News 13*, 14-19.

Jackson, M.L. 1971. *Soil Chemical Analysis.* Prentice Hall of India Pvt. Ltd., New Delhi.

Jasper, D.A. Abbott, L.K. and Robson, A.D. 1989. Acacias respond to additions of phosphorus availability and to the inoculation with VA mycorrhizal fungi in soils stockpiled during mineral sand mining. *Plant Soil 115*, 99-108.

Khan, A.G. 1974. The occurrence of mycorrhiza in halophytes, hydrophytes and xerophytes, and of *Endogone* spores in adjacent soils. *J. Gen. Microbiol. 81*, 7-14.

Koske, R.E. and Gemma, J.N. 1989. A modified procedure for staining roots to detect VA mycorrhizas. *Mycol. Res. 92*, 486-488.

Lovera, M. and Cuenca, G. 1996. Arbuscular mycorrhizal infection in Cyperaceae and Gramineae from natural disturbed and restored savannas in La Gran Sabana, Venzuela. *Mycorrhiza 6,* 111-118.

Matthew, K.M. 1991. An excursion Flora of Central Tamil Nadu, India. Oxford and IBH Publishing Co. Pvt. Ltd., New Delhi.

Mc Gonigle, T.P. Miller, M. H. Evens, D.G. Fairchild, G.L. and Swan, J.A. 1990. A new method, which gives an objective measure of colonization of roots by vesicular-arbuscular mycorrhizal fungi. *New Phytol. 115,* 495-501.

Menge, J.A. 1982. Utilization of vesicular arbuscular mycorrhizal fungi in agriculture. *Can. J. Bot. 61*, 1015-1024.

Mohanan, M. and Henry, A.N. 1994. *Flora of Thiruvananthapuram, Kerala.* Botanical Survey of India, Calcutta, India.

Mohankumar, V. and Mahadevan A., 1986. A. Survey of vesicular-arbuscular mycorrhizae in mangrove vegetation. *Current Science 55*, 936.

Morton, B. and Redecker, D. 2001. Two new families of Glomales, Archaeosporaceae and Paraglomaceae, with two new genera, *Archaeospora* and *Paraglomus*, based on concordant molecular and morphological characters. *Mycologia 93(1):* 181-195.

Morton, J. B. and Benny, G.L. 1990. Revised classification of arbuscular mycorrhizal fungi (*Zygomycetes*): A new order, *Glomales*, two new suborders, Glomineae and Gigasporineae, and two new families Acaulosporaceae and Gigasporaceae with an emendation of Glomaceae. *Mycotaxon 37*, 471-491.

Muthukumar, T.K. Udaiyan, N. and Manian, S. 2001. Vesicular arbuscular mycorrhizal association in the medicinal plants of Maruthamalai Hills, Western Ghats, Southern India. *J. Mycol. Pl. Pathol. 31*, 180-184.

Naithani, H.B., Sahni, K.C. and Bennet, S. S. R. 1997. *Forest Flora of Goa.* International Book Distributors, Dehradun.

Newman, E.I. and Reddell, P. 1987. The distribution of mycorrhizas among families of vascular plants. *New Phytol. 106,* 745-751.

Peat, H.J. and Fitter, A.H. 1993. The distribution of arbuscular mycorrhizas in the British Flora. *New Phytol. 125,* 845-854.

Powell, C.L. 1975. Rushes and sedges are non-mycotropic. *Plant Soil 42,* 481-484.

Rao, S.R. 1985. *Flora of Goa, Diu, Daman, Dadra and Nagarhaveli*, Volume I, Botanical Survey of India, Calcutta.

Rao, S.R. 1986. Flora of Goa, Diu, Daman, Dadra and Nagarhaveli, Volume II, Botanical Survey of India, Calcutta.

Read, D. J. 1984. The structure and function of the vegetative mycelium of mycorrhizal roots. In: *The ecology and physiology of the fungal mycelium* (Ed. Jennings, D.H. abd Rayner, A. D. M.). Cambridge University Press, Cambridge, 215-240.

Redecker, D. Morton, J.B. and Bruns, T.D. 2000. Ancestral lineages of arbuscular mycorrhizal fungi (*Glomales*). *Mol. Phylog. Evol. 14*, 276-284.

Schenck, N.C. and Perez, Y. 1990. Manual for the identification of VA mycorrhizal fungi. International Culture Collection of VA Mycorrhizal Fungi. Synergistic Publications, Gainesville, Florida, USA.

Setua, G.C., Kar, R., Ghosh, J. K., Das, K.K. and Sen, S.K. 1999. Influence of arbuscular mycorrhizae on growth, lead yield and phosphorus uptake in mulberry (*Morus alba* L.) under rainfed, lateritic soil conditions. *Biol. Fert. Soils 29*, 98-103.

Smith, S.E. and Read, D.J, 1997. *Mycorrhizal Symbiosis.* II Edition, Academic Press, Inc., San Diego.

Sonak, S., Kazi S. and Abraham M. 2005. *Khazans in Troubled Waters*. The Energy and Resources Institute, New Delhi.

Tester, M., Smith S.E. and Smith F.A. 1987. The phenomenon of "nonmycorrhizal" plants. *Can. J. Bot. 65,* 419-431.

Walker, C. and Trappe, J.M. 1993. Name and epithets in the Glomales and Endogonales *Mycol. Res. 97*, 339-344.

Walker, C. and Vestberg, M. 1998. Synonymy amongst the arbuscular mycorrhizal fungi: *Glomus claroideum, G. maculosum, G. multusubstensum* and *G. fistulosum*. *Ann. Bot. 82*, 601-624.

Wetzel, P.R. and van der Valk, A.G. 1995. Vesicular-arbuscular mycorrhizae in prairie pothole wetland vegetation in Iowa and North Dakota. *Can. J. Bot. 74,* 883-890.

Frontiers in Fungal Ecology, Diversity and Metabolites (edited by K.R. Sridhar), I.K. International Publishing House Pvt. Ltd., New Delhi, 2009.

10

Arbuscular Mycorrhizal Fungi and Salinity Stress in Crop Plants

D. Shobha[1*] and K.R. Chandrashekar[2]
[1]*Department of Botany, Mangalore University College, Mangalore 575 001, Karnataka, India*
[2]*Department of Applied Botany, Mangalore University, Mangalagangotri 574 199, Karnataka, India*
*(*E-mail: shobhadevadiga@yahoo.com)*

Introduction

Salinity is the most widespread soil constraint in agriculture, which causes imbalance in metabolism in crop plants (Reddy and Vora, 1986). Salt stress has a threefold effect: it reduces water potential, causes ion imbalance or disturbance in ion homeostasis, and is toxic. The altered water status leads to initial growth reduction and limitation in plant productivity. As salt stress involves both osmotic and ionic stress (Benlloch-Gonzales *et al.*, 2005; Hayashi and Murata, 1998; Munns, 2002), growth suppression is directly related to the total concentration of soluble salts or osmotic potential of soil water (Flowers, 2004; Katerji *et al.*, 2005; Pascal *et al.*, 2005). The detrimental effect (death of plants or decrease in productivity) is the result of salinity stress at the whole-plant level. Suppression of growth occurs in all plants due to salinity, but their tolerance levels and rates of growth reduction due to lethal concentrations vary widely among plant species. Salt stress is known to affect the major plant processes: growth, photosynthesis, protein synthesis, and lipid metabolism (Ramoliya *et al.*, 2004; Parida and Das, 2005). Some researchers showed that proline accumulation in plants may be a symptom of stress in less saline-tolerant species and suggested that it plays multiple roles in stress tolerance. Proline, a compatible solute, acts as a mediator of osmotic adjustment (Yoshiba *et al.*, 1997), protects macromolecules during dehydration (Sanchez *et al.*, 1998) and serves as a scavenger of hydroxyl radical (Alia *et al.*, 1995). Due to salinity stress, leaves become dark green and succulent. Such symptoms are also seen in those plants that suffer from phosphorus (P) deficiency (Maas and Nieman, 1978).

Salinity has been shown to reduce tissue P concentrations in of barley (*Hordeum vulgare* L.), corn (*Zea mays* L.), haricot (*Phaseolus sp.*), kidney bean (*Phaseolus vulgaris*), mustard (*Brassica sp.*), pea (*Pisum sativum* L.), onion (*Allium cepa* L.), millet (*Setaria sp.*), clover (*Trifolium* sp.), safflower (*Carthamis tinctoris* L.) and tomato (*Lycopersicon esculentum*) (Bernstein *et al.*, 1974; Champagnol, 1979; Hassan *et al.*, 1970a,b; Maas and Neiman, 1978; Ravikovitch and Yoles, 1971). Soil salinity may also decrease the uptake of other nutrients. Hassan *et al.* (1970a, b) have shown significant negative correlations between salinity and K, Ca, Cu and Fe uptake in barley, and K, Ca, Cu, Fe, and Mn uptake in corn. Detrimental effects of salinity on plant growth result from the direct influence of ion toxicity (Al-Karaki, 2000a; Ayers and Westcot, 1985; Hasegava *et al.*, 1986). Thus, in general terms, salinity inhibits plant growth as well as productivity. The current chapter emphasizes the importance of arbuscular mycorrhizal (AM) fungi on growth and mineral acquisition in crop plants.

Arbuscular Mycorrhizal Fungi

Many studies demonstrated that inoculation of AM fungi improves the growth of plants under a variety of salinity stress (Burke *et al.*, 2003; Tain *et al.*, 2004). Introduction of AM fungi to sites with saline soil resulted in the improvement of plant tolerance and growth (Al-Karaki 2000b; Jain *et al.*, 1989; Jindal *et al.*, 1993; Pond *et al.*, 1984; Poss *et al.*, 1985). The improved productivity of AM fungal plants has been attributed especially to an enhanced acquisition of low mobile nutrients such as P, Zn, and Cu (Al-Karaki and Al-Raddad, 1997; Al-Karaki and Clark, 1998; George *et al.*, 1994; Marschner and Dell, 1994) and improved water relations (Al-Karaki and Clark, 1998; Bethlenfalvay *et al.*, 1988; Sylvia *et al.*, 1993). Recently, Rabie (2005) suggested that AM fungi protect the host plants against the detrimental effects of salt. When mung bean (*Phaseolus radiatus* L. plants were irrigated with a different dilution of seawater, AM fungal plants showed higher growth than non-AM fungal plants at all levels of irrigation. In addition to stress tolerance and acquisition of nutrients by mycorrhizal plants (Zandavalli *et al.*, 2004), mycorrhizae modifies plant physiology (i.e., osmotic modifications) (Rao and Tak, 2002) as well as photosynthesis (Meroguiher *et al.*, 2002). Thus, the AM fungi have been considered as bio-ameliorators of saline soils (Tian *et al.*, 2004; Yano-Melo *et al.*, 2003) (see Chapter 9 for AM fungi in coastal 'Khazan Land' agroecosystem in Southwest India).

The AM fungi are ubiquitous symbionts belonging to the phylum Glomeromycota (formerly, Glomales within Zygomycotina). They have mutualistic associations with roots of a majority of higher plants including major crop species such as paddy, wheat, corn, sorghum and pasture plant species. The AM fungi exist in two different phases: inside the root (intraradicle) and in the soil (extraradicle). The intraradicle mycelia consist of hyphae, arbuscules (sites of nutrient and carbon exchange between the symbionts), and vesicles (sites of lipid storage for the fungus). The extraradicular mycelia form spores, explore soil and absorb nutrients. The intraradicular mycelia are connected to the extraradicle mycelia. The AM fungi typically penetrate roots through root hairs (Smith and Read, 1997). Some of the hyphae that grow along the roots spread AM fungi to new root tissues (Friese and Allen, 1991). Extraradicular hyphae extend away from roots into bulk soil, and also into the zones of root hair senescence and root turnover, thus complementing root uptake capacities and minimizing the impact of root exudation losses.

AM Fungi and Plant Growth in Saline Soils

The AM fungal colonization of roots as well as length of hyphae decreases with increase in soil salinity (Cantrell and Linderman, 2001; Shobha, 2003). In tomato plants, AM fungal colonization occurred irrespective of salt stress, but it was higher under control than in saline conditions. Salt-tolerant cultivar showed higher mycorrhizal colonization than salt-sensitive cultivar. Shoot dry matter yields and leaf areas were higher in mycorrhizal than non-mycorrhizal plants of both salt tolerant and salt sensitive cultivars (Al-Karaki *et al.*, 2001). Pre-inoculated lettuce plants grown up to 11 weeks in the saline soils showed greater shoot mass than non-mycorrhizal plants at a range of salinity (2, 4, 8, 12 ds m^{-1}). Leaves of AM lettuce at the highest salinity were significantly greener (more chlorophyll) than non-AM lettuce (Cantrell and Linderman, 2001). Mycorrhizal maize maintained high root and shoot dry weights and high concentrations of chlorophyll (Feng *et al.*, 2002). In *Strophostyles hevola*, the negative effects of salt concentrations were partially mitigated by the presence of mycorrhizal fungi. Tsang and Maun (1999) proved that *Strophostyles hevola* inoculated with *Glomus mossae* developed higher chlorophyll content and shoot dry weight at higher salt concentrations than non-mycorrhizal plants.

A major response to mycorrhizal colonization is massive synthesis of cytoplasm, which is proportional to increase in the area of plasmalemma arbuscular region (Jalali, 2001). Mycorrhizal symbiosis directs a significant portion of the host plant carbohydrate to the fungal partner (20-30%). The growth and development of fungal mass and reproductive structures directly influence the increased rate of plant photosynthesis (Nehls *et al.*, 2001). A positive correlation exists between soluble carbohydrate concentration within root and root length with arbuscules. But, mycorrhizal parameters were related more to plant nutrient levels and their ratios indicating that tissue nutrients have another level of control in addition to their effect on soluble carbohydrate concentration in roots. Increased AM colonization due to organic amendment significantly reduced nutrient imbalances. A strong relationship between colonization and root soluble carbohydrate concentration validates the basic assumption that mycorrhizal fungi act as a strong sink for photosynthesis (Muthukumar and Udaiyan, 2000). Fujiyoshi *et al.* (2000) indicated that AM fungi could be a large sink of the carbon assimilated by the host plants. The vesicles produced by AM fungi in large numbers act as stores filled with lipid and constitute a major carbon sink (Bago, 2000). Tomato plants inoculated with *Glomus mosseae* had significantly higher shoot and root dry mass and leaf area than non-mycorrhizal plants under medium salinity (4.9 ds m^{-1}) (Al-Karaki *et al.,* 2001). *Glomus monosporus* colonized onions possess significantly higher shoot, bulb and total dry weights. Similarly, *G. fasciculatus* colonized onions showed greater bulb and total weights at all salinity levels (Mpa solute potential -0.06, -0.20 and -0.41 $Ø_s$) (Ojala *et al.*, 1983). Cantrell and Linderman (2001) proved an increase in the onion bulb inoculated with AM fungal isolates from saline soils. Higher requirement of carbohydrates by AM fungi induces higher soluble sugar accumulation in host roots that enhances resistance to salt-induced osmotic stress in the mycorrhizal plant (Feng *et al.*, 2002). The increased salinity results in elevated proline in AM fungal faba plants. Proline concentration of non-AM fungal plants was higher than AM fungal plants at elevated salinity (Rabi and Almadini, 2005). Enhanced growth of mycorrhizal plants in saline environments has been attributed partly to mycorrhiza-mediated enhancement of host P nutrition (Al-Karaki, 2000b; Hirrel and Gerdemann, 1980; Pond *et al.*, 1985; Poss *et al.*, 1985). Several studies have indicated that AM fungi contribute to plant growth via enhancement of mineral nutrient uptake especially of immobile soil nutrients (P, Cu, Zn, Fe) (Al-Karaki and Al-Raddad, 1997; Al-Karaki and Clark, 1998; George *et al.*, 1994; Marschner and Dell, 1994).

AM Fungi and Mineral Acquisition

Minerals

K^+/Na^+ and Ca^{2+}/Na^+ ratios were significantly increased in AM than non-AM fungal faba plants. This may reflect the potential role of AM fungi in the increase of plant K^+ and Ca^{2+} uptake more than Na^+ under salinity stress (Rabi and Almadini, 2005). Under the salt stress, plants maintain high concentrations of K^+ and low concentrations of Na^+ in the root and shoot (Shobha, 2003) and in cytosol (Parida and Das, 2005). In the cytosol, they perform this by regulation of expression and activity of K^+ and Na^+ transporters and H^+ pumps as driving force for transport. In addition, externally supplied Ca^{2+} reduces the toxic effects of NaCl, presumably by facilitating higher K^+/Na^+ selectivity. High salinity also results in increased cytosolic Ca^{2+} that is transported from the apoplast and intracellular compartments. The resultant transient Ca^{2+} increase potentialities of stress signal transduction leading to salt adaptation. Based on the data of experiment on faba plants, Rabi and Almadini (2005) concluded that AM symbiosis may regulate the expression and activate K^+ and Na^+ transporters and H^+ pumps that generate the driving force for transport. Besides, it may increase the transient Ca^{2+} from apoplast and intracellular compartments. In tomato plants, P, K, Zn, Cu and Fe were higher in mycorrhizal than non-mycorrhizal plants under control and medium saline soils. The P, K, Zn, Cu and Fe acquisition was elevated due to AM fungal inoculation in salt-sensitive than salt-tolerant cultivars (Al-Karaki *et al.*, 2001). The AM fungi increased total uptake of P, K, Ca, Mg and Zn by 2.5-6 folds in sorghum and cowpea (Bagayoko *et al.*, 2000).

Phosphorus

Inoculation of AM fungi alleviated P deficiency and salinity effects in lettuce (Cantrell and Linderman, 2001). It has been postulated that the activity of AM hyphae extends beyond the zone of P depletion that develops in P deficient soils on absorption of P ions by roots faster than they diffuse through the soil to replenish the supply at the root surface (Hayman, 1983). The nutrients are transported directly from the soil to hyphae and arbuscules in the root cortex, where these are transferred to the host tissue. Mycorrhizae seem to absorb P from the soil as orthophosphate ions and translocate primarily by cytoplasmic streaming of polyphosphate granules in the vacuoles of mycorrhizal hyphae. Studies have shown that the uptake of P from low solubility sources particularly rock phosphate is due to the mycorrhizal fungi.

Release of phosphatases into the soil and the size of the soil microbial biomass (He *et al.*, 1997) are the factors influencing the availability of soil P to the plants. The AM fungi are capable of influencing these factors in addition to enhancing plant P extraction capabilities (Smith and Read, 1997). Koide and Kabir (2000) have demonstrated the production of extacellular phosphatase by AM fungi. In a compartmentalized *in vitro* system, AM fungi have also indicated that they can take up ^{32}P from AMP in the absence of heterotrophic organisms and translocate limited amounts of hydrolyzed P to a host root (Joner *et al.*, 2000). This clearly demonstrates that AM fungi have the potential to provide some amounts of P from organic sources to their host plants. Feng *et al.* (2003) reported that the ability of AM fungi to improve plant P uptake is greater when the P source is organic. The P uptake was stimulated by AM fungi in red clove after seven days of its growth in soil amended with organic substrates than inorganic fertilizer P or without fertilization. The results of the experiments carried out by Rabie and Almadini (2005) revealed the stimulation of phosphatases in saline conditions in faba plants inoculated with *Glomus clarum*. They also revealed that the protein concentration in AM plants was higher than non-AM plants at 1.5 and 6.0 dsm^{-1} salinity levels.

Nitrogen

It appears from the literature that AM fungi have the potential to improve plant N uptake from organic sources. The ability of mycorrhizal plants to exploit soil N sources might occur directly through uptake of organic molecules by AM fungi and increased ability to compete with heterotrophic microbes for mineral N or the enhancement of mineralization. According to Ames *et al.* (1983a), the AM fungi improve the exploitation of soil organic nitrogen by plants. An increase in the nitrogen level in the mycorrhizal plants indicates the availability of more soil N to mycorrhizal plants than to non-mycorrhizal plants (Feng *et al.*, 2002). The AM fungi increase the availability of organic N to plants through the enhanced mineralization and direct uptake of organic N (Ames *et al.*, 1983a). Recent investigations have revealed that AM fungi can enhance organic matter mineralization (Hodge, 2001). *Brassica napus* has been considered as non-mycorrhizal plant and could take up more N from an organic source in the presence of AM fungi (Hodge, 2003). The AM fungi absorb and transport NH_4^+-N to their host plants (Ames *et al.*, 1983b). The improved foraging capacity of a mycorrhizal root is likely to increase the uptake of NH_4^+ released from the mineralization of organic residues in soil. The AM fungi enhanced NH_4^+ uptake by plants and increased the rate of soil nitrification. Mycorrhizal association may increase plant N uptake by improving plant competitiveness against soil microorganisms. The impact of AM fungi on competition between plants and saprophytic microbes for nutrients released from mineralization was addressed by Moorhead *et al.* (1998). They hypothesized that plants compete with decomposers for mineral nutrients and such competition leads to reduce the litter decay in the presence of growing plants. They found that the growth of non-mycorrhizal blue grama (*Bouteloua gracilis* (H.B.K.) Lag Ex Steud) was reduced significantly by the presence of decaying litter in the N limited system. They hypothesized that mycorrhizal colonization might have increased the competitive ability of plants for the uptake of nutrients. Mycorrhizal colonization reduced the negative effect of the low N litter on blue grama growth and did not significantly reduce litter decay rate, but reduced the amount of N in the decaying litter. It has been suggested that AM fungi may have the potential to show litter decay and increase the C to N ratio of decaying litter, thus decreasing the ease of organic residue decomposition. It appears that the AM fungi have the potential to improve plant competitive ability for N uptake in N-limited systems. A delay in the decomposition of roots of *Pinus edulis* that were colonized by ectomycorrhizal fungi was also reported by Langley *et al.* (2006).

Conclusions and Outlook

Studies on AM fungi revealed that their colonization provides options and alternative mechanisms to satisfy plant's nutritive requirements and maintain physiological status in stressed and disturbed habitats. However, AM fungi differ in their symbiotic efficiencies. For instance, *Glomus*-colonized lettuce grew less and accumulated less N and P at highest salt levels (0.75 g NaCl/kg dry soil). Increase in salinity of soil decreased the hyphal growth and or viability of *Glomus* to a higher extent than those of *Glomus deserticola*. Therefore, mycorrhizal inoculation with suitable fungi has been proposed as a promising tool to improve salt tolerance of plants. Host-fungus specificity, interaction, and competitiveness between AM fungi and effectiveness of AM fungi under possible environmental stresses (e.g., salinity, drought, N deficiency, P deficiency) are the factors to be investigated in selecting AM fungi for mass multiplication.

References

Alia, K.V., Prasad, S.K. and Saradhi, P.P. 1995. Effect of zinc on free radicals and proline in *Brassica* and *Cajanus*. *Phytochem.* *39*, 45-47.

Al-Karaki, G.N. 2000a. Growth, water use efficiency and mineral acquisition by tomato cultivars grown under salt stress. *J. Plant Nutr.* *23*, 1-8.

Al-Karaki, G.N. 2000b. Growth and mineral acquisition by mycorrhizal tomato grown under salt stress. *Mycorrhiza 10*, 51-54.

Al-Karaki, G.N. and Al-Raddad, A.1997. Effects of arbuscular mycorrhizal fungi and drought stress on growth and nutrient uptake of two wheat genotypes differing in drought resistance. *Mycorrhiza 7*, 83-88.

Al-Karaki, G.N. and Clark, R.B. 1998. Growth, mineral acquisition, and water use by mycorrhizal wheat grown under salt stress. *J. Plant Nutr. 21*, 263-276.

Al-Karaki, N., Hammad, R. and Rusan, M. 2001. Response of two tomato cultivars differing in salt tolerance to inoculation with mycorrhizal fungi under salt stress. *Mycorrhiza 11*, 43-47.

Ames, R.N., Porter, L.K., St. John, T.V. and Reid, C.P.P. 1983a. Nitrogen sources and "A" values for vesicular arbuscular and non-mycorrhizal sorghum grown at three rates of ^{15}N-ammonium sulfate. *New Phytol. 96*, 555-563.

Ames, R.N., Reid, C.P.P. Porter, L.K. and Cambardella, C. 1983b. Hyphal uptake and transport of nitrogen from two ^{15}N labelled sources by *Glomus mosseae* a vesicular arbuscular mycorrhizal fungus. *New Phytol. 95*, 381-396.

Ayers, R.S. and Westcot, D.W. 1985. Water quality for agriculture. *FAO irrigation and drainage paper # 29*. Rome, 77-81.

Bagayoko, M., George, E., Romheld, V. and Buerkert, A.B. 2000. Effects of mycorrhizae and phosphorus on growth and nutrient uptake of millet, cowpea and sorghum on a West African soil. *J. Agric. Science 135*, 399-407.

Bago, B., Pffeffer, P.E. and Shachar-Hill, Y. 2000. Carbon transport and metabolism in arbuscular mycorrhiza. *Plant Physiol. 124*, 949-957.

Benlloch- Gonzallez, M., Fournier, J., Ramos, J. and Benlloch, M. 2005. Strategies underlying salt tolerance in halophytes are present in *Cynara cardunculus. Plant. Sci. 168*, 653-659.

Bernstein, L.F., Francois, F. and Clark, R.A. 1974. Interactive effect of salinity and fertility on yields of grains and vegetables. *Agron. J. 66*, 412-421.

Bethlenfalvay, G.J., Brown, M.S., Ames, R.N. and Thomas, R.S. 1988. Effects of drought on host and endophyte development in mycorrhizal soyabeans in relation to water use and phosphate uptake. *Physiol. Plant. 72*, 565-571.

Burke, D.J. Hamerlynck, E.P. and Hahn, D. 2003. Interaction between the salt marsh grass season. *Spartina patens*, arbuscular mycorrhizal fungi and sediment bacteria during the growing. *Soil Biol. Biochem. 35*, 501-511.

Cantrell, I.C. and Linderman, R.G. 2001. Preinoculation of lettuce and onion with mycorrhizal fungi reduces deleterious effects of soil salinity. *Plant Soil 233*, 269-281.

Champagnol, F. 1979. Relationships between phosphate metabolism of plants and salt toxicity. *Phosphorus Agric. 76*, 34-43.

Feng, G., Zhang, F.S., X. L. Li., Tian, C. Y., Tang, C. and Rengel, Z. 2002. Improved tolerance of maize plants to salt stress by arbuscular mycorrhizae is related to higher accumulation of soluble sugars in roots. *Mycorrhiza 12*, 185-190.

Flowers, T.J. 2004. Improving salt tolerance. *J. Exp. Bot. 55*, 307- 319.

Friese, C.F., Allen, M.F. 1991. The spread of VA mycorrhizal fungal hyphae in the soil: inoculum types and external hyphal architecture. *Mycologia 83*, 409-418.

Fujiyoshi, M., Nakatsubo, T., Ogura, S. and Horikoshi, T. 2000. Estimation of mycelial biomass of arbuscular mycorrhizal fungi associated with the annual legume *Kummerowia striata* (Thunb.) Schindler by ergosterol analysis. *Ecol. Res. 15*, 121-131.

George, E., Romheld, V. and Marschner, H. 1994. Contribution of mycorrhizal fungi to micronutrient uptake by plants. In: *Biochemistry of Metal Micronutrients in the Rhizosphere* (eds. Manthey, J A., Crowley, D.E. and Luster, D.G.). Lewis, Boca Raton, 93-109.

Hasegava, P.M., Bressan, R.A. and Hanada, A.K. 1986. Cellular mechanisms of salinity tolerance. *Hortic. Sci. 21*, 1317-1324.

Hassan, N.A.K., Drew, J.V., Knudsen, D. and Olsen, R. 1970a. Influence of soil salinity on production of dry matter and uptake and distribution of nutrients in barley and corn I. Barley. *Agron. J. 62*, 43-45.

Hassan, N.A.K., Drew, J.V., Knudsen, D. and Olsen, R. 1970b. Influence of soil salinity on production of dry matter and uptake and distribution of nutrients in barley and corn II. Corn. *Agron. J. 62*, 46-48.

Hayashi, H. and Murata, N. 1998. Genetically engineered enhancement of salt tolerance in higher plants. In: *Sterss Response of Photosynthetic Organisms. Molecular Mechanisms and Molecular Regulation* (Ed. Sato M.N.). Elsevier, Amsterdam, 133-148.

Haymann, D.S. 1983. The physiology of vesicular-arbuscular endomycorrhizal symbiosis. *Can. J. Bot. 61*, 944-963.

He, Z.L., Wu, J.,O'Donnell, A.G. and Syers, J.K. 1997. Seasonal responses in microbial biomass carbon, phosphorus, and sulphur in soils under pasture. *Biol. Fert. Soils. 24*, 421-428.

Hirrel, M.C. and Gerdemann, J.W. 1980. Improved growth of onion and bell pepper in saline soils by two vesicular arbuscular mycorrhizal fungi. *Soil Sci. Soc. Am. J. 44*, 654-655.

Hodge, A. 2001. Arbuscular mycorrhizal fungi influence decomposition of, but not plant nutrient capture from, *Glycine max* (L.) patches in soil. *New Phytol. 151*, 725-734.

Hodge, A. 2003. N capture by *Plantago lanceolata* and *Brassica napus* from organic material: the influence of spatial dispersion, plant competition and an arbuscular mycorrhizal fungus. *J. Exp. Bot. 54*, 2331-2342.

Jain, R.K., Paliwal, K., Dixon, R.K. and Gjerstad, D.H. 1989. Improving productivity of multipurpose trees on substandard soils in India. *J. Forestry 87*, 38-42.

Jalali, B.L. 2001. Mycorrhiza and plant health need for paradigm shift. *Indian Phytopath. 54*, 3-11.

Jindal, V., Atval, A., Sekhon, B.S. and Singh, R.1993. Effect of VAM on metabolism of mung bean plants under NaCl salinity. *Plant Physiol. Biochem. 31*, 475-481

Joner, E. J., Ravnskov, S. and Jakobsen, I. 2000. Arbuscular mycorrhizal phosphate transport under monoxenic conditions using radiolabelled inorganic and organic phosphate. *Biotechnol. Lett. 22*, 1705-1708.

Katerji, N., Hoorn, J. W., Hamdy, A., Mastrorilli, M., Nachit, M. and Oweis, T. 2005. Salt tolerance analysis of chickpea, faba bean and durum wheat varieties II. Durum wheat. *Agric. Water Manage. 45*, 59-68.

Koide, R.T. and Kabir, Z. 2000. Extraradicle hyphae of the mycorrhizal fungus *Glomus intraradices* can hydrolyse organic phosphate. *New Phytol. 148*, 511-517.

Langley, J.A., Chapman, S.K. and Hungate, B.A. 2006. Ectomycorrhizal colonization slows root decomposition: the post-mortemfungal legacy. *Ecology Lett. 9*, 955-959.

Maas, E.V., and Nieman, R.H. 1978. Physiology of plant tolerance to salinity. In: *Crop Tolerance to Suboptimal Land Conditions* (Ed. Jung, G.A.). ASA Spec., Madison, USA, 277-279.

Marschner, H. and Dell, B. 1994. Nutrient uptake in mycorrhizal symbiosis. *Plant Soil. 159*, 89-102.

Merguihur, A.E., Burity, H.A., Tabosa, J.N. and Maia, F.L. 2002. Salt stress response and proline accumulation in *Brachiaria humidicola. Revista Arg. Microbiol. A. 34*, 77-82.

Moorhead, D.L., Westerfield, M.M. and Zak, J.C. 1998. Plants retard litter decay in a nutrient limited soil: a case of exploitative competition? *Oecologia 113*, 530-536.

Munns, R. 2002. Comparative physiology of salt and water stress. *Plant Cell Environ. 20*, 239-250.

Muthukumar, T. and Udaiyan, K. 2000. Influence of organic manures on arbuscular mycorrhizal fungi associated with *Vigna unguiculata* (L.) Walp. in relation to tissue nutrients and soluble carbohydrate in roots under field conditions. *Biol. Fert. Soils 31*, 114-120.

Nehls, U., Mikolajeswski, S., Magel, E. and Hamp, R. 2001. Carbohydrate metabolism in ectomycorrhizas: gene expression, monosaccharide transport and metabolic control. *New Phytol. 150*, 533-541.

Ojala, J.C., Jarrell, W.M., Menge, J.A. and Johnson, E.L.V. 1983. Influence of mycorrhizal fungi on the mineral nutrition and yield of onion in saline soil. *Agron. J. 75*, 255-259.

Parida, S.K., and Das, A.B. 2005. Salt tolerance and salinity effects on plants. *Ecotoxicol. Environ. Safety 60*, 324-349.

Pascal, S.D., Maggio, A. and Barbieri, G. 2005. Soil salinization affects growth, yield and mineral compostion of cauliflower and broccoli. *Eur. Agron. 56*, 607-618.

Pond, E.C., Menge, J.A. and Jarrel, W.M. 1985. Improved growth of tomato in salinized soil by VAM fungi collected from saline soils. *Mycologia 76*, 74-84.

Poss, J.A., Pond, E.C., Menge, J.A. and Jarrel, W.M. 1985. Effect of salinity on mycorrhizal onion and tomato in soil with and without additional phosphorus. *Plant soil 88*, 307-319.

Rabie, G.H., and Almadini, A.M. 2005. Role of bio-inoculents in development of salt tolerance of Vicia faba plants under salinity stress. *African J. Biotech. 4*, 210-222.

Ramoliya, P., Patel, H. and Pandey, A.N. 2004. Effect of salinization of soil on growth and macro and micro nutrient accumulation in seedlings of *Salvadora persica* (Salvadoraceae). *Forest. Ecol. Manage. 202*, 181-193.

Rao, A.V. and Tak, R. 2002. Growth of different tree species and their nutrient uptake in limestone mine soil as influenced by arbuscular mycorrhizal fungi in India arid zone. *J. Arid Environ. 51*, 113-119.

Ravikovitch, S., and Yoles, D. 1971. The influence of phosphorus and nitrogen on millet and clover growing in soils affected by salinity. I plant development. *Plant Soil 35*, 555-567.

Reddy, M.P. and Vora, A.B. 1986. Salinity induced changes in pigment composition and chlorophyllase activity of wheat. *Indian J. Plant Physiol. 29*, 331-334.

Sanchez, F.J., Manzanares, M., De Andres, E.F., Tenorio, J.L. and Ayerbe, L. 1998. Turgor maintenance, osmotic adjustment and soluble sugar and proline accumulation in 49 pea cultivars in response to water stress. *Field Crop Res. 59*, 225-235.

Shobha, 2003. *Effect of Salinity stress on the Germination and Enzyme Activities in Different Varieties of Vigna unguiculata (L.) Walp. Seeds.* Ph.D. thesis, Mangalore University, India.

Smith, S.E. and Read, D.J. 1997. *Mycorrhizal symiosis.* 2nd Edition, Academic press, UK.

Sylvia, D.M., Hammond, L.C., Bennett, J.M., Haas, J.H. and Linda, S.B. 1993. Field response of maize to a VAM fungus and water management. *Agron. J. 85*, 193-198.

Tian, C.Y., Feng, G., Li X.L. and Zhang, F.S. 2004. Different effects of arbuscular mycorrhizal fungal isolates from saline or non-saline soil on salt tolerance of plants. *Appl. Soil. Ecol. 26*, 143-148.

Tsang, A. and Maun, M.A. 1999. Mycorrhizal fungi increase salt tolerance of *Strophostyles hevola* in coastal foredunes. *Plant Ecol. 144*, 159-166.

Yano-Melo, A.M., Saggin, O.J. and Maia, L.C. 2003. Tolerance of mycorrhizal banana (*Musa sp.* cv *pacovan*) plantlets to saline stress. *Agric. Ecosys. Environ. 95*, 343-348.

Yoshiba, Y., Kiyosue, T., Nakashima, K. and Yamaguchi- Shinozaki, K. 1997. Regulation of levels of proline as an osmolyte in plants under water stress. *Plant Cell Physiol. 38*, 1095-1102.

Zandavalli, R.B., Dillenburg, L.R. and Paulo, V.D. 2004. Growth responses of *Araucaria angustifolia* (Auraucariaceae) to inoculation with the mycorrhizal fungus *Glomus clarum*. *Appl. Soil. Ecol. 25*, 245-255.

Frontiers in Fungal Ecology, Diversity and Metabolites (edited by K.R. Sridhar), I.K. International Publishing House Pvt. Ltd., New Delhi, 2009.

11

Production of Indole Acetic Acid by Ectomycorrhizal Fungi

K. Natarajan* and Malathi Srinivasan
Centre for Advanced Studies in Botany, University of Madras, Guindy Campus, Chennai 600 025, Tamil Nadu, India
*(*E-mail: knraj42@yahoo.com)*

Introduction

The world's demand for forest products is increasing because of the increase in population and higher standards of living. Utilization of more land for wood production would increase, but land is also being increasingly required for other purposes. As a consequence, large proportions of forested areas must be devoted to manage fast-growing species intensively to meet the needs for wood for a wide range of purposes. A number of commercially important eucalypt species are suitable for this purpose. The relative ease with which this genus can be grown makes it reliable for afforestation practices. Apart from the well-known advantages of growing eucalypts (low nutrient requirements, adaptability to degraded and nutrient poor soils), this genus is amenable for two important strain improvement techniques: (1) easy protocols for tissue culture and micropropagation; and (2) easily tailored with growth-promoting mycorrhizal fungi, which makes it preferable in forestry.

Micropropagation has been tried on various genera and successful methods have been evolved for forest tree species including eucalypts. This enables stock improvement, wherein clones of selected trees can be produced, resulting in high vigor and tolerance to the changing edaphic factors. Coupled with this, is the wide range of natural associations that the tree exhibits. Eucalypts are known to have both ecto- and endo-mycorrhizal associations and it has been observed that these symbionts are the key factors in the normal growth of the phytobiont even on degraded regions. It would be moreadvantageous, if these mycorrhizal fungi are made to associate with the clones wherein the benefits of both tissue culture and mycorrhizal associations can be combined to get better outputs. Such types of studies, find more relevance in the recent years because of the inherent property of these fungi to secrete some compounds that are of great importance in micropropagation methods. Indole acetic acid (IAA) is one such compound. It is a phytohormone, which belong to the class auxin, and therefore, plays an important role in the formation of roots in plants. This hormone is available commercially as a synthetic powder and forms an indispensable constituent of any tissue culture medium. It has now been discovered that some of the ectomycorrhizal fungi also produce IAA when cultured on specific media and can, therefore, be employed directly or indirectly in tissue culture and micropropagation techniques. Use of ectomycorrhizal fungi even at the root initiation stage (juvenile stage) ensures early association with symbionts and hence, a successful association throughout the plant life. The current chapter deals with the importance of ectomycorrhizae in the production of IAA, its role and importance in micropropagation of commercially valued tree species with special reference to eucalypts.

IAA Synthesis by Ectomycorrhizal Fungi

Mycorrhizal fungi, by virtue of their symbiotic association with roots of all vascular plant systems, are among the most important microbes in the terrestrial ecosystem. Mycorrhizal symbiosis plays a significant role in the abatement of plant biological stress, and benefits their host plants by improving their growth and development. The complexity of ectomycorrhizal formation mechanisms is reflected by the characteristic morphology of these symbiotic organisms. Although, ectomycorrhizae vary to a great extent, they have some features in common, such as swellings, lack of root hairs on the swollen region and radially shaped cortical cells within these swellings. Melin, in 1925, demonstrated through his *in vitro* experiments that these characteristic structures developed only in the presence of the symbiotic fungus. Later, it was discovered that these profound morphological and anatomical changes in the short roots of the host are induced by auxin related metabolites exuded by the fungus.

Hitchcock and Zimmerman (1935), however, believed that substances like a-Naphthalene Acetic Acid (NAA), Indole Acetic Acid (IAA), Indole Butyric Acid (IBA), Indole Propionic Acid (IPA), and the methyl and ethyl esters of these acids were absorbed from the soil and transmitted to the distant parts of the shoots to promote growth. While conducting an experiment to confirm this theory, Hitchcock and Zimmerman (1935) discovered that the auxins were produced by the fungus and then translocated to the root because fungus-free root samples did not answer positive to the HCl-iron perchloride test for the presence of auxin. Subsequent experiments by MacDaugal and Dufrenoy (1944) also proved that auxin occurs abundantly in the fungal hyphae of pine mycorrhizae, and assumed that the coralloid branching of the short mycorrhizal roots in pine was only due to the translocation of the auxins from the hyphae. It has also been demonstrated experimentally that root structures strikingly similar to ectomycorrhizae can be synthesized in excised roots of *Pinus sylvestris* by adding to their nutrient solution, a cell-free filtrate in which the mycelium of a symbiotic fungus had grown for a certain period (Levisohn, 1953; Slankis, 1948; Turner, 1962). Similar morphogenesis has also been recorded by adding a nutrient solution of scots pine, white pine and synthetic auxins in concentrations that are supra-optimal for roots (Slankis, 1950, 1958). Depending upon the hormonal treatment, different mycorrhiza-like short roots were formed. Palmer (1954) confirmed these findings, but Barners and Naylor (1959) and Ulrich (1960a) were unable to induce mycorrhiza-like short roots with hormonal treatments. Slankis (1958) concluded that these metabolites were related to auxins. This was further supported by the findings of Subba Rao and Slankis (1959) on the occurrence of auxins within ectomycorrhiza-colonized roots of pine seedlings and not in non-mycorrhizal roots. Based on these findings, several researchers studied the ability of fungi to produce indole compounds (Gogala, 1971; Horak, 1963; Moser, 1959; Ulrich, 1960b).

Most of the fungi that were studied synthesized indole compounds, particularly IAA, but the amount of auxins produced, their concentrations, composition, and the duration necessary for their production in detectable amounts differed among species and strains. Experiments carried out by Shemakhanova (1962) proved unsuccessful as no trace of indole compounds was detected in the fungal culture filtrates. Later experiments revealed that most ectomycorrhizal fungi release IAA when cultivated on a medium containing an auxin-precursor, tryptophan (Ek *et al.*, 1983; Gay and Debaud, 1987; Tomaszewski and Wojciechowska, 1975; Ulrich, 1960a) or tryptophan precursors (Horak, 1963). Some fungi are even able to release IAA in the absence of precursors containing the indole ring (Tomaszewski and Wojciechowska, 1975), and a few fungi are able to produce IAA even when cultivated on a medium devoid of tryptophan (Rouillion *et al.*, 1986). Meister (1957) suggested that tryptophan can be converted by decarboxylation to tryptamine, which in turn could be oxidized via the corresponding aldehyde to IAA, or tryptophan can be deaminated by oxidation or transaminated to yield indole pyruvic acid, which on oxidative decarboxylation gives IAA.

Although there has been much speculation that fungal IAA might promote the establishment of the symbiotic relationship by stimulating the formation of short roots that are receptive to the fungus, the physiology of IAA release by ectomycorrhizal fungi and the role of this metabolite in the symbiotic association are yet poorly understood (Gay *et al.*, 1989). The release of IAA is not a characteristic of ectomycorrhizal fungi alone, as it has also been detected in saprophytic and pathogenic higher fungi, lower fungi, pathogenic bacteria (Kaper and Volstra, 1958; Kosuge and Wilson, 1965), soil bacteria (Azcon and Barea, 1975; Katznelson and Sirois, 1961), and rhizospheric bacteria (Brown, 1972). A correlative study of the morphogenetic effect of a fungus on its host roots, and its ability to release indole compounds should clarify the role of these metabolites in the establishment of the symbiotic relationship. Hacskaylo and

Bruchet (1972) reported that such correlative studies and their findings allowed the use of different fungi of the genus *Hebeloma* in forestry, apart from probing into the physiological aspects of these fungi.

Moser (1959) and Gay (1986) reported that IAA release by ectomycorrhizal fungi is greatly dependent on culture conditions, wherein high concentrations of the precursor (tryptophan) are required with a culture duration ranging from a few days to several months. Precursor availability in root exudates is probably one of the main limiting factors for fungal IAA synthesis in ectomycorrhzas. Although recent experiments implicate a tryptophan precursor rather than tryptophan itself as the predominant IAA precursor, other experiments continue to demonstrate conversion of tryptophan to IAA (Bartel, 1997). A report by Gay *et al.* (1989) also suggests that of all the precursors, aromatic amino acids, or intermediates of the shikimate pathway, the effect of 1mM tryptophan was the maximum on the IAA biosynthesis of *Hebeloma heimale*. The fungus did not produce detectable amounts of IAA when cultivated on a medium containing no precursor or supplemented with phenylalanine, tyrosine, or shikimic acid. Despite the evident theory of IAA biosynthesis by ectomycorrhizal fungi using tryptophan, the capability of these fungi to supply the auxin in concentrations that are required to induce the characteristic morphology of ectomycorrhizae remains a dubious factor (Ritter, 1968; Gay *et al*., 1989).

Role of IAA in Micropropagation

Lindermann and Call (1977) made a pioneering effort to hypothesize the production of growth-promoting substances by ectomycorrhizal fungi, thereby causing an enhancement in the rooting of cuttings of vegetatively propagated woody ornamentals. They found that the percentage of rooted cuttings and the root volumes of cuttings were greater on bearberry and huckleberry inoculated with ectomycorrhizal fungi than on uninoculated controls, even before mycorrhizal formation. Similar results were reported by Navratil and Rochon (1981) on poplar cuttings, and by Branzanti *et al.* (1984) and Cristoferi *et al.* (1985) on fruit root stock cuttings. The effect of inoculating an ectomycorrhizal fungus like *Pisolithus tinctorius* onto the roots of *Eucalyptus urophylla in vitro* was also reported by Malajcuzk *et al.* (1990), and the preferential colonization of the fungal isolate based on the host compatibility and its origin was explained. More recently, Reddy and Satyanarayana (1998) have tried to inoculate micropropagated plantlets of *E. tereticornis* with ectomycorrhizal fungi-like *Amanita murina*, *Hysterangium incarceratum*, *Laccaria laccata*, *Scleroderma cepa*, *S. flavidum* and three isolates of *Pisolithus tinctorius*. They have reported that plantlets with mycorrhizas formed by one of the local isolates of *P. tinctorius* showed significantly better growth than their non-mycorrhizal counterparts, when micropropagated plantlets of *Eucalyptus tereticornis* were inoculated *in vitro* with the fungus at the root initiation stage.

The ever increasing demand for eucalypt wood for industrial applications has necessitated better and faster methods of propagation on one hand and genetic improvement on the other. Plant tissue culture methods are useful in the rapid cloning of elite individuals. Micropropagation of *E. tereticornis* has been in vogue for a considerable period now. The name "micropropagation" has been derived from the miniature shoots or plantlets initially produced in this mode of plant propagation. The main advantages are that the cultures can be initiated in very small segments obtained from the mother plant, the space requirement is comparatively less, propagation is possible under aseptic conditions, extensive care is not required in between subcultures, and in most species, propagation is possible throughout the year (John *et al.*, 1997).

Eucalypts are fast-growing species capable of establishing under diverse environments and breeding programs have been developed for several species around the world (Eldridge *et al.*, 1993). An extensive body of literature concerning tissue culture in eucalypts exists currently (Gupta and Mascarenhas, 1987). Tissue culture work for large-scale production of planting stocks from superior mother trees was initiated in Australia and France (Cresswell and Fossard, 1974; Cresswell and Nitsch, 1975). Several types of micropropagation have been attempted with eucalypts. In 1970s, tissue culture-raised plants of *Eucalyptus dalrympleana* were successfully planted in the field followed by *in vitro* rooting of apical buds of *E. bacroftii*, *E. deglupta* and *E. grandis*. Fossard *et al.* (1977) tried to produce plantlets from seedling nodal bud explants, while Hartney (1980) tried this method with *E. globulus*, *E. grandis*, *E. curtisis*, *E. regnans* and *E. ficifolia*. Adventitious shoot formation from cultures of seedling explants has also been attempted (Bennet and McComb, 1982; Gupta and Mascarenhas, 1987; Kitahara and Caldas, 1975; Lakshmi Sita and Vidyanathan, 1979). Micropropagation of *Eucalyptus camaldulensis* from seedling explants has been reported by Hartney (1981), and Diallo and Duhoux (1984). Multiple shoot formation on nodal explants derived from 20-year-old mature tree has been studied by Gupta *et al.* (1983). Explants from mature trees of *E. marginata* and *E. sideroxylon* have been used to regenerate shoots and whole plants by Bennet and McComb (1982). Shoot formation from leaf explants of *E. camaldulensis* and *E. leichow* has been investigated by Muralidharan and Mascarenhas (1987), while Subbaiah and Minocha (1990) have reported shoot regeneration from stem and leaf callus of *E. tereticornis*. Plant regeneration of *Eucalyptus botryoides* has also been attempted using rotation nodule cultures and this system has been found to be beneficial for both mass propagation of selected elite clones and the creation of genetically engineered plants (Ito *et al.*, 1996).

Different hormonal concentrations and combinations have also been reported. Lakshmi Sita and Vaidyanathan (1979) used 2.2 to 8.8 μM BAP for multiple shoot formation in *E. citriodora*, while Lakshmi Sita *et al.* (1986) reported consistent multiple shoot formation in media supplemented with BAP (2.2-8.8 μM) and NAA (0.54-5.37 μM) using *E. grandis*. In *E. tereticornis*, Mascarenhas *et al.* (1982) have reported the induction of multiple shoots using a low concentration of cytokinins as 2.2 μM BAP and 0.92 μM kinetin. Later, Lakshmi

Sita and Chattopadhye (1987) used higher concentration of cytokinins while reporting multiple shoot formation and plantlet production.

Conclusions and Outlooks

In most of the tree breeding programs worldwide, increase in the growth rates of tree, stem volume, and shortening of rotation times are the important aims. Such trees would yield more biomass per unit area. A major disadvantage with traditional tree breeding, especially for forest tree species is the slow progress due to long generation periods (Eriksson *et al.*, 2000). Micropropagation protocols have advantages over traditional clonal propagation methods including high multiplication rates and production of disease-free plants (Amin and Jaiswal, 1988; Ghaffor and Alderson, 1994; Gowda and Muralikrishna, 1994; Sidiqqui and Farroq, 1995). Micropropagated plantlets that were inoculated with mycorrhiza enhanced plant quality, nutrient uptake and plant growth (Rapparini *et al.*, 1994). Mycorrhizal enhancement of rooting and root function has been reported in several species with increased pest resistance, enhanced crop production and a tolerance to environmental stress (Sbrana *et al.*, 1994). Recently, early mycorrhizal inoculation and colonization of tissue-cultured plantlets have been reported to reduce transplant shock during acclimatization, thereby increasing plant survival and establishment rates (Estrada-Luna *et al.*, 2000). Thus, it is evident from this review that the production of IAA by ectomycorrhizal fungi has an important role in the early initiation of mycorrhizal association with tree roots, and this aspect can be beneficially exploited in micropropagation of commercially important tree species.

References

Amin, M.N. and Jaiswal, V.S. 1988. Micropropagation as an aid to rapid cloning of a guava cultivar. *Sci. Hortic.* 36, 89-95.

Azcon, N.R. and Barea, J.M. 1975. Synthesis of auxins, gibberellins and cytokinins by *Azotobacter vinelandii* and *Azotobacter bejerinckii* related to effects produced on tomato plants. *Plant Soil 43*, 609-619.

Bartel, B. 1997. Auxin Biosynthesis. *Ann. Rev. Plant Physiol. Plant Mol. Biol.* 48, 51-66.

Bennet, I.J. and McComb, J.A. 1982. Propagation of jarrah (*Eucalyptus marginata*) by organ and tissue culture. *Aust. For. Res. 12*, 121-127.

Branzanti, B., Cristoferi, G., Zocca, A. and Zambonelli, A. 1984. Ectomycorrhizal fungi and IBA effects on fruit root stocking rooting. In: *Proceedings of the 6th North American Conference on Mycorrhizae* (Ed. Molina, R.). Forest Research Laboratory, Corvallis, OR.

Brown, M.R. 1972. Plant growth substances produced by microorganisms of soil and rhizosphere. *J. Bacteriol. 35*, 443- 451.

Burners, R.L. and Naylor, A.W. 1959. *In vitro* culture of pine roots and the use of *Pinus serotina* roots in metabolic studies. *For. Sci. 5*, 158-168.

Cresswell, R.J. and Fossard De, R.A. 1974. Organ culture of *Eucalyptus grandis. Aust. For. 37*, 55-69.

Cresswell, R.J. and Nitsch, C. 1975. Organ culture of *Eucalyptus grandis* L. *Planta* 125, 87-90.

Cristoferi, G., Branzanti, B. and Zocca, A. 1985. Role of mycorrhizae in fruit root stock rooting. In: *Proceedings of the 1st European Symposium on Mycorrhizae* (Ed. Gianinazzi-Pearson, V. and Gianinazzi, S.). Institut National de la Recherche Agronomique, Dijon, France, 483-488.

Diallo, N. and Duhoux, E. 1984. Organogenesis and multiplication *in vitro* of *Eucalyptus camaldulensis. J. Plant Physiol. 115*, 177-182.

Ek, M., Ljungquist, P.O. and Stenstrom, E. 1983. Indole-3-acetic acid production by *mycorrhizal* fungi determined by gas chromatography-mass spectrometry. *New Phytol. 94*, 401-407.

Eldridge, K., Davidson, J., Harwood, C. and van Wyk, G. 1993. *Eucalypt Domestication and Breeding.* Clarendon Press, Oxford.

Eriksson, M.E., Israelsson, M., Olsson, O. and Moritz, T. 2000. Increased gibberellin biosynthesis in transgenic trees promotes growth, biomass production and xylem fiber length. *Nature Biotech. 18*, 784-788.

Estrada-Luna, A.A., Davies Jr, F.T. and Egilla, J.N. 2000. Mycorrhizal fungi, enhancement of growth and gas exchange of micropropagated guava plantlets during *ex vitro* acclimatization and plant establishment. *Mycorrhiza 10*, 1-8.

Fossard De, R.A., Barker, P.K. and Bourne, R.A. 1977. The organ culture of nodes of four species of eucalypts. *Acta Hortic. 78*, 297-302.

Gay, G. 1986. Effect of glucose on indole-3-acetic acid production by the ectomycorrhizal fungus *Hebeloma hiemale* in pure culture. *Physiol. Veg. 24*, 185- 192.

Gay, G. and Debaud, J.C. 1987. Genetic study on indole-3-acetic acid production by ectomycorrhizal *Hebeloma* species, inter- and intra- specific variability in homo- and *di-karyotic mycelia. Appl. Microbiol. Biotechnol. 26*, 141-146.

Gay, G., Rouillon, R., Bernillon, J. and Favre- Bonvin. J. 1989. IAA biosynthesis by the *ectomycorrhizal* fungus *Hebeloma hiemale* as affected by different precursors. *Can. J. Bot.* 67, 2235-2239.

Ghaffoor, A. and Alderson, P.G. 1994. Somatic embryogenesis in guava. In: *Physiology, Growth and Development of Plants in Cultures* (Ed. Lumsden, P.J., Nicholas, J.R. and Davies, W.J.). Kluwer, Dordecht, 272-277.

Gogala, N. 1971. Growth substances in mycorrhiza of the fungus *Boletus pinicola* Vitt. and the pine tree *Pinus sylvestris* L. *Akad. Sci. Art Slov.* 123-202.

Gowda, A.N.S. and Muralikrishna, A. 1994. Micropropagation strategy for guava (*Psidium guajava* L.cv. Sardar). *In vitro 30*, 3.

Gupta, P.K. and Mascarenhas, A.F. 1987. *Eucalyptus.* In: *Cell and Tissue Culture in Forestry* (Ed. Bonga, J.M. and Durzan, D.J.). Volume 3, Martinus Bijhoff, Dordrecht, The Netherlands, 385-399.

Gupta, P.K., Mehta, U.J. and Mascarenhas, A.F. 1983. A tissue culture method for rapid propagation of *Eucalyptus camaldulensis* and *Eucalyptus torelliana* from mature trees. *Plant Cell Rep. 2*, 296-299.

Hacskaylo, E. and Bruchet, G. 1972. Hebelomas as mycorrhizal fungi. *Bull. Torrey Bot. Club 99*, 17-20.

Hartney, V.J. 1980. Vegetative propagation of the *Eucalyptus. Aust. For. Res. 10*, 191-211.

Hartney, V.J. 1981. Vegetative propagation of *Eucalyptus in vitro.* In: *Proceedings of IUFRO International Workshop on in vitro Cultivation of Forest Tree Species.* AFOCEL, Fountainebleau, France, 175-180.

Hitchcock, A.E. and Zimmerman, P.W. 1935. Absorption and movement of synthetic growth substances from soil as indicated by responses of aerial parts. *Contribution of Boyce Thompson Institute 7*, 447-476.

Horak, E. 1963. Untersuchungen zur Wuchsoffsynthese der Mykorrhiza Pilze. In: *Mycorrhiza* (Ed. Rawald, W. and Lyr, H.). Fischer Verlag, Jena, 147-163.

Ito, K., Doi, K., Tatemichi, Y. and Shibata, M. 1996. Plant regeneration of *Eucalyptus* from rotating nodule cultures. *Plant Cell Rep. 16*, 42-45.

John , C.K., Nadgauda, R.S. and Mascarenhas, A.F. 1997. Tissue culture of economic plants, including genetic engineering techniques. In: *Tissue Culture of Economic Plants.* Centre for Science and Technology of the Non-aligned and other Developing Countries and Commonwealth Science Council, London, 42-81.

Kaper, J.M. and Velstra, H. 1958. On the metabolism of tryptophan by *Agrobacterium tumefasciens. Biochim. Biophys. Acta 30*, 401-420.

Katznelson, H. and Sirois, J. C. 1961. Auxin production by species of *Arthrobacter. Nature 91*, 1323.

Kitahara, E.H. and Caldas, L.S. 1975. Shoot and root formation in hypocotyl callus cultures of *Eucalyptus. For. Sci. 21*, 242-243.

Kosuge, T. and Wilson, E.E. 1965. The conversion of L-*tryptophan* to indole acetic acid by a phytopathogenic bacterium *Pseudomonas savastoni. Plant Physiol. 40*, 59.

Lakshmi Sita, G. and Chattopadhye, S. 1987. Improvement of forest trees by tissue culture. In: *Plant Cell and Tissue Culture of Economically Important Plants* (Ed. Reddy, G.M.). Proceedings of the National Symposium, Hyderabad, India, 195-198.

Lakshmi Sita, G. and Vaidyanathan, C.S. 1979. Rapid multiplication of *Eucalyptus* by multiple shoot production. *Curr. Sci. 48*, 350-352.

Lakshmi Sita, G. and Shobharani, B. and Rao, K.S. 1986. Propagation of *Eucalyptus grandis* by tissue culture. In: *Eucalyptus in India - Past, Present and Future.* KFRI, Peechi, India, 318-321.

Levisohn, I. 1953. Growth response of tree seedlings to mycorrhizal mycelia in the absence of a *mycorrhizal* association. *Nature 172*, 316-317.

Lindermann, R.G. and Call, C.A. 1977. Enhanced rooting of woody plant cuttings by ectomycorrhizal fungi. *J. Am. Soc. Hortic. Sci. 102*, 629-632.

MacDougal, D.T. and Dufrenoy, J. 1944. Mycorrhizal symbiosis in *Aplectrum*, *Corallorhiza* and *Pinus. Plant Physiol. 19*, 440-465.

Malajczuk, N., Lapeyrie, F. and Garbaye, J. 1990. Infectivity of pine and eucalypt isolates of *P. tinctorius* on roots of *Eucalyptus urophylla in vitro* I - Mycorrhiza formation in model systems. *New Phytol. 114*, 627-631.

Mascarenhas, A.F., Hazra, S., Potdar, U., Kulkarni, D.K. and Gupta, P.K. 1982. Rapid clonal multiplication of mature forest trees through tissue culture. In: *Plant Tissue Culture* (Ed. Fujiwara, A.). Proceedings of the 5th International Congress on Plant tissue and cell culture, Tokyo, Japan, 719-720.

Meister, A. 1957. *Biochemistry of the Amino Acids.* Academic press Inc., New York.

Melin, E. 1925. *Untersuchunngen uber die Bedentung der Baummyhorriza.* Fisher, Jena, Germany.

Moser, M. 1959. Beitrage zur Kenntnis der Wuchsstoffbeziehungen im Bereich ectotropher Mycorrhizen. *Arch. Mikrobiol. 4*, 421-431.

Muralidharan, E.M. and Mascarenhas, A.F. 1987. *In vitro* plantlet formation by organogenesis in *E. camaldulensis* and by somatic embryogenesis in *E. citriodora. Plant Cell Rep. 6*, 256-259.

Navratil, S. and Rochon, G.C. 1981. Enhanced root and shoot development of poplar cuttings induced by *Pisolithus* inoculum. *Can. J. For. Res. 11*, 844-848.

Palmer, J.C. 1954. *Mycorrhizal* development in *Pinus virginiana* as influenced by growth regulators. Ph.D. Thesis, George Washington University, Washington DC.

Rapparini, F., Baraldi, R., Bertazza, G., Brazanti, B. and Predieri, S. 1994. Vesicular-arbuscular mycorrhizal inoculation of micropropagated fruit trees. *J. Hortic. Sci. 69*, 1101-1109.

Reddy, M.S. and Satyanarayana, T. 1998. Inoculation of micropropagated plantlets of *Eucalyptus tereticornis* with ectomycorrhizal fungi. *New Forests. 16*, 273-279.

Ritter, G. 1968. Auxin relations between mycorrhizal fungi and their partner trees. *Acta. Mycol. 4*, 421-431.

Rouillon, R., Gay, G., Bernillon, J., Favre-Brovin, J. and Bruchet, G. 1986. Analysis by HPLC-Mass spectrometry of the indole compounds released by the ectomycorrhizal fungus, *Hebeloma hiemale* in pure culture. *Can. J. Bot. 64*, 1893-1897.

Sbrana, C., Giovannetti, M. and Vitagliano, V. 1994. The effect of mycorrhizal infection on survival and growth renewal of micropropagated fruit rootstocks. *Mycorrhiza 5*, 153-156.

Shemakhanova, N.M. 1962. *Mycotrophy of Woody Plants.* Akademiya Nauk USSR, Moscow.

Sidiqqui, Z.M. and Farooq, S.A. 1995. Tissue culture studies on the nodal explants of *Psidium guajava. In vitro 31*, 3.

Slankis,V. 1948. Einfluss von exudaten von Boletus variegates auf die dictomische verzweigung isolierter kiefernwurzein. *Physiol. Plant. 1*, 390-400.

Slankis, V. 1950. Effect of a-naphthalene acetic acid on dichotomous branching of isolated roots of *Pinus sylvestris. Physiol. Plant. 3*, 40-44.

Slankis, V. 1958. The role of auxins and other exudates in mycorrhizal symbiosis of forest trees. In: *Physiology of Forest Trees* (Ed. Thimann, V.). Ronald Press, New York, 427-443.

Subbaiah, M.M. and Minocha, S.C. 1990. Shoot regeneration from stem and leaf callus of *Eucalyptus tereticornis. Plant Cell Rep. 9*, 370-373.

insect resistance, production of bioactive compounds, and future bioprospecting avenues.

Endophytic Fungi

Plants appear to be a reservoir of an innumerable numbers of endophytic organisms (Bacon and White, 2000). Endophytes are microbial entities that live within living tissues of plants and their relationship with the host plant is symbiotic and probably mutualistic. Fungal endophytes are those that colonize and grow asymptomatically within internal plant tissues without causing harm to their host. It is known that they help the hosts tolerate adverse abiotic and biotic factors. An enormous number of endophytic fungi have been isolated from a wide range of plants growing in their native habitat and their wide distribution with the host indicating an intimate beneficial mutualism. Almost all vascular plant species examined to date were found to harbor endophytic fungi (Sturz *et al*., 2000). In addition, endophytic fungi colonize mosses, ferns, marine algae, and seagrasses (Raviraja *et al*., 1996). Endophytes constitue an important component of microbial biodiversity and may provide at least a partial answer for some missing fungi out of the 1.5 million that are known. The environmental conditions under which the host grow, influences the endophytic fungal population (Hata *et al*., 1998) and the endophyte profile seems to be more diversified in tropical areas. Arnold *et al*. (2000) isolated 418 endophyte morphospecies (estimated 347 genetically distinct taxa) from 83 healthy leaves of *Heisteria concinna* and *Ouratea lucens* in a lowland tropical forest of Central Panama. He proposed that tropical endophytes themselves could be hyperdiverse with host preference and spatial heterogeneity. The genotypic diversity has been seen in single endophyte species originating from conifers (McCutcheon *et al*., 1993), birch (Lappalainen and Yli-Mattila, 1999) and grasses (Reddy *et al*., 1998). Accordingly, endophytes are ubiquitous in the plant kingdom with the population being dependent on host species and geographical location.

Mutualism

The plant-endophytic fungal interaction is a very complex phenomenon and needs to be understood explicitly. Endophytic fungi are unique among many plant microbial symbionts because they are the only microbial group that were thought to benefit their hosts through direct defense against pathogens or pests (Carroll, 1988; Clay, 1988; Faeth and Bultman, 2002). Each plant host has a range of physical, chemical, constitutive and induced control over the spread of fungi within its tissues. An enormous diversity of phenolic and other deterrent plant compounds are associated with the endophytic fungi, considerably more than those associated with potent pathogens in the same host. Endophytes upregulates plant responses to pathogens. In the absence of plant controls, a proliferation of endophytes throughout the tissues would be expected. The fungal endophytes possess the exoenzymes necessary to colonize their hosts and they grow well in the apoplastic washing fluid of the host. Endophytic fungi colonize the roots of larch (*Larix decidua*) and improve the host growth through sufficient nourishment (Paracer and Ahmadjian, 2000). Fungi associated with litter decompostion are involved in the initial stages of resource recycling.

Many aquatic fungi have an endophytic stage in their life cycle. There is an exchange of nutrients between the inhabiting fungus and the host plant species during such association. The fungus receives readymade nutrients within the host plant and in turn the host plants are benefited in several ways including the production of metabolites that inhibit insects and pests as well as the stimulation of defense mechanisms. For instance, the endophyte *Rhabdocline parkeri* produces a metabolite that reduces needle attack by borers. Similarly, metabolites of *Phomopsis* sp. in cotton deter the leaf feeding larvae of *Helicoverpa.* In another case, endophytic isolates of *Trichoderma* sp. and *Chaetomium globosum* have been shown to induce systemic resistance in the host against various phytopathogens.

Endophytic Fungi and Plant Growth

Endophytic fungi inhabiting plants have often shown the promotion of health as well as growth. Plants that are colonized by endophytes often grow faster than non-colonized plants. This impact may be due to the production of phytohormones such as indole-3-acetic acid (IAA), cytokines, other plant growth-promoting substances (Zou and Tan, 1999) and this may due partly to the fact that endophytes had enhanced the hosts' uptake of nutritional elements such as nitrogen and phosphorus (Gasoni and De Gurfinkel, 1997; Malinowski and Belesky, 1999). Such growth stimulation is best understood in grasses. Endophyte colonized grasses tend to be comparatively vigorous, particularly under conditions of minimal fertilization and irrigation. Such plants produce a high number of tillers and roots, making them more drought-tolerant, more competitive with weed species, able to recover more rapidly from injury, and generally more persistent in the field. Such grasses enhance their performance under conditions of stress, such as high temperature, nutrient and water deficiency, salinity, insects and diseases.

Many of the endophytes are capable of producing plant hormones. The growth-promoting phytohormone, IAA was re-isolated from cultures of a root associated bacterial endophyte, *Azospirillum brasilense* (Zimmer *et al*., 1998), the tall fescue fungal endophyte *Acremonium coenophialum* (De Battista *et al*., 1990), *Aureobasidium pullulans, Epicoccum purpurascens* (Petrini *et al*., 1992) and *Colletotrichum* sp. (Zou *et al*., 2000). In addition to IAA and indole-3-acetonitrile (Dew *et al*., 1990), cytokinins were also produced by an endophytic strain, *Hypoxylon serpens* isolated from tobacco (Petrini *et al*., 1992). The alternative production of endophytic phytohormones is assumed to be related to the growth-promoting feature of the plant. On the other hand, a new indole derivative, 6-isoprenylindole-3-carboxylic acid was characterized recently from the endophyte *Colletotrichum* sp. of *Artemisia annua* (Bush *et al*., 1997). It shows moderate antibacterial activity against the Gram-positive bacteria (*Bacillus subtilis*, *Staphylococcus aureus* and *Sarcina lutea*) and the Gram-negative bacterium (*Pseudomonas* sp.). This new product is also inhibitory to the growth of some phytopathogenic fungi (*Phytophthora capisici, Rhizoctonia cerealis* and *Gaeumannomyces graminis* var. *tritici*) (Lu *et al*., 2000).

Endophytic Fungi and Pests

Endophytic fungi are capable of imparting other important strengths to plants such as greater resistance to stress conditions (e.g., water), alteration in physiological properties, production of phytohormones and other compounds of biotechnological value (e.g., enzymes, pharmaceuticals). The endophytic fungus, *Phomopsis oblonga* protected elm trees against the beetle (*Physocnemum brevilineum*), which in turn resulted in the control of the Dutch elm disease caused by *Ceratocystis ulmi* (Webber, 1981). Kanda *et al*. (1994) reported the preference of larvae of bluegrass webworm, *Parapediasia teterrella* for diets with endophyte-free plants of *Lolium perenne* and *Festuca arundinacea* to a point that the larvae would starve to death if plants colonized only with *Acremonium* were available. In the flour beetle (*Tribolium castaneum*), diets composed of ground seeds of *F. arundinacea* and *L. perenne* colonized by endophytic fungi resulted in a reduction of infested seeds and insect growth more notably in *F. arundinacea* (Cheplick and Clay, 1988). The alcoholic extracts from *Poa ampla* colonized or non-colonized with *Neotyphodium typhnium* were assayed against mosquito larvae. It was found out that only the extracts from the colonized plants were effective against larvae and not the extract obtained directly from the endophytic fungi (Ju *et al.,* 1998).

Certain endophytes improve the ecological adaptability of hosts by enhancing their tolerance to environmental stresses and resistance to phytopathogens and herbivores, including some insects. Endophyte colonized grasses usually possess high tolerance to drought (Arachevaleta *et al*., 1989; Ravel *et al*., 1997) and aluminium toxicity (Malinowski and Belesky, 1999). Some endophytes are able to provide protection to host plants against some nematodes (Hallman and Sikora, 1996), mammals (Bacon *et al*., 1977), insect herbivores (Wilkinson *et al*., 2000), and bacterial and fungal pathogens (Christensen, 1996). Some endophytes enhance the hosts' allelopathic effects on other species that are co-occurring in the vicinity, usually being competitor(s) for the nutrition and space (Sturz *et al*., 1998). This could be the reason why some plants with specific endophytes are usually competitive and become dominant species in successional fields (Clay and Holah, 1999). Endophytic grasses have shown high resistance to foliar-feeding insects (e.g., billbugs, chinch bugs, sod webworms, fall army-worms, and argentine stem weevils). Biologically active alkaloids are found only in endophyte colonized grasses. The insecticidal effects of these compounds deter insect infestations, resulting in a population decline. As the alkaloid levels in the roots are low, endophytes are not effective against root feeders such as white grubs. It is shown that significant reduction in proportions of cocoa pods lost to *Phytophthora* sp. on treating with two of the endophytic fungi and these results suggest that endophytes may be used as biocontrol agents to benefit cacao plants (Holmes *et al.*, 2002).

There are several endophytic fungi, which exhibit strong insect repellent and insecticidal properties. The best known endophytic fungal product with insecticidal properties is the nodulisporic acid, which is an indole diterpenes, isolated from *Nodulisporium* sp. of *Bontia daphnoides*. Insect toxins have also been isolated from an unidentified endophytic fungus of wintergreen (*Gaultheria procumbens*). The two new compounds, 5-hydroxy-2-(1'-hydroxy-5'-methyl-4'-hexenyl)benzofuran and 5-hydroxy-2-(1'-oxo-5'-methyl-4'-hexenyl)benzofuran showed toxicity to spruce budworm (Findlay *et al.*, 1997). Another endophytic fungus, *Muscodor vitigenus* isolated from a liana (*Paullina paullinioides*) yielded naphthalene as its major product. Naphthalene is an active ingredient of common mothballs, a widely exploited insect repellant. The *M. vitigenus* showed promising preliminary results as an insect deterrent and has exhibited potent insect repellence against the wheat stem sawfly (*Cephus cinctus*) (Daisy *et al.*, 2002). As the world has become wary of ecological damage by synthetic insecticides, endophytic research should continue to discover novel, powerful, selective and safe alternatives.

Bioactive Metabolites

Endophytic fungi of medicinal plants are a potential source of a diverse array of bioactive metabolites, which can be used for the development of drugs and act as powerful research tools. The potential role of the endophytes and their biologically active metabolites, in association with hosts has been investigated substantially. The spectrum of isolated bioactive secondary metabolites originated from endophytic fungi belong to diverse structural groups like terpenoids, steroids, xanthones, chinones, phenols, isocumarines, benzopyranones, tetralones, cytochalasines and enniatines.

Volatile antimicrobials

Endophytic fungus, *Muscodor vitigenus* isolated from the liana (*Paullinia paullinioides*) in rainforests of the Peruvian Amazon produces naphthalene, which effectively repells the adult wheat stem sawfly (*Cephus cinctus*). *Muscodor albus* from *Cinnamomum zeylanicum* of the Honduras (Worapong *et al.*, 2001) and *Muscodor roseus* from two monsoonal rainforest trees in Northern Australia (Worapong *et al.*, 2002). It produced a mixture of volatile antimicrobials that effectively inhibit and kill a wide range of plant-associated fungi and bacteria (Strobel *et al.*, 2001).

Alkaloids and steroids

The fungal alkaloids include amines and amides, indole derivatives and pyrrolizidines. Peramine 1, a pyrrolopyrazine alkaloid toxic to insects was characterized from *Neotyphodium coenophialum, N. lolli, Epichloë festucae* and *E. typhina* present in the stem and leaf of tall fescue, ryegrass and other grasses (Schardl and Phillips, 1997). Ergot alkaloids that are neurotoxic to insects and mammal herbivores were also discovered in cultures of *Neotyphodium* endophytes (Tan and Zou, 2001). The novel amide, ras-farnesyltransferase inhibitor from the culture broth of an endophyte *Phoma* sp., *cytotoxic* cytochalasans and cytochalasin E were purified from the culture broth of *Rhinocladiella* sp. endophytic to the perennial twining vine of *Tripterygium wilfordii. Phomopsichalasin* has been characterized from the culture of an endophyte, *Phomopsis* sp. that originated from the twigs of *Salix gracilostyla* var. *melanostachys.* Most recently, a tetramic acid analog, cryptocin, which is a potent antimycotic against *Pyricularia oryzae* and other phytopathogens has been characterized from the culture of endophytic fungus *Cryptosporiopsis* cf. *quercina* isolated from the inner bark of the stems of *Tripterygium wilfordii.* Indole alkaloids such as chanoclavine (Belesky and Malinowski, 1999), agroclavine (Arachevaleta *et al.*, 1989) and elymoclavine (Ravel *et al.*, 1997) were isolated from the culture of *Neotyphodium* endophytes (Petroski and Powel, 1992). Four new neurotoxic metabolites (lolitrem N, lolitriol, lolicine A and B) were isolated from *L. perenne* infected with *N. lolii. Pyrrolizidines* like lolines and saturated *aminopyrrolizidine* alkaloids were also exclusively found in endophyte colonized grasses: *Festuca arundinacea* (colonized by *N. coenophialum*) and *F. pratensis* (colonized by *N. unicinatum*) (Tan and Zou, 2001).

Steroids like ergosterol, 5α,6β-trihydroxyergosta-diene, β-hydroxyergosta-5-ene, 3-oxoergosta-4,6,8(14), 22-tetraene, 3β-hydroxy-5α,8α-epidioxyergosta-6,22-diene 57, 3β-hydroxy-5α,8α-epidioxyergosta-6,9(11),22-triene,3-oxoergosta-4-ene,3β,5α-dihydroxy-6β-acetoxyergosta-7,22-dieneand3β,5α-dihydroxy-6β-phenyl- acetoxyergosta-7,22-diene were characterized from the liquid culture of endophyte *Colletotrichum* sp. of *A. annua.* Many of these metabolites were shown to be antifungal against some crop pathogens: *Gaeumannomyces graminis* var. *tritici, Rhizoctonia cerealis, Helminthosporium sativum* and *Phytophthora capisici* (Tan and Zou, 2001).

Terpenoids and flavonoids

Terpenoids of endophytic fungi include mainly sesqui- and diterpenes. 2α-*Hydroxydimeninol* and *pestalotiopsins* A–C are the sesquiterpenes characterized from endophytic *Pestalotiopsis* spp. associated with *Taxus brevifolia* (Pulici *et al.*, 1997). Heptelidic and hydroheptelidic acids isolated from encophytic fungus *Phyllosticta* sp. of *Abies balsamea* are toxic to insects (Calhoun *et al.*, 1992). Some of the sesquiterpenes reported for the first time by fungal endophytes were chokols A-G, isolated from *Epichloë typhina* on *Phleum pratense* and were found to be fungitoxic to the leaf spot disease pathogen, *Cladosporium phlei* (Koshino *et al.*, 1989). Tricin and related flavone glycosides toxic to mosquito larvae have also been isolated from endophyte colonized blue grass (*Poa ampla*) (Ju *et al.*, 1998).

Phenols, phenylpropanoids and quinones

Phenol and phenolic acids were frequently detected in cultures of endophytes often with pronounced biological activities. The 2-Hydroxy-6-methylbenzoic acid isolated from endophytic *Phoma* sp. was shown to be antibacterial (Yang *et al.*, 1994) and 2-Methoxy-4- hydroxy-6-methoxymethylbenzaldehyde produced by a tree endophyte *Pezicula* sp. was shown to have antifungal (Schulz *et al.*, 1995) properties. Five known antifungal phenolic acids (*p*-hydroxybenzoic acid, *p*-hydroxyphenylacetic acid, tyrosol, *cis*- and *trans-p*-coumaric acids) were isolated from the stromata of *Epichloe typhina* on *Phleum pratense* (Brady and Clardy, 2000; Horn *et al.*, 1996; Koshino *et al.*, 1988; Koshino *et al.*, 1989), phenylpropanoids (Horn *et al.*, 1996; Koshino *et al.*, 1989;) and lignan (Brady and Clardy, 2000) were characterized from the stromata of *Epichloë typhina* on *Phleum pratense* (Koshino *et al.*, 1988). Coniferin (Bouarab *et al.*, 1999) and syringin (Stierle *et al.*, 1993), two monolignol glucosides were isolated from endophytic *Xylariaceae* sp. as chemical signals during the establishment of fungus-plant interactions (Chapela *et al.*, 1991). Mellein (Li *et al.*, 1996), an isocoumarin isolated from *Pezicula* spp. is strongly fungicidal, herbicidal, and algicidal (Schulz *et al.*, 1995). Gamahorin (Li *et al.*, 1998) is a new isocoumarin isolated from the stromata of *E. typhina* on *P. pratense*. Insecticidal rugulosin (Koshino *et al.*, 1988) was characterized from *Hormonema dematioides*, an endophytic fungus of balsam fir (Calhoun *et al.*, 1992). A highly hydroxylated quinone altersolanol A (Arai *et al.*, 1973) characterized from phytopathogenic *Alternaria* spp. was re-isolated from an endophytic *Phoma* sp. with its antibacterial activity (Yang *et al.*, 1994). Preussomerin *N1* (Strobel *et al.*, 1997), palmarumycin CP4a (Strobel and Hess, 1997) and palmarumycin CP5 (Benz *et al.*, 1974) were the new farnesyl-protein transferase inhibitors produced by an endophytic *Coniothyrium* sp.

Other metabolites

Endophytic fungi are also known to produce a wide array of important metabolites like peptides, aliphatic compounds and enzymes. Leucinostatin A, an oligopeptide with phytotoxic, anticancer and antifungal properties isolated from culture of *Acremonium* sp. is endophytic to *Taxus baccata* (Strobel *et al.*, 1997). Different antimicrobial cyclopeptides, echinocandins were isolated from endophytic *Cryptosporiopsis* sp. and *Pezicula* sp. of *Pinus sylvestris,* and *Fagus sylvatica.* From an endophyte of the eastern larch, a novel ester metabolite was isolated as an antibacterial agent (Yang *et al.*, 1994). Three chlorinated metabolites from endophytic fungi were reported including an insect-toxic heptelidic acid, chlorohydrin, and two antimicrobial and algicidal compounds (-)-mycorrhizin A and (+)-cryptosporiopsin from the cultures of balsam fir needle endophyte *Phyllosticta* sp. (Calhoun *et al*., 1992), tree endophytes *Pezicula* sp. and *P. livida* (Schulz *et al*., 1995), respectively. Two novel antitumor metabolites, sequoiatones A and B were isolated from endophytic fungus, *Aspergillus parasiticus* of redwood (Stierle *et al*., 1993). Three new monoterpenes and C-methylated acetogenins were produced by *Pestalotiopsis* spp. endophytic to *Taxus brevfolia*, which were shown to be of chemotaxonomic significance (Pulici *et al*., 1997). Two new metabolites (ethyl 2,4-dihydroxy-5,6-dimethylbenzoate and phomopsilactone) were isolated from *Phomopsis cassiae*, endophytic to *Cassia spectabilis*, which displayed strong antifungal activity against the phytopatogenic fungi, *Cladosporium cladosporioides* and *C. sphaerospermum*. They were cytotoxic against human cervical tumor cell line (HeLa) in *in vitro* assays as well (Silva *et al*., 2005). Lumyong *et al*. (2002) reported that endophytic fungi isolated from rainforest seedlings of Thailand showed the ability to produce a variety of enzymes (cellulase, hydrolytic enzyme, mannanase, protease and xylanase).

Bioprospecting

In recent years, there has been a great demand for new highly effective antibiotics, chemotherapeutic agents and agrochemicals that have low toxicity and have the least environmental impact. The development of resistance by the pathogens to the existing drugs, manifestation of diseases (e.g., AIDS), and severe acute respiratory syndrome have necessistated the discovery and development of new drugs. Endophytic fungi serve as a novel natural source for this purpose.

Antibiotics

Endophytic fungi are a rich source of antibiotics, which inhibit or kill a wide range of harmful phytopathogens as well as animal and human pathogens. Cryptocandin, a unique peptide antimycotic showed antifungal activity against important human fungal pathogens (*Candida albicans* and *Trichophyton*) and plant pathogenic fungi (*Sclerotinia sclerotiorum* and *Botrytis cinerea*) and was isolated from an endophytic fungus *Cryptosporiopsis quercina* of the tree *Tripterigeum wilfordii* (a native medicinal plant of Eurasia) (Strobel *et al*., 1999). Cryptocin, a unique tetramic acid is also produced by *C. quercina*, which is potent against *Pyricularia oryzae* as well as a number of other plant pathogenic fungi (Li *et al*., 2000). This compound is being examined as a natural biocontrol agent for rice blast and is used as a base model to synthesize other antifungal compounds. *Pestalotiopsis* produces a secondary metabolite, ambuic acid, which is an antifungal agent (Li *et al*., 2001). *Pestalotiopsis microspora* isolated from *Torreya taxifolia* trees, delivers several antifungal compounds like pestaloside, pestalopyrone and hydroxypestalopyrone (Lee *et al*., 1995). Other newly isolated secondary products obtained from *P. microspora* (endophytic to *T. brevifolia*) include new caryophyllene sesquiterpenes (pestalotiopsins A and B) (Pulici *et al*., 1996).

A newly described fungus, *Estalotiopsis jesteri* (from the Sepik River area of Papua New Guinea) produces jesterone and hydroxy-jesterone, which exhibited antifungal activity against a variety of fungal plant pathogens (Li and Strobel, 2001). *Phomopsichalasin*, a metabolite of endophytic *Phomopsis* sp. represents the first cytochalasin-like compound with a three ring system replacing the cytochalasin macrolide ring. This metabolite mainly exhibits antibacterial activity against *Bacillus subtilis, Salmonella enterica* and *Staphylococcus aureus.* It also displays a moderate activity against *Candida tropicalis* (Horn *et al.*, 1995). On screening the endophytic *Fusarium* sp. of plant *Selaginella pallescens* in the Guanacaste Conservation Area of Costa Rica revealed antifungal activity: a new pentaketide CR377 isolated from the culture broth exhibited potent activity against *C. albicans* (Brady and Clardy, 2000). Colletotric acid is a metabolite of endopohytic *Colletotrichum gloeosporioides* of *Artemisia mongolica* and displayed an inhibitory activity against bacteria as well as the fungus (*Helminthsporium sativum*) (Zou *et al.*, 2000). Another endopohytic *Colletotrichum* sp. isolated from *Artemisia annua* produced metabolites with a varied antimicrobial activity. The traditional Chinese herb, *A. annua* is well recognized for its synthesis of artemisinin and it has the ability to serve as anti-malarial drug in many geographically different areas. The *Colletotrichum* sp. of *A. annua* produced not only metabolites with activity against human pathogenic fungi and bacteria, but also fungistatic compounds against plant pathogenic fungi (Lu *et al.*, 2000).

Muscodor albus, endophytic to *Cinnamomum zeylanicum* effectively inhibits and kills certain fungi and bacteria through a mixture of volatile compounds (Strobel *et al.*, 2001), which collectively act synergistically to cause death to a wide range of plant and human pathogenic fungi and bacteria. The most effective class of inhibitory compounds was the esters and amongthem isoamyl acetate was the most biologically potent. The ecological implications and potential benefits of the 'mycofumigation' effects of *M. albus* are highly promising. Endophytic *Gliocladium* sp. is also a producer of a volatile antibiotic compound. In fact, the most abundant volatile inhibitor is annulene, formerly used as a rocket fuel, which was discovered for the first time as a natural product of endophytic fungus (Stinson *et al.*, 2003).

Anticancer agents

Some of the most commonly found endophytic fungi of yews (*Taxus wallichiana*) are *Pestalotiopsis* spp. and *P. microspora* that are most abundant. The discovery of taxol-producing fungus, *Taxomyces andeanae*, an endophyte of *Taxus brevifolia* by Stierle *et al.* (1993) paved the way to produce this drug by employing large-scale fungal fermentation to generate taxol at a low cost with a high yield. An examination of the endophytes of *T. wallichiana* yielded *P. microspora* and a preliminary monoclonal antibody test revealed that it might produce paclitaxel (Strobel *et al.*, 1996). Paclitaxel, a highly functionalized diterpenoid found in yew trees is useful to treat a number of human tissue-proliferating diseases like cancer. After several years of effort, a novel paclitaxel producing endophytic fungus, *T. andreanae* was discovered in *T. brevifolia* (Strobel *et al*., 1993). The most critical line of evidence for the presence of paclitaxel in the culture fluids of this fungus was the electrospray mass spectrum of the putative paclitaxel isolated from *T. andreanae.* From the extremely rare and previously thought to be extinct wollemi pine (*Wollemia nobilis*), *Pestalotiopsis guepini* was isolated and shown to produce paclitaxel (Strobel *et al*., 1997). A rubiaceous plant, *Maguireothamnus speciosus* yielded a novel endophytic fungus, *Seimatoantlerium tepuiense* that also produces paclitaxel. Further, fungal paclitaxel production has also been noted in a *Periconia* sp. and in *Seimatoantlerium nepalense* (Li *et al*., 1998).

Torreyanic acid, a selectively cytotoxic quinone dimer (anticancer agent) was isolated from a strain of *P. microspora* endophyte of an endangered tree *T. taxifolia* (Lee *et al*., 1995). Torreyanic acid was tested on several cancer cell lines and demonstrated that it is 5-10 times more potent on those cell lines that are sensitive to protein kinase C agonists and cause cell death by apoptosis. Fungi like *Xylaria*, *Phoma*, *Hypoxylon* and *Chalara* produce cytochalasins, which possesses antitumor and antibiotic activities (Wagenaar *et al*., 2000). Three novel cytochalasins with antitumor activity have recently been reported from a *Rhinocladiella* sp. endophytic to *Tripterygium wilfordii* (Wagenaar *et al*., 2000).

Pharmaceutically-valued compounds

Although fungal secondary metabolites are a good source of antiviral compounds, research on these lines is limited. The fungus, *Cytonaema* sp. produces two important antiviral compounds, which are protease inhibitors (inhibit the human cytomegalovirus) (Guo *et al*., 2000). A non-peptide fungal metabolite (L-783,281) isolated from an endophytic fungus (*Pseudomassaria* sp.) and isolated from an African rainforest acts as an insulin (Zhang *et al*., 1999). The *P. microspora* endophytic to *Terminalia morobensis* in Papua New Guinea produces pestacin and isopestacin that have antimicrobial and antioxidant properties. Isopestacin scavenges superoxide and hydroxyl free radicals in a solution (Strobel *et al*., 2002). Pestacin also possesses potent antioxidant activity at least one order magnitude greater than trolox, a vitamin E derivative (Harper *et al*., 2003). The endophytic fungus, *Fusarium subglutinans* isolated from *T. wilfordii* produced immunosuppressive noncytotoxic diterpene pyrones, subglutinol A and B (Lee *et al*., 1995).

Genetic Engineering

Genetic engineering and genetic modification of endophytes is one of the interesting areas gaining importance. There are a few examples of the application of genetic engineering to the area of endophytic fungi. By transplanting symbiotic fungi from drought-tolerant or heat-tolerant plants into crops, researchers have turned wimpy species into hardy ones. It has been shown that fungi 'can have a big impact in adapting the plant to a particular environment'. Exposing some grasses, specially those that lacked the normal complement of endophytic fungi and those devoid of it to hot soils resulted in the survival of only the grass with endophytic fungi (Redman *et al.,* 2002). It has been shown that isolated fungi have the ability to overcome heat-stress suggesting that fungi and its host deal better with stressful environment. Other investigations have indicated that fungi help plants wardoff invading pathogens. With fungi, plants mount a defense response within 24 hours, while plants devoid of endophytes require three days. It has been demonstrated that the same fungi can protect other plants on transferring fungal spores on watermelon, tomato, wheat, and other seedlings. On comparing the stamina of infected and uninfected plants in places with heat and water stress where they would not typically survive, watermelon and tomatoes, whose roots would barely withstand 38°C were not fine at 50°C, on cool off at night, some of the plants survived even up to 70°C in the day. Uninfected wheat succumbed within 10 dry days, while wheat carrying the fungi lasted for 18 days without water.

Conclusions and Outlook

Great diversity has been found among endophytic fungi that are isolated from various plant species. The role of endophytes in the protection of plants against insects and diseases has been well studied. Recent evidences indicate that in addition to increasing plant resistance to herbivores, endophytic fungi may also enhance plant competitiveness, germination success, resistance to drought and water stress. However, research is based mainly on endophytes from temperate hosts and studies have been concentrated on some genera of grasses. Apart from isolated studies, more recently, efforts are being directed to endophytes of tropical plant hosts. Endophytes play important roles such as protecting plants against pests and environmental stresses. Even though enormous studies have been carried out in discovering the endophytes and thereby, exploiting them for the production of secondary metabolites, this important resource still remains largely untapped. Endophytic microorganisms are a major source of compounds of biotechnological potential as antibiotics and antitumor drugs. In recent years, there is a great demand for new antibiotics, chemotherapeutic agents and agrochemicals those are highly effective, less toxic, and have a minimal environmental impact. The development of resistance by the pathogens to the existing drugs and the evolution of novel diseases have necessitated the discovery and development of new drugs. It is expected that new drugs of biotechnological potential produced by endophytes will be discovered with enhanced studies focusing on this group of microorganisms. Further research at the molecular level in this area is clearly required for a better understanding of the host-endophyte interaction and to judiciously harness this rich and reliable source of bioactive and chemically novel compounds with vast medicinal and agricultural potential.

References

Arai, T., Mikami, Y., Fukushima, T., Utsumi, T. and Yazawa, K. 1973. A new antibiotic, leucinostatin, derived from *Penicillium lilacium*. *J. Antibiot. 28*, 157-161.

Arechavaleta, M., Bacon, C.W., Hoveland, C.S. and Radclife, D.E. 1989. Effects of tall fescue endophyte on plant response to environmental stress. *Agron. J. 81*, 83-90.

Arnold, A.E., Maynard, Z., Gilbert, G.S., Coley, P.D. and Kursar, T.A. 2000. Are tropical endophytes hyperdiverse? *Ecol. Lett. 3*, 267-274.

Bacon, C.W. and White, J.F. 2000. *Microbial endophytes*. Marcel Dekker Inc., New York.

Bacon, C.W., Porter, J.K., Robins, J.D. and Luttrell, E.S. 1977. *Epichloë typhi* from toxic tall fescue grasses. *Appl. Environ. Microbiol. 34*, 576-581.

Benz, F., Knüsel, F., Nüesch, J., Treichler, H., Voser, W., Nyfeler, R. and Keller-Schierlein, W. 1974. Stoffwechselprodukte von mikroorganismen echinocandin B, ein neuartiges polypeptid-antibioticum aus *Aspergillus nidulans* var. *echinulatus*: Isolierung und bausteine. *Helvetica Chimica Acta 57*, 2459-2477.

Bouarab, K., Potin, P., Correa, J. and Kloareg, B. 1999. Sulfated oligosaccharides mediate the interaction between a marine red alga and its green algal pathogenic endophyte. *Plant Cell 11*, 1635-50.

Brady, S. F. and Clardy, J. 2000. CR377, a new pentaketide antifungal agent isolated from an endophytic fungus. *J. Nat. Prod. 63*, 1447-1448.

Bush, L.P., Wilkinson, H.W. and Schardl, C.L. 1997. Bioprotective alkaloids of grass-fungal endophyte symbioses. *Plant Physiol. 114*, 1-7.

Calhoun, L.A., Findrlay, J.A., Miller, J.D. and Whitney, N.J. 1992. Metabolites toxic to spruce budworm from balsam fir needle endophytes. *Mycol. Res. 96*, 281-286.

Carroll, G. 1988. Fungal endophytes in stems and leaves: from latent pathogens to mutualistic symbionts. *Ecology 69*, 2-9.

Chapela, I.H., Petrini,O. and Hagmann, L. 1991. Monolignol glucosides as specific recognition messengers in fungi and plant symbiosis. *Physiol. Mol. Pl. Pathol. 39*, 289-298.

Cheplick, G.P. and Clay, K. 1988. Acquired chemical defenses in grasses: The role of fungal endophytes. *Oikos 52*, 309-318.

Christensen, M. J. 1996. Antifungal activity in grasses infected with *Acremonium* and *Epichloe* endophytes. *Aust. Pl. Pathol. 25*, 186-191.

Clay, K. 1988. Fungal endophytes of grasses: A defensive mutualism between plants and fungi. *Ecology 69*, 10-16.

Clay, K. and Holah, J. 1999. Fungal endophyte symbiosis and plant diversity in successional fields. *Science 285*, 1742-1744.

Daisy, B. H., Strobel, G.A., Castillo, U., Ezra, D., Sears, J., Weaver, D.K. and Runyon, J.B. 2002. Naphthalene, an insect repellent, is produced by *Muscodor vitigenus*, a novel endophytic fungus. *Microbiol. 148*, 3737-3741.

De Battista, J.P., Bouton, J.H., Bacon, C.W. and Siegel, M.R. 1990. Rhizome and herbage production of endophyte removed tall fescue clones and populations. *Agron. J. 82*, 6521-6654.

Dew, R.K., Boissonneault, G.A., Gay, N., Boling, J.A., Cross, R.J. and Cohen, D.A. 1990. The effect of the endophyte *(Acremonium coenophialum)* and associated toxin(s) of tall fescue on serum titer response to immunization and spleen cell flow, cytometry analysis and response to mitogens. *Vet. Immunol. Immunopathol. 26*, 285-295.

Faeth, S.H. and Bultman, T.L. 2002. Endophytic fungi and interactions among host plants, herbivores and natural enemies. In: *Multitrophic Level Interactions* (Ed. Tscharntke, T. and Hawkins, B.A.). Cambridge University Press, 89-123.

Findlay, J.A., Bethelezi, S., Li, G. and Sevek, M. 1997. Insect toxins from an endophyte fungus from wintergreen. *J. Nat. Prod. 60*, 1214-1215.

Gasoni, L. and Stegman de Gurfinkel, B. 1997. The endophyte *Cladorrhinum forcundissimum* in cotton roots: phosphorus uptake and host growth. *Mycol. Res. 101*, 867-870.

Guo, B., Dai, J., Ng, S., Huang, Y., Leong, C., Ong, W. and Carte, B.K. 2000. Cytonic acids A and B: novel tridepside inhibitors of hCMV protease from the endophytic fungus *Cytonaema* species. *J. Nat. Prod. 63*, 602-604.

Hallmann, J. and Sikora, R.A. 1996. Toxicity of fungal endophyte secondary metabolites to plant-parasitic nematodes and soil-borne plant-pathogenic fungi. *Eur. J. Plant Pathol. 102,* 155-162.

Harper, J.K., Ford, E.J., Strobel, G.A., Arif, A., Grant, D.M., Porco, J., Tomer, D.P. and Oneill, K. 2003. Pestacin: a 1,3-dihydro isobenzofuran from *Pestalotiopsis microspora* possessing antioxidant and antimycotic activities. *Tetrahedron 59,* 2471-2476.

Hata, K, Futai K. and Tsuda, M. 1998. Seasonal and needle age-dependent changes of the endophytic mycobiota in *Pinus thunbergii* and *Pinus densiflora* needles. *Can. J. Bot. 76,* 245-250.

Holmes, K.A., Evans, H. C. and Pomella, A. 2002. Insider knowledge: Endophytes for cacao disease biocontrol. *Biocontrol News and Information 23,* 86-88.

Horn, C.P., Dalal, R.C., Birch, C.J. and Doughton, J.A. 1996. Sowing time and tillage practice affect chickpea yield and nitrogen fixation. II. Nitrogen accumulation, nitrogen fixation and soil nitrogen balance. *Aust. J. Exp. Agric. 36,* 701-706.

Horn, W.S., Simmonds, M.S.J., Schwartz, R.E. and Blaney, W.M. 1995. Phomopsichalasin, a novel antimicrobial agent from an endophytic *Phomopsis* sp. *Tetrahedron 14,* 3969-3978.

Ju, Y., Sacalis, J.N., and Still, C.C. 1988. Bioactive flavonoids from endophyte-infected blue grass (*Poa ampla*). *J. Agric. Food Chem. 46,* 3785-3788.

Kanda, K., Hirai, Y., Koga, H. and Hasegawa, K. 1994. Endophyte-enhanced resistance in perennial ryegrass and tall fescue to bluegrass webworm. *Jap. J. Appl. Entomol. Zool. 38,* 141-145.

Koshino, H., Yoshihara, T., Sakamura, S., Shimanuki, Y., Sato, T. and Tajimi, A. 1989. A ring B aromatic sterol from stromata of *Epichloe typhina. Phytochem. 28,* 771-772.

Koshino, H., Terada, S., Yoshihara, T., Sakamura, S., Shimanuki, T., Sato, T. and Tajimi, A. 1988. Three phenolic acid derivates from stromata of *Epichloë typhina* on *Phleum pratense. Phytochem. 27,* 1333-1338.

Lappalainen, J.H. and Yli-Mattila, T. 1999. Genetic diversity in Finland of the birch endophyte *Gnomonia setacea* as determined by RAPD-PCR markers. *Mycol. Res. 103,* 328-332.

Lee, J.C., Yang, X., Schwartz, M., Strobel, G. and Clardy, J. 1995. The relationship between an endangered North American tree and an endophytic fungus. *Chem. Biol. 2,* 721–727.

Lee, J., Lobkovsky, E., Pliam, N.B., Strobel, G.A. and Clardy, J. 1995.Subglutinols A and B: immunosuppressive compounds from the endophytic fungus *Fusarium subglutinans. J. Org. Chem. 60,* 7076-7077.

Li, J. Y. and Strobel, G.A. 2001. Jesterone and hydroxy-jesterone *antioomycetcyclohexenenone* epoxides from the endophytic fungus *Pestalotiopsis jesteri. Phytochem. 57,* 261-265.

Li, J.Y., Harper, J.K., Grant, D.M., Tombe, B.O., Bashyal, B., Hess, W.M. and Strobel, G. A. 2001. Ambuic acid, a highly functionalized cyclohexenone with antifungal activity from *Pestalotiopsis* spp. and *Monochaeti*a sp. *Phytochem. 56,* 463-468.

Li, J. Y., Sidhu, R.S., Ford, E., Hess, W.M. and Strobel, G. A. 1998. The induction of taxol production in the endophytic fungus *Periconia sp.* from *Torreya grandifolia. J. Ind. Microbiol. Biotechnol. 20,* 259-264.

Li, J.Y., Strobel, G.A., Harper, J.K., Lobkovsky, E. and Clardy, J. 2000. Cryptocin, a potent tetramic acid antimycotic from the endophytic fungus *Cryptosporiopsis cf. quercina. Org. Lett. 2,* 767-770.

Li, J.Y., Strobel, G.A., Sidhu, R., Hess, W.M. and Ford, E. 1996. Endophytic taxol producing fungi from Bald Cypress *Taxodium distichum. Microbiol. 142,* 2223-2226.

Lu, H., Zou, W.X., Meng, J.C., Hu, J., and Tan, R.X. 2000. New bioactive metabolites produced by *Colletotrichum* sp., an endophytic fungus in *Artemisia annua. Pl. Sci. 151,* 67-73.

Lumyong, S., Lumyong, P., McKenzie, E.H.C., and Hyde, K.D. 2002: Enzymatic activity of endophytic fungi of six native seedling species from Doi Suthep-Pui National Park, Thailand. *Can. J. Microbiol. 48,* 1109-1112.

Malinowski, D.P. and Belesky, D.P. 1999. *Neotyphodium coenophialum*-endophyte infection affects the ability of tall fescue to use sparingly available phosphorus. *J. Pl. Nutr. 22,* 835-853.

McCutcheon, T.L., Caroll, G.C. and Schwab, S. 1993. Genotypic diversity in populations of a fungal endophyte from Douglas fir. *Mycologia 85,* 180-186.

Paracer, S., and Ahmadjian, V. 2000. Symbiosis: *An Introduction to Biological Associations.* 2nd Edition, Oxford University Press, Oxford.

Frontiers in Fungal Ecology, Diversity and Metabolites (edited by K.R. Sridhar), I.K. International Publishing House Pvt. Ltd., New Delhi, 2009.

13

Identity and Toxigenic Potentiality of *Fusarium* spp. Colonizing *Sorghum*

A.H. Rajasab* and S. Mohammed Shabbir
Mycology and Plant Pathology Laboratory, Department of Botany, Gulbarga University, Gulbarga, 585 106, Karnataka, India
*(*E-mail: rajasab55@rediffmail.com)*

Introduction

Grain mold, ergot, loose smut, grain smut and head smut are common diseases on many varieties of sorghum. *Fusarium* species are found to cause diseases from seed to senescence and recognized as one of the major grain mold pathogens of sorghum. They cause economic losses and affect the health of animals including that of human beings. Disease surveys have shown that the hybrid varieties of sorghum although bred for disease resistance, sometimes become more susceptible to grain molds, smut and ergot (Rajasab and Frederiksen, 2001). Sundaram (1977) and Ravindranath (1980) reviewed sorghum disease situation in India, while Anahosur (1992) reported regional distribution of diseases and their causal organisms. In all, thirteen *Fusarium* spp. are reported to colonize sorghum grown in different parts of the world, whereas in India eight species are reported. *Fusarium* colonization of grain sorghum induces a variety of symptoms resulting in different phases of diseases (Table 1). *Fusarium*-induced diseases include: seed rot, seedling blight, root rot, stalk rot, Pokkah boeng, pre-emergence ear rot, head blight and grain mold (Frederiksen *et al.*, 1982; Rajasab and Maribhat, 1989; Singh and Bandyopadhyay, 2000; Williams and Rao, 1981). This review presents information on taxonomic identity and toxigenic potency of *Fusarium* spp. colonizing sorghum. Information on the identity of *Fusarium moniliforme*, which does not exist any more, is reviewed. The existing biological species within *Gibberella fujikuroi* species complex (anamorphs and teleomorphs) are discussed including a commentary on toxins produced by fusaria.

Table 1. List of *Fusarium* species, their sexual stages and the parts of sorghum that they are associated with (Source: Leslie, 2000; Marasas *et al.*, 2001a)

***Fusarium* spp.**	**Sexual stage**	**Comments**
F. equiseti	*Gibberella intricans*	Associated with roots and stalk rots.
F. graminearum	*G. zeae*	Homothallic, associated with seedling blight, root and stalk rot, grain mold, storage mold and mycotoxicoses. More common in cool, wet sorghum growing areas.
F. verticillioides	*G. fujikuroi* Mating population A	The most commonly identified *Fusarium* species from sorghum. Associated with seedling blight, root and stalk rot, grain mold, storage rots and mycotoxicoses. Distributed throughout sorghum growing areas.
F. napiforme	None known	Newly described species commonly associated with sorghum. Distribution limits are not yet known.
F. nygamai	*G. nygamai* Mating population G	Newly described species, commonly associated with sorghum. Distribution limits are not yet known, but widespread in Africa and Australia.
F. oxysporum	None known	A very common and widely dispersed soil fungus. Probably a complex; not a single species. Reported to colonize sorghum root and stalk tissue, but pathogenic association with sorghum has not been proven.
F. proliferatum	*G. fujikuroi* Mating population D	Associated with seedling blight, Pokkah boeng, grain mold, root and stalk rot, storage diseases and mycotoxicoses.
F. semitectum (*F. pallidoroseum*)	None known	It has not been studied as extensively as many of the other *Fusarium* spp. listed here. Associated with grain mould, root and stalk rot.
F. solani	*Nectria haematococea*	A very common soil fungus. A complex; not a single species. Reported to colonize sorghum root and stalk tissue, but pathogenic association with sorghum has not been proven.
F. subglutinans	*G. fujikuroi* Mating population B, E and H.	A species complex containing at least three and probably more different biological species. Associated with root and stalk rots.
F. thapsinum	*G. thapsina* Mating population F	Recently distinguished from *F. moniliforme sensu lato,* and perhaps the most common *Fusarium* spp. on sorghum. Associated with seedling blight, root and stalk rot, grain mold, storage mold and *mycotoxicoses.*
F. tricinctum	*G. indica*	Reported to colonize sorghum root and stalk tissue, but pathogenic association with sorghum has not been proven.
F. andiyazi	None known	Recently distinguished from *F. moniliforme sensu lato,* through molecular technique (AFLP), found to cause seedling blight.

Fusaria and Sorghum

Fusarium moniliforme Sheldon is the name given to all isolates obtained from different hosts based on the production of microconidia in chains. However, they belong to diverse species because many species of *Fusarium* produce microconidia in chains (Leslie and Marasas, 2002) (Table 2). Based on the data of microconidial production in chains by fusaria, Leslie and Marasas (2002) and Summerell *et al.* (2003) suggested that most isolates obtained from sorghum with the characteristic microconidial chains would belong to one of the five species (*Fusarium andiyazi, F. nygamai, F. proliferatum, F. thapsinum* and *F. verticillioides*). A researcher involved in fusarial study on sorghum has to resolve to respective biological species instead of clumping whole isolates under a common name as *F. moniliforme*. If such an assignment is impossible, the strain should be termed *F. moniliforme sensu lato*, which clearly indicates that an accurate diagnosis to species level was not made thereby alerting others that the strains described may belong to one of the several species of fusaria (Summerell *et al.*, 2003).

Table 2. *Fusarium* species that form microconidia in chain and might be identified as *F. moniliforme* (Source: Leslie and Marasas, 2002)

Anamorph *Fusarium* spp.	Teleomorph *Gibberella* spp.
F. andiyazi	None
F. annulatum Bugnicourt	None
F. brevicatenulatum	None
F. fujikuroi	*G. fujikuroi*; *G. fujikuroi* mating population C
F. globosum	None
F. lactis	None
F. napiforme	None
F. nygamai	*G. nygamai*; *G. fujikuroi* mating population G
F. phyllophilum	None
F. proliferatum	*G. intermedia*; *G. fujikuroi* mating population D
F. pseudocircinatum	None
F. pseudonygamai	None
F. thapsinum Klittich *et al.*	*G. thapsina* Klittich *et al.*; *G. fujikuroi* mating population F
F. verticillioides (Sacc.) Nirenberg	*G. moniliformis*; *G. fujikuroi* mating population A

Fusarium spp. are known to be associated with sorghum from seed to senescence. Among them *F. moniliforme* is the name that has traditionally been used for the various isolates of *Fusarium* from the *Liseola* section. These isolates were recovered from the sorghum stalk and grain causing stalk rot and grain mold. Booth (1971) recognized *F. moniliforme* based on conidiogenesis and the presence of polyphialides. Hsieh *et al.* (1977) depending on the vegetative compatibility grouping, divided *F. moniliforme* into three different heterothallic mating groups. During subsequent years, based on biological and phylogenetic species concept, various investigators divided *Gibberella fujikuroi* into nine distinct mating populations (Britz *et al.*, 1999; Kathariou, 1981; Klaasen and Nelson, 1996; Klittich and Leslie, 1992; Kuhlman, 1982; Leslie, 1991; Nelson *et. al.*, 1983; Nirenberg and O'Donnell, 1998; Wineland, 1924; Zeller *et al.*, 2003). These mating populations were designated with alphabetic epithets A-I and anamorphic and teleomorphic names were given for a majority of these populations (Table 3). During the past ten years, tremendous progress has been made to resolve the identity of *F. moniliforme*. The name *F. moniliforme* was given to the isolates belonging to the mating population A of *G. fujikuroi* species complex have been retired for conformity with ICBN and *F. verticillioides* Nirenberg has been assigned (Leslie and Marasas, 2002). Mating population A is common on maize and mating population F in common on sorghum and termed *F. thapsinum* (Klittich *et al.*, 1997). They differ in their ability to produce mycotoxins.

Table 3. Biological species within *Gibberella fujikuroi* species complex.

Mating Population	Anamorph	Teleomorph	Reference (teleomorph)
A	*F. verticillioides*	*G. moniliformis*	Wineland, 1924
B	*F. sacchari*	**G. subglutinans*	Nelson *et al.*, 1983
C	*F. fujikuroi*	*G. fujikuroi*	Nirenberg, 1976
D	*F. proliferatum*	*G. intermedia*	Kuhlman, 1982
E	*F. subglutinans*	*G. subglutinans*	Nelson *et al.*, 1983
F	*F. thapsinum*	*G. thapsina*	Klittich *et al.*, 1997
G	*F. nygamai*	*G. nygamai*	Klaasen and Nelson, 1996
H	*F. circinatum*	*G. circinata*	Nirenberg and O'Donnell, 1998
I	*F. konzum*	*G. konza*	Zeller *et al.*, 2003

* - Indicates that name has not been given.

Fusarium spp. cause a range of diseases in rice, sugarcane, maize and sorghum (Booth, 1971). In sorghum, seedling blight, Pokkah boeng, grain mold, seed, root and stalk rot, and storage diseases are common (Frederiksen and Odvody, 2000). Presently, 13 species of *Fusarium* are reported on sorghum (Leslie, 2000; Marasas *et al.*, 2001a) (Table 1). Extensive surveys were carried out by Onyike *et al.* (1991) and Onyike and Nelson (1992, 1993) on *Fusarium* species associated with sorghum and millets as well as on soil from Nigeria, Lesotho and Zimbabwe. Isolates of *F. nygamai* obtained from millet grain and soil planted with millets and sorghum showed differences in the length of microconidial chains. Klaasen and Nelson (1998) resolved them into two groups based on mating tests by genetic studies. The first group with short microconidial chain and smooth or rough-walled chlamydospores were identical to *F. nygamai* as described by Burgess and Trimboli (1986). The second group, with long microconidial chain resembled *F. moniliforme*, mating population A and production of smooth-walled thickened hyphae were described as over-wintering structures by Nyvall and Kommedahl (1968) in corn stalks.

Burgess and Trimboli (1986) described *F. nygamai* in association with sorghum grain in Australia. Later, Marasas *et al.* (1987, 1988) described *F. napiforme* on millet and sorghum and *F. nygamai* on millet from South Africa. The *F. napiforme* is distinguished from *F. nygamai* by the presence of napiform to globose microconidia in chains. Taxonomically, *F. nygamai* and *F. napiforme* are of considerable interest because these are intermediate between the sections

Eligans and *Liseola* as defined by Wollenweber and Reinking (1935). They cannot be included in *Liseola* as they produce chlamydospores and are excluded from *Eligans* as they produce microconidia in chains (Klaasen and Nelson, 1998). Leslie *et al.* (1990) surveyed large areas of central and eastern US and reported eight species of fusaria from plant tissue, soil debris, and soil collected from 18 sorghum fields. Among them, *F. moniliforme* was predominant in tissues and soil debris and was absent in soil samples. Mansuetus *et al.* (1997) reported seven mating populations (A-F), the biological species in *G. fujikuroi* species complex belonging to section *Liseola* from sorghum grain of three sites in Tanzania. The frequency of the different mating populations was similar on all three sites with A and F the most frequent, C and E the least, and F was predominant mating population. Similarly, Klittich and Leslie (1992) and Leslie (1991) recovered all the six mating populations of *G. fujikuroi* from corn as well as sorghum in the US. They found A population prevalent on corn, while F on sorghum. The predominance of population A in sorghum and millet grains in the African countries have been reported by Klaasen and Nelson (1998).

Klittich *et al.* (1997) erected a new species, *F. thapsinum* in section *Liseola* belonging to the mating population F in *G. fujikuroi* species complex recovered from sorghum in Egypt, South Africa, Philippines, Thailand, and nine states of the US. It produces characteristic diffusing yellow pigment on PDA and is distinguishable by other characters such as isozyme polymorphism, electrophoretic karyotype, production of mycotoxin, Benomyl sensitivity and difference in the sequence of the internally transcribed spacer (ITS) region of the rDNA repeat. Marasas *et al.* (2001a) described a new species *F. andyazi* in section *Liseola* based on AFLP technique from sorghum in Africa and the US. This species is distinguished morphologically by the production of unique pseudochlamydospores in carnation leaf pieces. Sexual stages were not observed in laboratory conditions. From India, Bhale and Khare (1982), Gopinath and Shetty (1985) and Mahalinga *et al.* (1988) have observed four species of fusaria. Later, Mohammed and Rajasab (2004) isolated additional four species of *fusaria* (*F. compactum, F. nygamai, F. oxysporum, F. proliferatum, F. scirpi, F. semitectum, F. solani* and *F. verticillioides*) from sorghum grain. Among these, perhaps *F. scirpi* on sorghum seeds was not reported elsewhere.

Toxins

Toxins of *Fusarium* have been studied for over 60 years with the first isolation of low molecular weight wilt toxins. By the 1960s, naphthazarin toxins, cyclic peptide toxins, *trichothecenes* and *zearalenones* have been isolated from *Fusarium* and elucidated their chemical structure. The development of new analytical tools in 1960s and 1970s (NMR and MS) facilitated rapid and efficient identification of the compounds of fungi. As a result, a large and extraordinarily diverse group of toxins and biologically active secondary metabolites have been identified from fusaria. Different species of *Fusarium* have the potential to produce a variety of sets of toxins (profile of mycotoxins) and these have been reviewed by De Nijs *et al.* (1996), Marasas *et al.* (1984), Stoessl (1981), Summerell *et al.* (2001) and Thrane (1989). Eleven classes of toxins produced by thirty species of fusaria are presented in Table 4 (Desjardins and Proctor, 2001). Fusarial toxins are grouped in sections according to their biogenetic origins (Desjardins and Proctor, 2001; Stoessl, 1981).

Table 4. Selected toxins produced by *Fusarium* spp. affecting different crops (Source: Desjardins and Proctor, 2001)

Fusarium spp.	Fumonisin	Fusaric acid	Fusarin	Moniliformin	Naphthazarin	Sambutoxin	Zearalenone	Fusaproliferin	Trichothecenes	Cyclic peptides	Fusarochrome
F. moniliforme MP-A	●	●	●	◉	●	○		○		○	○
F. napiforme	◉	●		●							
F. thapsinum MP-F	◉	●		●				○		○	
F. nygamai MP-G	●	●		●							
F. sacchari MP-B	◉	●						○		●	
F. fujikuroi MP-C	●	●		●				○		●	
F. proliferatum MP-D	●	●	●	●				●		●	○
F. dlamini	◉										
F. subglutinans MP-E	◉	●	○	●							
F. anthophilum	◉			●							
F. oxysporum complex	◉	●		●							
F. beomiforme	○			●							
F. lateritium	○								◉	●	
F. sambucinum	○	●	●			●	○		●	●	
F. venenatum									●	○	
F. camptoceras	○								●		
F. poae	○		●			○			●		
F. sporotrichioides	○		●			○			●		○
F. acuminatum	○		○	●		○			●	●	
F. graminearum	○		●			○	●		●		○
F. culmorum			●			○	●		●		
F. crookwellense		●	●				●		●	○	
F. heterosporum		●	◉				●				
F. avenaceum	○		●	●						●	○
F. semitectum	○		○	●		●					○
F. equiseti	○		○	●		○	●		●	○	●
F. scirpi	○									●	
F. tricintum			●						○	●	○
F. torulosum									○	●	
F. solani complex		●	○	○	●	○	○		◉	○	

● – Majority of strains within a species produce toxin
◉ – Minority of strains within a species produce toxin
○ – Strains within a species have been tested, but toxin producers have not been found

Polyketides

Fumonisins: The fumonisins are a group of recently discovered mycotoxins commonly found in corn worldwide. It was first reported in 1988 by South

Africa while investigating the association between human oesophageal cancer and the consumption of maize infected with *F. moniliforme* (Bezuidenhout *et al.*, 1988). Fumonisin B_1 (FB_1) is most abundant in maize, whereas FB_2, FB_3 and FB_4 occur at lower levels (Desjardins and Proctor, 2001). *Fusarium anthophilum, F. dlamini, F. globosum, F. moniliforme* MP-A, *F. napiforme, F. nygamai, F. proliferatum* and *F. thapsinum* have been reported to produce fumonisins in cultures (Desjardins and Proctor, 2001; Marasas *et al.*, 2001b). Along with these, unconfirmed reports on *Alternaria alternata* (Chen *et al.*, 1992), *F. oxysporum* (Abbas and Ocamb, 1995b) and *F. polyphialidicum* (Abbas and Ocamb, 1995a) for the production of fumonisin are also available. *Fusarium moniliforme* and *F. proliferatum* are the most common fumonisin B_1 producing fungi associated with maize (Shephard *et al.*, 1996a). Fumonisins have been associated with a number of mycotoxicoses including equine leucoencephalomalacia, porcine pulmonary edema, and experimental liver cancer in rats (Bhat *et al.*, 1997; Riley *et al.*, 1996). Fumonisins (B_1+FB_2+FB_3) with mean concentrations ranging from 11-140 and 13-130mg/kg feed are associated with equine leucoencephalomalacia and porcine pulmonary edema respectively (Shephard *et al.*, 1996b). See Chapter 14 for detailed discussion on fumonisins produced by fusaria.

Fusaric acid: It is one of the most common mycotoxins produced by several species of fusaria including *F. crookwellense, F. heterosporum, F. napiforme, F. nygamai, F. oxysporum, F. proliferatum, F. sambucinum F. solani F. subglutinans* and *F. verticillioides*. Although fusaric acids were first discovered as wilt-inducing toxins in the 1930s, their importance in the development of wilt disease is still unclear (Scheffer, 1983). It is the major toxin associated with plants and grains (Bacon *et al.*, 1996; Porter *et al.*, 1995). *Fusaric* acid was originally reported as a phytotoxin (Gaumann, 1957) and other investigations established it as a hypotensive agent on several mammals (Hidaka *et al.*, 1969). Although it is not acutely toxic, it enhances the activity of fusarial toxins with other toxins synergistically (Bryden *et al.*, 2001).

Fusarins: Fusarins were identified in 1981 from *F. moniliforme* as a result of its mutagenic activity to *Salmonella typhimurium* (Wiebe and Bjeldanes, 1981). A wide range of fusaria produce fusarin C and the related fusarins *A, D* and F (Desjardins and Proctor, 2001). Among them, fusarin C has been found to be elaborated by *F. avenacum, F. crookwellense, F. culmorum, F. graminearum, F. oxysporum, F. poae, F. sambucinum, F. sporotrichioides, F. tricintum* and *F. verticillioides* (Bryden *et al.*, 2001). Fusarin C is a potential contaminant of food. Its effect on animal health and productivity may be diminished because of the instability and the need for the host to activate fusarin *C* to more potent mutagenic compounds (fusarin X and Z) (Bryden *et al.*, 2001).

Moniliformin: It was first isolated from *F. moniliforme* (Cole *et al.*, 1973). Subsequesntly, the moniliformin production has been reported from *F. acuminatum, F. anthophilum, F. avenaceum, F. concolor, F. culmorum, F. equiseti, F. fusarioides (F. chlamydosporum), F. graminearum, F. moniliforme* var. *subglutinans* (*F. subglutinans*), *F. nygamai*, *F. oxysporum*, *F. proliferatum*, *F. reticulatum F. semitectum*, *F. sporotrichioides*, and *G. fujikuroi* (Bryden *et al.*, 2001). *Gibberella fujikuroi* mating types A and F shared the *F. moniliforme* anamorph and can be differentiated based on the production of moniliformin. The *G. fujikuroi* produces moniliformin more often at higher levels than the A mating population (Leslie *et al.*, 1996). Moniliformin is known to cause toxicity and death in poultry and rodents (Engelhardt *et al.*, 1989). Keshan, a serious endemic disease in some regions of China, especially in the mountainous regions and marshy lands is attributed to *F. moniliforme* (Chen *et al.*, 1990; Kriek *et al.*, 1977; Zang and Li, 1989).

Zearalenones: Zearalenones are a group of estrogenic metabolites first identified and characterized during 1966 (Hidy *et al.*, 1966). Subsequently, extensive studies have been conducted and many reviews have been published (Betina, 1984, 1989; Hagler *et al.*, 2001; Hidy *et al.*, 1977; Mirocha and Christensen, 1974; Mirocha and Pathre, 1979; Pathre and Mirocha, 1976). Zearalenone, a non-steroidal oestrogen and its major metabolites, α-zearalenone and β-zearalenone elicit significant oesterogenic activity in animals, corresponding to their binding affinities for hepatic, uterine, mammary and hypothalamus oestrogen receptors (Coulombe, 1993). In addition, over hundreds of compounds of the zearalenone family were screened that exhibited a wide range of pharmacological properties (e.g., anabolic, corticoid, lipodiatic, mineralcorticoid, antibacterial, antifungal, antioestrogenic, antigonadotrophic, antiosteoporetic, antifertility) (Hurd, 1977). Zearalenone and its derivatives are the only naturally occurring fungal non-steroidal oestrogen, which have been shown to be oestrogenic in humans because the amounts of zearalenone to which humans are likely to be naturally exposed are several orders of magnitude lower than the amounts necessary for biological effects (Hagler *et al.*, 2001).

Terpenoides

Fusaproliferin: It was discovered in 1993 in cultures and subsequently in maize naturally contaminated with *F. proliferatum* (Moretti *et al.*, 1996; Randazzo *et al.*, 1993). A difference was found in the production of fusaproliferin (Moretti *et al.*, 1996) with regard to the teleomorph of *Fusarium* species belonging to the section *Liseola*, *G. fujikuroi* mating populations from A-F (Leslie, 1991). Fusaproliferin was produced by isolates from D and E, but not by the isolates from A, B, C and F mating populations.

Trichothecenes: Trichothecenes are a large family of tricyclic sesquiterpene toxins. It is named after the fungus, *Trichothecium roseum* from which the first member of this toxin class was isolated (Freeman and Morrison, 1948). Many species of *Fusarium* potentially produce a wide variety of trichothecenes (e.g., diacetoxyscirpenol, deoxynivalenol, nivalenol, fusarenone X, T-2 toxin) (Desjardins and Proctor, 2001; Miller *et al.*, 2001). Diacetoxyscirpenol (DAS) was reported for the first time from *F. equiseti* (Brian *et al.*, 1961). It is acutely toxic to animals (Ueno, 1983) and perhaps the only mycotoxin for which experimental data on toxicity in humans is available (Miller *et al.,* 2001). Several reviews are available on trichothecenes (Miller *et al.*, 2001; Rotter *et al.*, 1996). The T-2 toxin was the second toxin among the trichothecene family reported from *F. sporotrichioides* (Ueno, 1983). The third important acutely toxic trichothecene is nivalenol that was isolated first from grain contaminated by *F. sporotrichioides* (misidentified as *F. nivale*) (Marasas *et al.*, 1984; Tatsuno *et al.*, 1968). The trichothecenes (DAS, T-2 toxin and nivalenol) do not occur in adequate concentration reported from different parts of the world (IARC, 1993; Müller *et al.*, 1997; Scott, 1989). However, it was the discovery of deoxynivalenol (DON) in Eastern Canada (in 1979-81) that focused the spotlight for the first time on a toxin from fusaria (Vesonder and Hesseltine, 1980). Eventually, DON was discovered by Japanese researchers from grain that had made humans ill (Morooka *et al.*, 1972). The DON is an important mycotoxin in terms of human exposure worldwide (Miller, 1995; IARC, 1993). A large-scale human toxicoses ascribed to DON have been reported from India and other parts of the world (Bhat *et al.*, 1989; Miller, 1990). There are many reports on DON, which have been reviewed by Miller *et al.* (2001).

Amino Acid Derivatives

Enniatins, including beauvericin comprise a family of cyclic depsi-peptides produced by many fusaria and by other filamentous fungi (Desjardins and Proctor, 2001). Species of *Fusarium* producing these toxins under the common cyclic peptides are given in Table 4. Reports regarding investigations to demonstrate toxicity to mammals or poultry are very few. However, there is an indirect evidence that beauvericin is toxic to mammals or poultry (Logrieco *et al.*, 1997). The plant pathogenic species of the genus *Fusarium* are known to be capable of producing beauvericin and also an array of other mycotoxins (e.g., fumonisin, fusaric acid, fusarins, moniliformin) (Bryden *et al.*, 2001). Beauvericin was found to occur along with either FB_1 or moniliformin in culture extracts of *F. verticillioides* and *F. proliferatum* isolates (Plattner and Nelson, 1994).

Shikimic Acid Derivatives

Fusarochromes were identified in 1985 as a result of its ability to cause dyschondroplasia, an abnormal cartilage formation in experimental poultry (Lee *et al.*, 1985). Although, the structural analysis suggested that it has been derived from the shikimic acid pathway (Pathre *et al.*, 1986), it is yet to be confirmed by biochemical studies (Desjardins and Proctor, 2001). Fusarochromanone production appears to be rare among fusaria and has been reported only in *F. equiseti* (Thrane, 1989; Wu *et al.*, 1990). The natural occurrence of the toxins has been reported only in Danish poultry feed associated with field cases of tibial dyschondrioplasia in broiler chickens (Krogh *et al.*, 1989).

Conclusions and Outlook

Thirteen species of *Fusarium* colonizing sorghum have been reported worldwide and eight of them exist in India. Identification of *Fusarium* species based on their morphological characters becomes very difficult even for highly skilled mycologists. Many of these species heavily colonize post harvest agricultural produce and are potent producers of mycotoxins. This warrants the development of modern diagnostic techniques and tools to identify *Fusarium* strains and their toxins before visible symptoms to develop strategies to save the agricultural produce.

Acknowledgements

Authors are grateful to the authorities of Gulbarga University, Gulbarga for the facilities to conduct research on *Fusarium* species inciting sorghum. One of us (SMS) is indebted to Gulbarga University for the award of research fellowship.

References

Abbas, H.K. and Ocamb, C.M. 1995a. First report on production of fumonisin B_1 by *Fusarium polyphialidicum* collected from seeds of *Pinus strobus*. *Plant Dis. 79*, 642.

Abbas, H.K. and Ocamb, C. M. 1995b. First report of fumonisin B_1, B_2, and B_3 production by *Fusarium oxysporum* var. *redolens*. *Plant Dis. 79*, 968.

Anahosur, K.H. 1992. Sorghum diseases in India: Knowledge and research needs. In: *Sorghum and Millets Diseases*: *A Second World Review* (Ed. de Milliano, W.A.J., Frederiksen, R.A. and Bengston, J.D.). International Crops Research Institute for Semiarid Tropics, Patancheru, Andhra Pradesh, 45-56.

Bacon, C.W., Porter, J.K., Norred, W.P. and Leslie, J.F. 1996. Production of fusaric acid by *Fusarium* species. *Appl. Environ. Microbiol. 62*, 4039-4043.

Betina, V. 1984. Zearalenone and Brefeldin A. In: *Mycotoxins – Production, Isolation, Separation and Purification* (Ed. Betina, V.). Elsevier, Amsterdam, 237-257.

Betina, V. 1989. Zearalenone and Brefeldin A. In: *Mycotoxins – Production, Isolation, Separation and Purification* (Ed. Betina, V.). Elsevier, Amsterdam, 271-284.

Bezuidenhout, S.C., Gelderbolm, W.C.A., Gorst-Allman, C.P., Horak, R.M., Marasas, W.F.O. Spiteller, G. and Vleggaar, R. 1988. Structure elucidation of the fumonisins, a mycotoxin from *Fusarium moniliforme*. *J. Chem. Soc. Chem. Comm.* 743-745.

Bhale, M.S. and Khare, M.N. 1982. Seed borne fungi of sorghum in Madhya Pradesh and their significance. *Indian Phytopath. 35*, 676-678.

Bhat, R.V., Beedu, S.R., Ramakrishna, Y. and Munshi, K.L. 1989. Outbreak of trichothecene mycotoxicosis associated with the consumption of mould-damaged wheat products in Kashmir valley, India. *Lancet 1*, 35-37.

Bhat, R.V., Shetty, P.H., Amrith, R.P. and Sudershan, R.V. 1997. A food-borne disease outbreak due to consumption of moldy sorghum and maize containing fumonisin mycotoxins. *J. Toxicol. Clin. Toxicol. 35*, 249-255.

Booth, C. 1971. *The Genus Fusarium*. Commonwealth Mycological Institute, Kew, Surrey, UK.

Brian, P.W., Dawkins, A.W., Groue, J.F., Hemming, H.G., Lowe, D. and Norris, G.L.F. 1961. Phytotoxic compounds produced by *Fusarium equiseti*. *J. Exp. Bot. 12*, 1-12.

Britz, H., Coutinho, T.A., Wingfield M.J., Marasas, W.F.O., Gordon, T.R. and Leslie, J.F. 1999. *Fusarium subglutinans* f. sp. *pini* represents a distinct mating population in the *Gibberella fujikuroi* species complex. *Appl. Environ. Microbiol. 65*, 1198-1201.

Bryden, W.L., Logrieco, A., Abbas, H.K., Porter, J.K., Vesonder, R.F., Richard, J.L. and Cole, R.J. 2001. Other significant *Fusarium* mycotoxins. In: *Fusarium : Paul E. Nelson Memorial Symposium* (Ed. Summerell, B.A. Leslie, J.F., Backhouse, D., Bryden, W.L. and Burgess, L.W.). American Phytopathological Society, St. Paul, MN, 360-392.

Burgess L.W and D Trimboli. 1986. Characterization and distribution of *Fusarium nygamai* sp. nov. *Mycologia 78*, 223-229 .

Chen, J., Mirocha, C.J., Xie, W., Hogge, L. and Olson, D. 1992. Production of the mycotoxin fumonisin B_1 by *Alternaria alternata* f. sp. *Lycopersici. Appl. Eniviron. Micribiol. 58*, 3928-3931.

Chen, L., Tian, X., and Yang, B. 1990. A study on the inhibition of rat myocardium glutathione peroxidase and glutathione reductase by moniliformin. *Mycopathologia 110*, 119-124.

Cole, R.J., Kirksey, J.W., Cutler, H.G., Doupnik, B.L. and Peckham, J.C. 1973. Toxin from *Fusarium moniliforme*, effect on plants and animals. *Science* 179, 1324-1326.

Coulombe, R.A. 1993. Biological action of mycotoxins. *J. Dairy Sci. 76*, 880-891.

De Nijs, M., Rombout, F. and Notermans, S. 1996. *Fusarium* molds and their toxins. *J. Food Safety 16*, 15-58.

Desjardins, A.E. and Proctor, R.H. 2001. Biochemistry and genetics of *Fusarium* toxins. In: *Fusarium : Paul E. Nelson Memorial Symposium* (Ed. Summerell, B.A. Leslie, J.F., Backhouse, D., Bryden, W.L. and Burgess, L.W.). American Phytopathological Society, St. Paul, MN, 50-69.

Engelhardt, J.A., Carlton, W.W. and Tuite, J.F. 1989. Toxicity of *Fusarium moniliforme* var. *subglutinans* for chicks, ducklings, and turkey poults. *Avian Dis. 33*, 357-360.

Frederiksen, R.A. and Odvody, G.N. 2000. *Compendium of Sorghum Diseases*. II Edition, APS Press, 30-31.

Frederiksen, R.A., Castor, L.L. and Rosenow, D.T. 1982. Grain mold, small seed and head blight: The *Fusarium* connection in sorghum. *Proceedings of the thirty-seventh Annual Corn and Sorghum Industry Research Conference 37*, 26-36.

Freeman, G.G. and Morrison, R.I. 1948. Trichothecin-an antifungal metabolic product of *Trichothecium roseum. Nature 162*, 30.

Gaumann, F. 1957. Fusaric acid as wilt toxin. *Phytopath. 47*, 342.

Gopinath, A. and Shetty, H.S. 1985. Occurrence and location of *Fusarium* species in Indian sorghum seed. *Seed Sci. Technol. 13*, 521-528.

Hagler, Jr., M.H., Towers, N.P., Mirocha, J.C., Eppley, R.M. and Bryden, W.L. 2001. Zearalenone: Mycotoxins or mycoestrogen? In: *Fusarium : Paul E. Nelson Memorial Symposium* (Ed. Summerell, B.A. Leslie, J.F., Backhouse, D., Bryden, W.L. and Burgess, L.W.). American Phytopathological Society, St. Paul, MN, 321-331,

Hidaka, H., Nagatsu, T., Takeya, K., Takeuchi, T, Suda, H., Kojiri, K., Matsuzaki, M. and Umezawa, H. 1969. Fusaric acid, a hypotension agent produced by fungi. *J. Antibiot. 22*, 228.

Hidy, P.H., Baldwin, R.S. Greasham, R.L., Keith, C.L. and Mcmullen, J.R. 1977. Zearalenone and some derivatives: production and biological activities. *Adv. Appl. Microbiol. 22*, 59-82.

Hidy, P.H., Hodge, E.B., Urry, W.H. and Wehrmeister, H.L. 1966. The structure of zearalenone. *Tetrahedron Lett. 27*, 3109-3114.

Hsieh, W.H., Smith, S.N. and Snyder, W.C. 1977. Mating groups in *Fusarium moniliforme. Phytopath. 67*, 1041-1043.

Hurd, R.N. 1977. Structure activity relationships in zearalenones. In *Mycotoxins in Human and Animal Health* (Ed. Rodricks, J.V., Hesseltine C.W. and Mehlman, M.A.). Pathotox, USA, 379-391.

IARC, 1993. *Monograph 56: Some Naturally-Occuring Substances: Some Food Items and Constituents, Heterocyclic Amines and Mycotoxins*. International Agency for Research on Cancer, Lyon, France.

Kathariou, S. 1981. *Gene pool organization in Fusarium moniliforme*. Ph.D. Dissertation, University of California, Berkeley.

Klaasen, J.A. and Nelson, P.E. 1996. Identification of a mating population, *Gibberella nygamai* sp. nov., within the *Fusarium nygamai* anamorph. *Mycologia 88*, 965-969.

Klaasen, J.A. and Nelson, E. 1998. Identity of *F. nygamai* isolates with long and short microconidial chains from millet, sorghum, and soil in Africa. *Mycopathol. 140*, 171-176.

Klittich, C.J.R. and Leslie, J.F. 1992. Identification of a second mating population within the *Fusarium moniliforme* anamorph of *Gibberella fujikuroi. Mycologia 84*, 541-547.

Klittich, C.J.R., Leslie, J.F., Nelson, P.E. and Marasas, W.F.O. 1997. *Fusarium thapsinum* (*Gibberella thapsina*): A new species in section *Liseola* from sorghum. *Mycologia 89*, 643-652.

Kriek, N.P.J., Marasas, W.F.O., Steyn, P.S., Van Rensburg, P.J. and Steyn, M. 1977. Toxicity of a moniliformin-producing strain of *Fusarium moniliforme* var. *subglutinans* isolated from Maize. *Food Cosm. Toxicol. 15*, 579-587.

Krogh, P., Christensen, D.H., Hald, B., Harlou, B., Laresen, C., Pedersen, E.J. and Thrane, U. 1989. Natural occurrence of the mycotoxin fusarochromanone, a metabolite of *Fusarium equiseti*, in cereal feed associated with tibial dischondroplasia. *Appl. Environ. Microbiol. 55*, 3184-3188.

Kuhlman, E.G. 1982. Varieties of *Gibberella fujikuroi* with anamorphs in *Fusarium* section *Liseola*. *Mycologia 74*,759-768.

Lee, Y.W., Mirocha, C.J., Shroeder D.J. and Walser M.M. 1985. TDP-1, a toxic component causing tibial dyschondroplasia in broiler chickens and trichothecenes from *Fusarium roseum* "Graminearum". *Appl. Eniviron. Microbiol. 50*, 102-107.

Leslie, J.F. 1991. Mating population in *Gibberella fujikuroi* (*Fusarium* section *Liseola*). *Phytopath. 81*, 1058-1060.

Leslie, J.F., Pearson, C.A.S., Nelson, P.E. and Toussoun, T.A. 1990. *Fusarium* spp. from corn, sorghum and Soybean fields in the central and eastern United States. *Phytopath. 80*, 343-350.

Leslie, J.F. 2000. *Fusarium* species associated with sorghum. In: *Compendium of Sorghum Diseases*. II Edition (Ed. Frederiksen, R.A. and Odvody, G.N.), APS Press, 30-31.

Leslie, J.F. and Marasas, W.F.O. 2002. Will the real "*Fusarium moniliforme*" please stand up! In: *Sorghum and Millets Diseases* (Ed. Leslie, J.F.), Lowa State Press, Ames, Lowa, 201-210.

Leslie, J.F., Marasas, W.F.O., Shephard, G.S., Sydenham, E.W., Stockenstrom, S. and Thiel, P.G. 1996. Duckling toxicity and the production of *fumonisin* and *moniliformin* by isolates in the A and F mating populations of *Gibberella fujikuroi* (*Fusarium moniliforme*). *Appl. Environ. Microbiol. 62*, 1182-1187.

Logrieco, A., Ritieni, A., Moretti, A, Randazo, G. and Bottalico, A. 1997. *Beauvericin* and *fusaproliferin*, new emerging *Fusarium* toxins. *Cereal Res. Comm. 25*, 407-413.

Mahalinga, D.M., Anahosur, K.H. and Hegde, R.K. 1988. *Fusarium* species associated grain mold and stalk rot of sorghum and their effect on seed germination and growth of seedlings. *Curr. Sci. 57*, 177-178.

Mansuetus, A.S.B., Odvody, G.N., Frederiksen, R.A. and Leslie, J.F. 1997. Biological species in the *Gibberella fujikuroi* species complex (*Fusarium* section *Liseola*) recovered from sorghum in Tanzania. *Mycol. Res. 101*, 815-820.

Marasas, W.F.O., Nelson, P.E. and Toussoun, T.A. 1984. *Toxigenic Fusarium Species: Identity and Mycotoxicology*. Pennsylvania State University Press, Pennsylvania.

Marasas, W.F.O., Rabie, C.J., Lubben, A., Nelson, P.E., Toussoun, T.A. and Van Wyk, P.S. 1987. *Fusarium napiforme*, a new species from millet and sorghum in Southern Africa. *Mycologia 79*, 910-914.

Marasas, W.F.O., Burgess, L.W., Anelich, R.Y., Lamprecht, S.C. and Van Schalkwyk, D.J. 1988. Survey of *Fusarium* species associated with plant debris in South African soils. *S. Afr. J. Bot. 54*, 63-71.

Marasas W.F.O., Rheeder, J.P., Lamprecht, S.C., Zeller, K.A. and Leslie, J.F. 2001a. *Fusarium andiyazi* sp. nov., a new species from Sorghum. *Mycologia 93*, 1203-1210.

Marasas, W.F.O., Miller, J.D., Riley, R.T. and Visconti, A. 2001b. Fumonisins – Occurrence, Toxicology, Metabolism and Risk Assessment. In: *Fusarium : Paul E. Nelson Memorial Symposium* (Ed. Summerell, B.A. Leslie, J.F., Backhouse, D., Bryden, W.L. and Burgess, L.W.). American Phytopathological Society, St. Paul, MN, 332-359

Miller, J. D. 1990. Contamination of food by *Fusarium* toxins: studies from Austria – Asia. *Proc. Jap. Assoc. Mycotoxicol. 32*, 17-24.

Miller, J. D. 1995. Fungi and mycotoxins in grain: Implications for stored product research. *J. Stored Prod. Res. 31*, 1-6.

Miller, D.J., Apsimon, J.W., Blackwell, B.A., Greenhalgh, R. and Taylor, A. 2001. Deoxynivalenol: A 25-year perspective on a trichothecene of agricultural importance.

In: *Fusarium : Paul E. Nelson Memorial Symposium* (Ed. Summerell, B.A., Leslie, J.F., Backhouse, D., Bryden, W.L. and Burgess, L.W.). American Phytopathological Society, St. Paul, MN, 310-320.

Mirocha, C.J. and Christensen, C.M. 1974. Oesterogenic mycotoxins synthesized by *Fusarium*. In: *Mycotoxins*. Elsevier Science Publishers, Amsterdam, 129-148.

Mirocha, C.J. and Pathre, S.V. 1979. Mycotoxin – their biosynthesis in fungi *zearalenone* biosynthesis. *J. Food Prot. 42*, 821-824.

Mohammed, S. and Rajasab, A.H. 2004. Diversity of *Fusarium* species on sorghum grain. *Indian Phytopath. 57*, 450-453.

Moretti, A., Logrieco, A., Bottalico, A., Fogliano, R.A. and Randazzo, G. 1996. Diversity in *beauvericin* and *fusaproliferin* production by different population of *Gibberella fujikuroi* (*Fusarium* section *Liseola*). *Sydowia 48*, 44-56.

Morooka, N., Uratsuji, N., Yoshizawa, T. and Yamamoto, H. 1972. Studies on the toxic substances in barley infected with *Fusarium* species. *J. Food Hygiene Soc. Japan 13*, 368-375.

Müller, H.-M., Reimann, J., Schumacher, U. and Schwadorf, K. 1997. *Fusarium* toxins in wheat harvested during six years in an area of Southwestern Germany. *Nat. Toxins 5*, 24-30.

Nirenberg, H.I. and O'Donnell, K. 1998. New *Fusarium* species and combinations within the *Gibberella fujikuroi* species complex. *Mycologia 90*, 434-458.

Nelson, P.E., Toussoun, T.A., and Marasas, W.F.O. 1983. *Fusarium* species: *A Illustrated Manual for Identification*. The Pennsylvania State University Press, Pennsylvania.

Nirenberg, H.I. 1976. Untersuchungen uber die morphologischi and Biologische differenzierung in der *Fusarium* – section Liseola. *Mitteilungen Aus der biologischen bunderan stalt fur land – Forlstwir fschft Berlin: Dahlem 169*, 1-117.

Nyvall, R.F. and Kommedahl, T. 1968. Individual thickened hyphae as survival structures of *Fusarium moniliforme* in corn. *Phytopath. 58*, 1704-1707.

Onyike, N.B.N. and Nelson, P.E. 1992. *Fusarium* species associated with sorghum grain from Nigeria, Lesotho and Zimbabwe. *Mycologia 84*, 452-458.

Onyike, N.B.N. and Paul E. Nelson 1993. The distribution of *Fusarium* species in soils planted to millet and sorghum in Lesotho, Nigeria and Zimbabwe. *Mycopathol. 121*, 105-114.

Onyike, N.B.N., Nelson, P.E. and Marasas, W.F.O. 1991. *Fusarium* species associated with millet grain from Nigeria, Lesotho, and Zimbabwe. *Mycologia 83*, 708-712.

Pathre, S.V. and Mirocha, C.J. 1976. Zearalenone and related compounds. In: *Mycotoxins and Other Fungal Related Food problems* (Ed. Rodericks J.V.). Chemical Series # 149, American Chemical Society, Washington DC.

Pathre, S.V., Gleason, W.B., Lee, Y.W. and Mirocha, C.J. 1986. The structure of *fusarochromanone*: new mycotoxin from *Fusarium roseum* "Graminearum". *Can. J. Chem. 64*, 1308-1311.

Plattner, R.D. and Nelson P.E. 1994. Production of *beauvericin* by a strain of *Fusarium proliferatum* isolated from corn fodder for swine. *Appl. Eniviron. Microbiol. 60*, 3894-3896.

Porter, J.K., Bacon, C.W., Wray, E.M. and Hagler, W.M. Jr. 1995. Fusaric acid in *Fusarium moniliforme* cultures, corn and feed toxis to livestock and the neurochemical effects in the brain and pineal gland of rats. *Natural Tox. 3*, 91-100.

Rajasab A.H. and Maribhat, M. 1989. Pre-emergence ear rot, a new phase of the fusarial disease of sorghum in India. *Geobios New Rep. 8*, 74-76.

Rajasab, A.H. and R.A. Frederiksen, 2001. Prevalence of sorghum diseases in Karnataka and Maharastra states, India during 2001 rainy season. *Int. Sorghum News Lett. 42*, 40-44.

Randazzo, G., Fogliano, V., Ritieni, A., Mannina, L., Rossi, E., Scarallo A. and Segre, A. 1993. Proliferin, a new sesterterpene from *Fusarium proliferatum*. *Tetrahedron 49*, 10883-10896.

Ravindranath, V. 1980. Sorghum diseases in India. In: *Sorghum and Millets Diseases: A Second World Review* (Ed. de Milliano, W.A.J., Frederiksen, R.A. and Bengston, J.D.). International Crops Research Institute for Semiarid Tropics, Patancheru, Andhra Pradesh, 57-66.

Riley, R.T., Wang, E., Schroeder, J.J., Smith, E.R., Plattner, R.D., Abbas, H., Yoo, H.S. and Merril, A.H.Jr. 1996. Evidence for disruption of sphingolipid metabolism as a contributing factor in the toxicity and carcinogenecity of *fuminosins*. *Nat. Toxins 4*, 3-15.

Rotter, B.A., Prelusky D.B. and Petska, J.J. 1996. Toxicology of deoxynivalenol (vomitoxin). *J. Toxicol. Eniviron. Health 48*, 101-134.

Scheffer, R.P. 1983. Toxins as chemical determinants of plant disease. In: *Toxins in Plant Pathogenesis* (Ed. Daly, J.M. and Deverall, B.J.). Academic Press, New York, 1-40.

Scott, P.M. 1989. The natural occurrence of *trichothecenes*. In: *Trichothecene Mycotoxin: Pathophysiologic Effects*. Volume 1 (Ed. Beasley, V.R.). CRC Press, Boca Raton, FL, 1-26.

Shephard, G.S., Thiel, P.G., Stockenstrom, S. and Sydenham, E. W. 1996a. Worldwide survey of fumonisin contamination of corn and corn-based products. *J. AOAC Int. 79*, 671-687.

Shephard, G.S., vander Westhuizen, L., Thiel, P.G., Gelderblom, W.C.A., Marasas, W.F.O., and van Schalkwyk, D.J. 1996b. Disruption of sphingolipid metabolism in non-Human primate consuming diets of fumonisin-containing *Fusarium moniliforme* culture material. *Toxicon 34*, 527-534.

Singh, S.D., and Bandyopadhyay, R. 2000. Panicle and Seed Diseases. In: *Compendium of Sorghum Diseases*. II Edition (Ed. Frederiksen, R. and Odvody, G.N.). American Phytopathological Society, St. Paul, MN, 38-40.

Stoessl, A. 1981. Structure and biogenetic relations: fungal non-host specific. In:*Toxins in Plant Disease* (Ed. Durbin, R.D.). Academic Press, New York, 110-220.

Summerell, B.A., Salleh, B. and Leslie, J.F. 2003. A utilitarian approach to *Fusarium* identification. *Plant Dis. 87*, 117-128.

Summerell, B.A., Leslie, J.F., Backhouse, D., Bryden, W.L. and Burgess, L.W. 2001. *Fusarium: Paul E. Nelson Memorial Symposium* (Ed. Summerell, B.A. Leslie, J.F., Backhouse, D., Bryden, W.L. and Burgess, L.W.). American Phytopathological Society, St. Paul, MN.

Sundaram, N.V. 1977. Pathological research in India. International sorghum workshop, March 6-13, ICRISAT, Hyderabad, Patancheru, Andhra Pradesh, 1-19.

Tatsuno, T., Saito, M. Enomoto, M. and Tasunoda, H. 1968. Nivalenol, a toxic principle of *Fusarium nivale. Chem. Pharm. Bull. 16*, 2519-2520.

Thrane, U. 1989. *Fusarium* species and their specific profile of secondary metabolites. In: *Fusarium: Mycotoxin, Taxonomy and Pathogenicity* (Ed. Chelkowski, J.). Elsevier Science Publishers, Amsterdam, 199-225

Ueno, Y. 1983. *Trichothecenes*. Elsevier, Amsterdam.

Vesonder, R.F. and Hesseltine, C.W. 1980. *Vomitoxin*: natural occurrence on cereal grains and significance as a refusal and emetic factor to swine. *Process Biochem. 44*, 12-15.

Wiebe, L.A. and Bjeldanes, L.F. 1981. Fusarin C, a mutagen from *Fusarium moniliforme* grown on corn. *J. Food Sci. 46*, 1424-1426.

Williams, R.J. and Rao, K.N. 1981. A review of sorghum grain mold. *Trop. Pest Manage.* 27, 200-211.

Wineland, G.O. 1924. An ascigenous stage and synonym for *Fusarium moniliforme. J. Agric. Res. 28*, 909-922.

Wollenweber, H.W. and Reinking, O.A. 1935. *Die Fusarien, inhre Beschreibung, Schadwirkung und Bekampfung*. Verlag Paul Parey, Berlin, Germany.

Wu, W., Nelson, P.E., Cook, M.E. and Smalley, E.B. 1990. Fusarochromanone production by *Fusarium* isolates. *Appl. Eniviron. Microbiol. 56*, 2989-2993.

Zang, H. and Li, L. 1989. Mechanism of toxicity of moniliformin. *Acta. Microbiologica Sinica. 29*, 93-100.

Zeller, K., Summerell, B.A., Bullock, S. and Leslie, J.F. 2003. *Gibberella konza* (*Fusarium konzum*) sp. nov., a new biological species within the *Gibberella fujikuroi* species complex from Prarie grasses. *Mycologia 95*, 943-954.

Frontiers in Fungal Ecology, Diversity and Metabolites (edited by K.R. Sridhar), I.K. International Publishing House Pvt. Ltd., New Delhi, 2009.

14

Fumonisins–A New Class of Fusarial Mycotoxins

G.R. Janardhana*, Regina Sharmila Dass and M.Y. Sreenivasa
Mycology and Plant Pathology Laboratory, Department of Studies in Botany and Microbiology, University of Mysore, Manasagangotri, Mysore 570 006, Karnataka, India
*(*E-mail: janardhanaplantclinic@india.com)*

Introduction

Fumonisins are the new class of mycotoxins identified from the cultures of mitosporic fungus *Fusarium moniliforme* having cancer-promoting properties (Gelderblom *et al.*, 1988). It is a major breakthrough close to a century of investigation of animal and human diseases associated with consumption of maize contaminated with *F. moniliforme* (Synonym: *F. verticillioides)*. This was the starting point for worldwide efforts to describe the structure, properties and toxicology of this new group of toxins. Fumonisins are synthesized by different species of the genus *Fusarium*. The most significant producers of fumonisins are strains of *F. moniliforme* and *F. proliferatum.* These moulds and their fumonisins, are under intense scientific investigation as they are the causative agents of oesophageal cancer in humans, toxic syndromes in domestic animals including equine leucoencephalomalacia, porcine pulmonary edema, hepatotoxicity and hepatocarcinogenicity in mice and toxicity in poultry, fish sheep, cattle and plants (Lumlertdacha *et al.*, 1995). See Chapter 13 for a variety of toxins produced by fusaira in sorghum.

A large body of literature on *F. moniliforme* and fumonisins is available (Marasas, 1995; Nelson, 1992; Nelson *et al.*, 1993). In spite of more than 90 years of research on the biology, pathology, toxicology and genetics of this fungus, there are still a few options for the effective management of the diseases it causes in cereals like maize. Recent international development in mycotoxin regulations has increased pressure to find strategies to manage the contamination of maize by fumonisins. This problem is unique because of close association of *F. moniliforme* with maize plants and implications of maize disease on animal and human health. The main focus of this chapter is to describe the problems posed by *F. moniliforme* and other fumonisin-producing fungi and the impact of fumonisins on human and livestock health.

Occurrence of Fumonisins

Fumonisins are toxic secondary metabolites produced by *Fusarium* spp. (*F. anthophilum, F. moniliforme* and *F. proliferatum*) (Fig. 1). They are ubiquitous in distribution and frequently present on freshly harvested and stored agricultural commodities such as cereals, particularly during pre- and post-harvest phases of production, processing and storage of grains. In order to assess human and animal exposure to fumonisins accurately, surveys on the natural occurrence of fumonisins in foods and feeds are necessary. A range of methods are available for detecting fumonisins at the µg to ng levels at which they commonly occur in naturally contaminated foods and feeds. Liquid chromatography of fluorescent

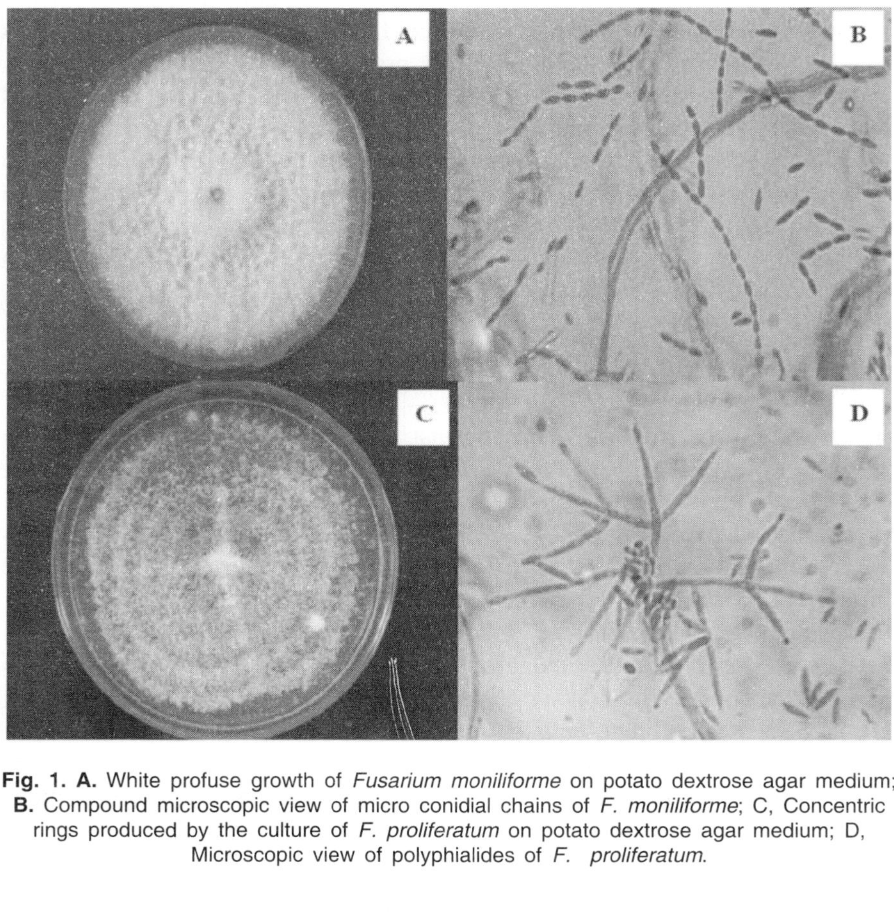

Fig. 1. A. White profuse growth of *Fusarium moniliforme* on potato dextrose agar medium; **B.** Compound microscopic view of micro conidial chains of *F. moniliforme*; C, Concentric rings produced by the culture of *F. proliferatum* on potato dextrose agar medium; D, Microscopic view of polyphialides of *F. proliferatum*.

derivatized fumonisins is the most widely used method for detecting fumonisins in food and feeds. Fumonisin detection limits for liquid chromatography are generally in the range of 0.01-0.05 μg/g dry weight (Gelderblom *et al.*, 1993). Because *F. moniliforme* infects cereals (e.g., maize, sorghum) worldwide, it is not surprising to find fumonisin-contaminated cereals from every geographic region tested to date. Two surveys of maize-based feed samples associated with animal diseases [equine leucoencephalomalacia (ELEM) and porcine pulmonary edema (PPE)] present the highest fumonisin B1 (FB_1) contamination up to 87-88% of the samples above detection limits, 1 μg/g (maximum, 126-330 μg/g; mean, 29-63 μg/g). Corn screening from the Central US comprises the third most contaminated sample type (maximum, 239μg/g; mean, 21 μg/g). The next most contaminated group comprises maize collected from high oesophageal cancer areas of China and South Africa, which shows a high incidence of FB_1 (maximum, 47-155 μg/g; mean, 13-54 μg/g) (Munkvold and Desjardins, 1997).

Fumonisins are commonly detected in symptom-less maize kernels. However, fumonisin levels in randomly selected good quality maize are generally much lower than maize samples associated with human and animal health problems. Surveys of 1,300 maize samples collected from the Central US from 1988 through 1995 indicate that although FB_1 is present in most samples, the levels are generally low (maximum, of 5-38μg/g; mean, 1-3 μg/g). Survey of good quality maize and maize-based foods from other countries have generally been more limited but indicate that a majority of samples contain fumonisins although the levels are less than 1 μg/g. These limited surveys indicate that human food samples from a number of countries (e.g., Brazil, Italy, Kenya, US) may occasionally be contaminated with FB_1 at 5-10 μg/g or more, which would generally be considered a level of concern (Munkvold and Desjardins, 1997).

Chemistry of Fumonisins

The core structures of fumonisins consist of a 20-carbon backbone and an amino group of C-2, which is derivatized in some fumonisins (Fig. 2). There are also methyl groups of C-12 and C-16 and as many as five hydroxyl groups. The hydroxyl at C-14 and C-15 are usually esterified with propane-1, 2, 3 tricarboxylic acid moieties. The FB_1 is a long fatty acid chain ester, stable during grain processing. Dry mixing of maize results in the distribution of FB_1 into the bran, gram and flour. The FB_1 is stable in polenta (maize porridge), where as it hydrolyzed by nixtamalization (calcium hydroxide processing). As FB_1 is water soluble, its concentration is reduced during the manufacture of cornstarch by wet-milling. A number of factors made it difficult to extract FB_1 from processed food. Nixtamalization and ammoniation lead to hydrolyzed FB_1 and aminopentol, respectively. These treatments reduce the fumonisin, while increasing the concentration of hydrolyzed fumonisins without deaminating the toxic product.

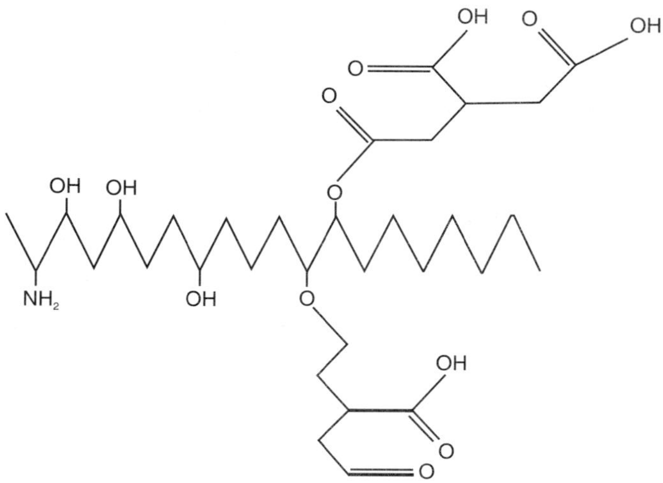

Fig. 2. Fumonisin B1 (FB_1) 1, 2, 3 – propanetricarboxylic acid, 1, 1 – [(12 – amino – 4, 9, 11 – trihydroxy – 2- methyl (tridecyl) – 2 – (1 – methyl pentyl) – 1, 2´ ethane – dicyl] ester, $C_{34} H_{59} NO_{15}$. MW: 721.838.

Fumonisins in India

India with its varied agroclimatic conditions produces a variety of food crops throughout the year. Non-scientific methods of agricultural practices, poor storage facilities and unfavorable environmental conditions during pre- and post-harvest handling of cereal grains are mainly responsible for contamination, infection and colonization by fumonisin producing fungi (Bhat *et al.*, 1997). In a limited epidemiological survey in India, association between high levels of fumonisins (not other mycotoxins) in moldy sorghum and corn and gastrointestinal symptoms (e.g., cramping and diarrhoea) was evident (Bhat *et al.*, 1997). However, factors difficult to extrapolate results of these studies are: differences in agricultural and nutritional conditions between US and other countries. For instance, the US corn supply generally contains significantly low levels of fumonisins than corn from rural areas of South Africa. In some instances, the study population will be significantly malnourished in comparison with the US population.

During October 1995, non-seasonal rains in certain parts of the Deccan Plateau damaged much of the rainy season (Kharif) sorghum crop. The sorghum crop harvested at that time was left as heaps in the field. The high moisture prevailing inside the heaps provided conducive environment for mould growth and FB_1 production. Most of the crops harvested during this period became visibly moldy. The disease was manifested only in adults who were consuming the moldy grains. The affected people mostly belonged to the low socioeconomic groups such as marginal farmers and landless agricultural laborers who did not have access to other staple foods. The disease was self-limiting, non-contagious and non-fatal. It did not affect the population consuming grains harvested during the previous year, harvested before or after the rains and those who had access to other foods. Children and pregnant women were not affected, as they did not include sorghum in their diet. Moldy nature of the grains affected the rheological properties of flour. Difficulties were experienced in the preparation of unleavened bread from moldy sorghum flour; the dough became sticky and unevenly cooked resulting in dark sticky bread.

Mould affected grains frequently get channelized into animal feeding systems. An outbreak similar to the one that occurred in humans was also found in some poultry farms during the same period due to fumonisin contaminated feed ingredients (Prathapkumar *et al.*, 1997). The disease was characterized by black sticky diarrhea, reduced weight, decline in egg production, lameness, and mortality. Antibiotics and vitamin supplements have no effect, while feed withdrawal showed improvements. The disease could be reproduced in one-day-old-cockerels with FB_1 contaminated sorghum. The feed upon analysis revealed contamination of FB_1 (8.5 mg/g).

Toxicity of Fumonisins

Although the fumonisins were discovered recently, the toxicity of corn contaminated by *F. moniliforme* has been well documented for more than a century. A disease of farm animals known as 'moldy-corn disease' or 'blind staggers' was first described in the US in 1850 (Graham, 1913). The causative agent remained unknown until Sheldon (1904) identified *F. moniliforme* and associated outbreaks of moldy corn disease of horses, cattle, mules, hogs and chicken in Nebraska. The most dramatic manifestation of livestock disease is known as ELEM, a fatal brain disease of horses, donkeys, mules and rabbits (Marasas, 1995; Nelson *et al.*, 1993; Wilson and Marsonport, 1971). In horses, this disease resulted in the death within a few hours to one week after horses were fed with fumonisin-contaminated feed. In 1971, ELEM was resulted by feeding horses material contaminated with a pure culture of *F. moniliforme* (Wilson, 1971), which firmly established this fungus as the causative agent. The South African research group was the first to identify and characterize the fumonisins and demonstrate that pure FB_1 produce ELEM in horse, as horses are unusually sensitive to fumonisins (Marasas *et al.*, 1998). Surveys of feed samples associated with outbreaks of ELEM in North America, South America and Africa found FB_1 levels ranging between 0.2 and 126µg/g feed. Experiments on the minimum toxic dose of fumonisins indicate that ponies consuming naturally contaminated feeds containing FB_1 at levels as low as 8 µg/g are at risk for developing ELEM (Wilson *et al.*, 1992). The ELEM may be the most dramatic, but certainly not the only animal disease associated with the consumption of feed contaminated with *F. moniliforme*. A study in 1981 regarding the oral toxicity of *F. moniliforme* culture to various animal species reported the deaths of two of the treated pigs (Kriek *et al.*, 1981). The principal lesions were not in the lungs where a fatal edema developed. Subsequently, consumption of corn contaminated with *F. moniliforme* and fumonisins was associated with numerous outbreaks of PPE in the Central US and Georgia (Gilchrist *et al.*, 1992). Survey of feed samples associated with outbreaks of PPE in North and South America found FB_1 levels ranging from 2-330 μg/g feed (Nelson *et al.*, 1993). Although pigs are unusually sensitive to trichothecene mycotoxins produced by *Fusarium* species, they appear to be less sensitive than horses to fumonisins. Survey and feeding experiments suggest that pigs consuming approximately 100 μg fumonisins (FB_1 and FB_2) per gram are at risk for PPE (Nelson *et al.*, 1993).

The first report of fumonisin toxicity to poultry from India reported 1997 (Bhat *et al.*, 1997). Poultry are apparently more resistant to fumonisins than swine and equines. Relatively high doses are required to induce measurable effects. Chicks fed with 450-525 ppm of fumonisin for 21 days, exhibited low feed consumption and weight loss. At 75 ppm, free sphingosine levels were elevated and disruption of sphingolipid metabolism through inhibition of ceramide synthetase activity is thought to be the mechanism of action. In another study, it was found that one-day-old poults fed with rations containing 199 and 200 ppm FB_1 for 21 days showed lower body weight gains and feeding efficiency compared to the controls (Weibking *et al.*, 1993a). There were also differences in organ weights and blood parameters. Weibking *et al.* (1993a, 1993b) concluded that *F. moniliforme* culture containing fumonisin is toxic to young turkey poults and the poultry appears to be more sensitive to the toxin than the broiler chick. Dairy cattle (Holsteins and Jerseys) fed diets containing 100 ppm fumonisin for approximately seven days prior to freshening and for seventy days thereafter demonstrated lower milk production (6 kg/cow/day), which was explained primarily by reduced feed consumption, increases in serum enzymes concentration and mild liver disease (Diaz *et al.*, 2000). Dairy cattle may be more sensitive to fumonisins than beef cattle, perhaps due to greater stress. Fumonisin carry over from feed to milk is thought to be negligible (Richard *et al.*, 1996; Scott *et al.*, 1994). Thus, fumonisin contamination has implications on animal health, at least from a regulatory perspective. Although the concentrations of fumonisins associated with adverse health effects are variable, the American Association of Veterinary Laboratory Diagnosticians has recommended maximum safe concentrations for livestock (Table 1). Although the role of fumonisins in some moldy corn diseases of livestock has been well established, their role in human diseases most particularly their carcinogenic potential in humans is much more difficult to determine. The International Agency for Research on Cancer (IARC), the South African Medical Research Council, the US Department of Agriculture (USDA) and many other agencies are in the process of evaluating the carcinogenic potential of fumonisins. These efforts include studies on the mechanism of action of fumonisins in various animal models and in cell culture systems and epidemiological studies of humans.

Table 1. Recommended levels of fumonisin concentration in animal feed for different animal species (Source: Miller *et al.*, 1996)

Animal	Recommended maximum concentration
Horses and other equine species	5 µg/g
Porcine species	10 µg/g
Beef cattle	50 µg/g
Poultry	50 µg/g

During the last one and a half decade, fumonisins have received worldwide attention. These mycotoxins have been shown to exhibit unique toxicological properties that differed with different animal species (Norred, 1993). The structural similarity of fumonisins to the long-chain base backbones of sphingolipids demonstrated that fumonisins affect sphingolipid metabolism (Wang *et al.*, 1996). In cultured cells from rat liver and pig kidney, FB_1 inhibited the activity of sphingosine N-acetyl transferase (ceramide synthase). This inhibition led to the rapid accumulation of high levels of the sphingoid base, sphinganine, an increase in the level of the sphingoid base, sphingosine and the depletion of complex sphingolipids. In fact, high levels of free sphinganine and or sphingosine in serum and urine may be useful for dietary exposure to fumonisins in humans and animals. Gelderblom *et al.*, (1991) in South Africa demonstrated that pure FB_1 could cause an increased incidence of hepatocellular carcinoma when fed to rats for two years at 50 µg/g of feed. Subsequent studies with FB_1, FB_2 and FB_3 in short-term initiation promotion bioassays demonstrated hepatocarcinogenesis in the rat (Gelderblom *et al.*, 1993). Fumonisins are also acutely hepatotoxic and it has been proposed that liver cell necrosis and regeneration play a role in carcinogenecity of fumonisins. Although current models link the biological activities of fumonisins to sphingolipid biosynthesis, there is still much to understand about the mechanisms by which fumonisins are toxic and carcinogenic. Sphingolipids are structural components of eukaryotic cell membranes, however, there is increasing evidences that sphingolipids also affect cellular proliferation, differentiation and induce apoptosis in several types of cultured human and animal cells (Miller *et al.*, 1996; Scott *et al.*, 1994). The role of sphingolipids and the sphingolipid-analog mycotoxins in apoptosis is a fast developing field that may provide insights into the diseases caused by fumonisins.

The FB_1 exhibits a variety of biological activities including phytotoxicity, which is of interest for its potential role as a virulence factor to facilitate invasion of plant tissues by the fungi. Droplets of FB_1 solution applied to the leaf surface of jimson weed, black nightshade and tomatoes caused necrosis, growth inhibition and death. With *Arabidopsis thaliana* grown on agar plates, an IC_{50} was less than 1 ppm (Marasas, 1996). When (3H) FB_1 was prepared by biosynthetic incorporation of commercially available radiolabelled precursors into the toxin in rice medium solid cultures of *F. moniliforme*, the toxin produced by incorporation of (9,10-3H) palmitate-induced phytotoxic symptoms were identical to unlabelled material indicating its biological activity. The area of necrosis on treated leaves was similar in light- and dark-treated plants. When (3H) FB_1 was applied to a wound site on target plants, severe damage occurred at the site of FB_1 application and in tissue above the site. These results reveal that FB_1 applied to intact surfaces of target plants exhibits contact activity (Marasas, 1996). In addition, purified fumonisins at low concentrations (10^{-6} M) have been shown to cause necrosis and other symptoms in maize and tomato seedlings (Gelderblom *et al.*, 1991; Gilchrist *et al.*, 1992; Wang *et al.*, 1991).

Assessment of Risk

Determination of risks due to the presence of fumonisins in cereals (e.g., maize, sorghum, paddy), animals and humans involves many factors. Firstly, toxicity tests indicate that fumonisins cause adverse effects in a wide range of animal species with disease symptoms usually observed at FB_1 (5-10μg/g feed) although physiological changes may occur at lower concentrations. There are species differences in fumonisin sensitivity and target organ specificity that makes it difficult to extrapolate results from animals to humans. Human consumption of maize-based foods varies widely among populations around the world and within the Indian subcontinent. It is especially difficult to estimate the risks of long-term chronic exposure to low levels of fumonisins in humans. Secondly, it appears reasonable to conclude from surveys conducted earlier that maize usually contains fumonisins. Because of the close relationship of *F. moniliforme* with maize, fumonisins cannot be completely eliminated without banning maize as a food and feed ingredient. Maize is one of the most important agricultural commodities not only in India, but also in different parts of the world largely because it is the major ingredient of animal feeds. In some regions (Africa, Asia, Central and South America) humans consume maize meal as a staple diet. In countries like North America and Europe, maize products are important components of many processed foods including breakfast cereals, snacks, soft drinks and beer. Fumonisins are not destroyed by the methods used for food processing. Thus, technology is not yet available to ensure that all maize-based foods and feeds are completely devoid of fumonisins. Although fumonisins are natural contaminants of maize, they are not inherent components of maize and are thus considered to be 'added substances' by the criteria of US Food and Drug Administration (FDA). As fumonisins cannot be completely eliminated from the foods, the goal of the FDA is to determine an acceptable daily intake: a level that would result in negligible risk or a reasonable certainty of no harm. The goal of FDA and other agencies is to protect public health while minimizing the costs of risk assessment and risk management (Troxell, 1996).

Methods of Reducing Fumonisins

A number of strategies for reducing fumonisin concentrations in maize are currently in progress. Some approaches are directed towards resistance to infection or reduction of fumonisins in the grain, while others are aimed at detoxification of contaminated maize. Among the possibilities, genetically engineered resistance to *Fusarium* infection or genetic engineering approaches of detoxification of fumonisins in plants has more potential. Plants could also be engineered to produce antifungal proteins, which is a viable approach to enhance resistance to fungi. Various proteins identified from microorganisms or plants have been shown to have antifungal activities. Another possible approach would be to identify naturally occurring components in maize, sorghum and paddy, which inhibit fumonisin synthesis. Concentrations of these components could be enhanced through selection or genetic engineering, but so far no such compounds have been identified. Two species of saprophytic fungi isolated from moldy corn ears were shown to be capable of utilizing FB_1 as sole-carbon-source and these fungi were shown to possess enzymes hydrolyzing and metabolizing fumonisins (Duvick *et al*., 1994). A cDNA encoding fumonisin esterase (tricarboxylate ester hydrolase) has been cloned from one of these fungi to black yeast (*Exophiala spinifera*) (Duvick *et al*., 1994) and research is in progress to express this gene in transgenic maize in order to evaluate the effects on toxin accumulation.

A general understanding of different studies on reduction of fumonisins is that the reduction in detectable toxins does not always result in reduced toxicity. The fumonisin molecules may be altered by various treatments to make them non-toxic. Therefore, it is crucial that toxicity assays of end products of these treatments accompany the fumonisin detection data. The effects of heat on fumonisin toxicity are not completely clear. Some reports indicate significant reductions in fumonisin concentration as a result of heating aqueous solutions to 150°C or above (Jackson, 1996) or heating moist maize kernels (Murphy *et al*., 1996). However, heating apparently causes hydrolysis of the primary amine group of the fumonisins, leaving the backbone of the molecule intact. Toxicity of the hydrolyzed products has been demonstrated in some experimental systems. Other reports found that fumonisin concentrations were not reduced by heat treatment (Sydenham *et al*., 1995). When maize is made into tortilla flour, it is subjected to a process known as nixtamalization. During this process, the maize is treated with calcium hydroxide and heated. Research has shown that calcium hydroxide treatment can reduce detectable FB_1 concentrations (Murphy *et al*., 1996; Park *et al*., 1992; Sydenham *et al*., 1995). However, there are conflicting reports about the toxicity of the hydrolyzed end products (Murphy *et al.*, 1996). It has also been found that the hydrolyzed fumonisins may be nearly as toxic as the unaltered FB_1 (Murphy *et al*., 1996). Similar results have been reported by Park *et al*. (1992), but they found that modified nixtamalization resulted in detoxification up to some extent.

Chemical methods to detoxify fumonisins have met with limited success. Several commercially available enzymes have been tested for their ability to detoxify fumonisins (Wang *et al*., 1991). None of these products significantly reduced the recovery of FB_1. Ammoniation as a detoxification method for other mycotoxins (particularly aflatoxins), may successfully detoxify fumonisins when combined with high temperature (Park *et al.*, 1992), but low temperature ammoniation has not been successful. A promising method for detoxification is also reported: non-enzymatic browning is a reaction that occurs in the presence of a primary amine, a reducing sugar and water at pH 7. This reaction results in the removal of the primary amine group from the fumonisin molecule. Treatment of FB_1 with fructose under the above conditions resulted in a significant reduction in detectable FB_1. More importantly, when a diet containing the product of the FB_1-fructose reaction was fed to rats and did not result in cancer induction. Another reaction that has been invented is, treatment of maize with a combination of H_2O_2 and $NaHCO_3$. It was reported that this reaction reduced fumonisin concentration up to 100% in contaminated maize. Toxicity of end products of this reaction was greatly reduced compared with untreated maize (Park *et al.*, 1992).

Conclusions and Outlook

Currently there are no proven practical methods to significantly reduce fumonisin concentrations in maize and there is a debate whether efforts in this direction are warranted. Fumonisins in corn result in potential human and animal health risk and it decreases the value of corn to the producers. Structural modification of fumonisins destroys the ability to cause the adverse effect that leads to toxicity, but contradictory results are available. Based on chemical and biological properties of fumonisins, efforts should be directed to develop methods to decontaminate fumonisin contaminated corn products. Biological control of *F. moniliforme* by competitive exclusion is untested under field conditions. There will be many obstacles to successful reduction of fumonisin by biological control due to numerous pathways for infection by the fungi. The most feasible approach may be to reduce infection by fumonisin-producing *Fusarium* strains through competition with non-producing strains. This approach has been successfully used with *Aspergillus flavus* to reduce aflatoxin concentrations in cottonseed in small field experiments. It is difficult to evaluate the success of traditional breeding in control of fumonisin levels. It is possible that currently used selection methods have prevented higher levels of fumonisins, nevertheless substantial amounts were found in maize and other cereals of many countries. Unless specific sources of resistance are identified and incorporated in modern hybrids, breeding will not contribute further to the management of fumonisins although genetic engineering approaches are the most attractive methods to reduce fumonisin levels.

References

Bhat, R.V., Shetty, P.H., Amruth, R.P. and Sudershan, R.V. 1997. A Food-borne disease outbreak due to the consumption of moldy sorghum and maize containing fumonisin mycotoxins. *Clinical Toxicol.* *35*, 249-255.

Diaz, D.E., Hopkins, B.A., Leonard, L.M., Hagler, W.M.Jr. and Whitlow, L.W. 2000. Effect of Fumonisin on Lactating dairy cattle. *J. Dairy. Sci. 83*, 1171.

Duvick, J., Rood, T. and Grant, S. 1994. Isolation of Fumonisin-metabolizing fungi from maize seed. In: *Proceedings of 5th International Mycological Congress*, August 14-21, Vancouver, Canada, 56.

Gelderblom, W.C.A., Cawood, M.E., Sayman, S.D., Vleggaar, R. and Marasas, W.F.O. 1993. Structure-activity relationship of fumonisins in short-term carcinogenesis and cytotoxicity assays. *Food Chem. Toxicol. 31*, 407-414.

Gelderblom, W.C.A., Jaskiewicz, K., Marasas, W.F.O., Thiel, P.G., Horak, R.M., Vleggar, R. and Kriek, N.P.J. 1988. Fumonisins-novel mycotoxins with cancer-promoting activity produced by *Fusarium moniliforme. J. Appl. Environ. Microbiol. 54*, 1806-1811.

Gelderblom, W.C.A. Krick, N.P.J., Marasas, W.F.O. and Thiel, P.G. 1991. Toxicity and Carcinogenity of the *Fusarium moniliforme* metabolite, fumonisin B_1, in rats. *Carcinogenesis 12*, 1247-1251.

Gilchrist, D.G., Ward, B., Moussato, V. and Mirocha C.J. 1992. Genetic and physiological response to fumonisin and AAl - toxin by intact tissue of a higher plant. *Mycopathol. 117*, 57-64.

Graham, R. 1913. Forage poison in horses, cattle and mules so-called cerebrospinal meningitis and commonly termed "staggers". *Kentucky Agricultural Experiment Station Bulletin # 17*, 369-383.

Jackson, L.S., Hlwyka, J.J., Senthil, K.R., and Bullerman, L.B. 1996. Effect of thermal processing on the stability of fumonisins. In: *Fumonisins in Food* (Ed. Jackson, L.S., Devries, J.W. and Bullerman, L.B.). Plenum Press, New York, 345-353.

Kreik, N.P.J., Kellerman, T.S. and Marasas, W.F.O. 1981. A comparative study of the study of the toxicity of *Fusarium moniliforme* to horses, primates, pigs, sheep and rats.Onderstepoort. *J. Vet. Res. 48*,129-131.

Lumlertdacha, R.T., Lovell, R. A., Shelby, S., Lenz, D. and Kempainen, B.W. 1995. Growth, hematology and histopathology of channel catfish *Ictalurus punctatus* fed toxins from *Fusarium moniliforme. Aquaculture 130*, 201-218.

Marasas, W.F.O. 1995. Fumonisins: Their implications for human and animal health. *Natural Toxins 3*, 193-198.

Marasas, W.F.O. 1996. Fumonisins: History, worldwide occurrence and impact. In: *Fumonisins in Food* (Ed. Jackson, L.S., Devries, J.W. and Bullerman, L.B.). Plenum press, New York, 1-18.

Marasas, W.F.O., Kellerman, T.S., Gelderblom, W.C.A., Coetzer, J.A.W., Thiel, P.G. and van der Lugt, J.J. 1988. Leukoencephalomalacia in a horse induce by fumonisin B1 isolated form *Fusarium moniliforme*. Onderstepoort *J. Vet. Res. 55*, 197-203.

Miller, M.A., Hanstead, J.P. and Lovell, R.A. 1996. Regulatory aspects of fumonisins with respect to animal feed. In: *Fumonisins in Food* (Ed. Jackson, L.S., Devries, J.W. and Bullerman, L.B.). Plenum Press, New York, 363-368.

Munkvold, G. P. and Desjardins, A. E. 1997. Fumonisins in maize: Can we reduce their occurrence? *Plant Disease 81*, 556-565.

Murphy, P.A. Hendrich, S., Hopmans, E.C., Hauck, C.C., Lu, Z., Buseman, G. and Munkvold, G. 1996. Effect of processing on fumonisin content of corn. In: *Fumonisins in Food* (Ed. Jackson, L.S., Devries, J.W. and Bullerman, L.B.). Plenum Press, New York, 323-334.

Nelson,P.E. 1992. Taxonomy and Biology of *Fusarium moniliforme. Mycopathologia 117*, 29-36.

Nelson, P.E., Desjardins, A.E. and Plattner, R.D., 1993. Fumonisins, Mycotoxins produced by *Fusarium* species: Biology Chemistry and significance. *Ann. Rev. Phytopathol. 31*, 233-252.

Norred, W.P. 1993. Fumonisins: mycotoxins produced by *Fusarium moniliforme. J. Toxicol. Environ. Health, 38*, 309-328.

Park, D.L., Rau, S.M., Jr. Mirocha, C.J., Abd-Alla, E.S.A.M. and Weng. C.Y. 1992. Mutagenic Potential of Fumonisin contaminated corn following ammonia decontamination procedure. *Mycopathol. 117*, 105-108.

Prathapkumar, S.H., Rao, V.S., Paramkishan, R.J. and Bhat, R.V. 1997. Disease outbreak in laying hens arising from the consumption of fumonisin-contaminated food. *Br. Poultry Sci. 38*, 475-479.

Richard, J.L., Meerdink, G., Maragos, C.M., Tumbleson, M., Bordson, G., Rice, L.G. and Ross,P.F. 1996. Absence of detectable fumonisins in the milk of cows fed *Fusarium proliferatum* (Matsushima) Nirenberg Culture Material. *Mycopathol. 133*, 123-126.

Scott, P.M., T. Delgado, D.B. Prelusky, H. L. Trenholm and Miller, J.D. 1994. Determination of Fumonisins in milk. *J. Environ. Sci. Health B 29*, 989-998.

Sheldon, J.L. 1904. *A corn mold.* Agricultural Experimental Station, Nebraska, 17th Annual Report, 23-32.

Sydenham, E.W., Stockenstrom, S., Thiel, P.G., Shephard, G.S., Koch, K.R. and Marasas, W.F.O. 1995. Potential of alkaline hydrolysis for removal of fumonisins from contaminated corn. *J. Agric. Food chem. 43*, 1198-1201.

Troxell, T.C. 1996. Regulatory aspects of Fumonisins in the United States, In: *Fumonisins in Food* (Ed. Jackson, L.S., Devries, J.W. and Bullerman, L.B.). Plenum Press, New York, 355-362.

Wang, E., Norred, W.P. Bacon, C.W., Riley, R.T. and Merrill, A.H.Jr. 1991. Inhibition of Sphingolipid Biosynthesis by Fumonisins. *J. Biol. Chem. 266*, 14486-14490.

Wang, H., Jones, C., Ciacci-Zanella, J., Holt, T., Gilchrist. D.G. and Dickman, M.B. 1996. Fumonisins and AAL toxins: Sphinganine Analog Mycotoxins induce apoptosis in monkey kidney cells. *Proc. Nat. Acad. Sci. 93*, 3461-3465.

Weibking, T.S., Ledoux, D.R., Bermudez, A.J., Tusk, J.R., Rottingghaus, G.E. Wang, E. and Merrill, A.H. 1993a. Effects of feeding *Fusarium moniliforme culture* material, containing known levels of fumonisins B1 on young boiler chicks. *Poultry Sci. 72*, 456-465.

Weibking, T.S., Ledoux, D.R., Brown, T.P. and Rottinghaus, G.E. 1993b. Fumonisin toxicity in turkey poults. *J. Vet. Diag. Invest. 5*, 75-83.

Wilson, B.J. and Marsonport, R.R. 1971. Causative fungal agent of leucoencephalomalacia in equine animals. *Vet. Res. 88*, 484-486.

Wilson, T.M., Ross, P.F., Owens, D.L., Rice, L.G., Green, S.A., Jenkins, S.J. and Nelson, H.A. 1992. Experimental reproduction of ELEM - A study to determine the minimum toxic dose in ponies. *Mycopathol. 117*, 115-120.

Frontiers in Fungal Ecology, Diversity and Metabolites (edited by K.R. Sridhar), I.K. International Publishing House Pvt. Ltd., New Delhi, 2009.

15

Improvement of Coffee Quality through Prevention of Moulds

K. Velmourougane[1] and Rajeev Bhat[2*]

[1]*Coffee Research Sub-Station, Coffee Board, Chettalli Post, North Coorg 571 248, Karnataka, India*

[2]*Microbiology and Biotechnology, Department of Biosciences, Mangalore University, Mangalagangotri 574 199, Mangalore, Karnataka, India*

(E-mail: rajeevbhat76@rediffmail.com)*

Introduction

Coffee is one of the most popular and widely preferred non-alcoholic beverages consumed worldwide for its unique taste, flavour, aroma, and stimulation. Today, coffee is the second largest traded commodity in the world market after petroleum. Among the most valuable agricultural commodities, coffee holds a significant status in the livelihoods of a large population of the tropical developing countries. It is estimated that 25 million families in 52 exporting countries depend on coffee. The major coffee producing countries in the world are Brazil, Vietnam, Colombia, Indonesia, Mexico, Ethiopia, Guatemala, India, Mexico, and Uganda. Out of the total global coffee production, nearly 70% belongs to the Arabica variety, while the rest 30% belongs to Robusta variety. During the past decade, Vietnam has emerged as a major contributor of the Robusta variety and has become the second largest producer of coffee replacing Colombia.

Coffee beans are obtained from an evergreen perennial woody plant that grows up to a height of 10 meters (if not pruned regularly) and belongs to the family Rubiaceae. Worldwide, only two species are popular and cultivated: *Coffea arabica*, known as Arabica and *Coffea canephora*, known as Robusta. However, Robusta coffee is known to produce an inferior tasting beverage. Another variety, *Coffea liberica* (Liberian coffee) is also grown in some parts of the world and accounts for 1% of the world's production.

Coffee beans are obtained from the coffee fruit (berry), wherein each fruit contains two beans. After harvesting, the beans are separated from the fruit, processed, fermented and roasted before trade. The coffee plant is widely believed to be a native of East Africa and was first used in Ethiopia. The Arabs were the first race who brewed coffee and later on this spread to Egypt and to Europe. This chapter deals with coffee quality, safety, trade, importance of mould proliferation, mycotoxins, prevention of mould contaminants at different stages and positive implications of coffee moulds in natural fermentation.

The Indian Coffee

The coffee industry in India has a rich heritage that dates back to the 19th century when the British introduced coffee plantation on a commercial scale in the Western Ghats. In India, coffee is cultivated mainly in the Southern States: Karnataka (area, 58%; production, 72%), Kerala (area, 25%; production, 20%) and Tamil Nadu (area, 9%; production, 7%) and to a lesser extent in Andhra Pradesh, Orissa and North Eastern states. The state of Karnataka produces both Arabica and Robusta in an equal proportion, while Kerala specializes in the production of Robusta and Tamil Nadu specializes in Arabica. Overall, in India, Arabica and Robusta are produced in equal quantities.

Coffee plantations are instrumental in preserving the precious forest ecosystem in the traditional areas (Muralidhara *et al.*, 2004), while in non-traditional areas, coffee development serves the twin objectives of socioeconomic development of the tribals as well as the preservation of forest ecosystem by checking 'podu' or shifting cultivation. The average productivity of coffee in the country is about 800 kg/ha for Arabica and 1000 kg/ha for Robusta. The productivity of smallholdings is often below the national average, while in large holdings it is up to 2.0-2.50 tons/ha.

Coffee is processed by two methods in India, *viz.*, wet and dry processing (see Fig. 1 for steps involved). Arabica is wet processed to produce parchment coffee, while the major portion of the Robusta is converted into cherry coffee. India's share of coffee is about 4% (300,000 tons) in world production and 4.5% (246,000 tons) of international trade. Over 80% are exported and 20% used for internal consumption. Up to 70% of coffee produced in India is exported to European countries (Italy, Germany, Spain, Belgium, Slovenia, Greece, Netherlands and France). Over the past few decades, India has become a predominantly Robusta producer which accounts for 60% of the total production and 68% of exports. The most preferred coffee from India is the 'monsooned coffee', which finds a niche in the international market and has high demand in Scandinavian countries (Bhushan *et al.,* 2003; Variyar *et al.,* 2003). The structure of coffee marketing chain in India has been depicted in Figure 2.

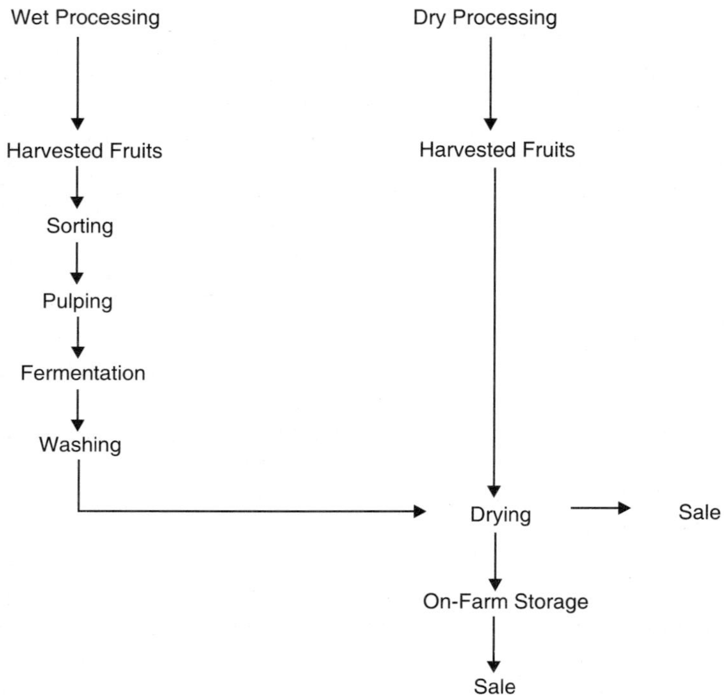

Fig. 1. Types of coffee processing in India.

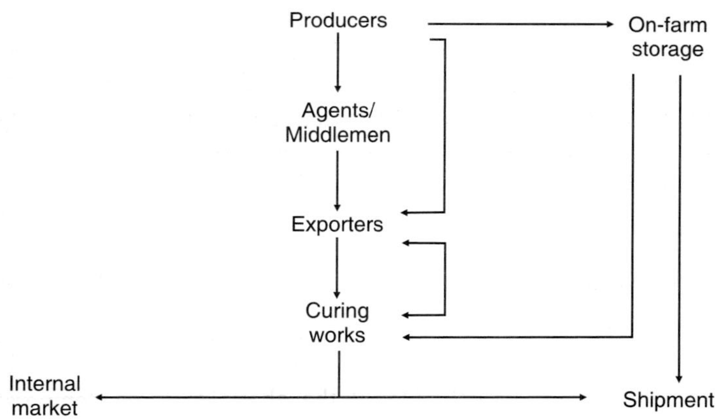

Fig. 2. Structure of coffee marketing chain in India.

Quality, Processing and Safety

The quality and safety aspects of a raw plant produce is a major global challenge. These features have assumed importance throughout the world especially with the establishment of World Trade Organization (WTO), which aims at the harmonization of international trade rules for natural products and services. Food safety is an assurance that a produce will not cause harm to the consumers when it is prepared or consumed according to its intended use. As a member of WTO, India is a signatory to the Sanitary and Phyto-Sanitary (SPS) agreement, which primarily aims to ensure food safety, which directly or indirectly affects International trade. Some of the common quality criteria that influence the coffee quality are the presence of extraneous matters, defects, pesticide residues, hydrocarbon contamination and the presence of toxic trace elements (e.g., arsenic and cadmium). However, the presence of spoilage moulds and their capacity to produce toxic secondary metabolites (mainly mycotoxins) is of principle concern to minimize the economic losses and reduce the potential risks to human and livestock (Ueno, 2000).

In the recent past, the presence of mycotoxins (mainly Ochratoxin A, OTA) in coffee has become a major barrier in international trade, thereby affecting the economy of the producing country. The European Commission (EU) has not yet fixed the maximum levels for OTA in coffee, but has given a recommendation of a reference level of 3 μg/kg (Romani *et al.,* 2000). The occurrence of OTA has attracted worldwide attention ever since the Hungarian health authorities rejected a coffee consignment during 1994 for the first time citing excess levels of OTA. Nine countries have specific regulations for OTA and the legislative limits range from 5 to 50 μg/kg (Boutrif, 1999). According to the estimates of the Institute for Scientific Information on Coffee (ISIC) implemented by FAO, if the EC establishes a regulation for OTA on the proposed level (5 μg/kg), 7% of coffee produce worldwide would exceed this quantity (Boutrif and Bessy, 2001). In addition, the EU has implemented harmonized tolerance limits for OTA in coffee. Imposition of such limits for OTA could have serious repercussions for major coffee producing countries including India. Ultimately, due to rejection of consignments, the farmers or producers will suffer, as the cost for testing coffee lots for OTA levels would be passed on to the producer.

Mould Growth and Mycotoxins

The main factors that favour the mould growth are temperature, water activity (a_w), relative humidity, toxigenic fungal stock (propagules) and the occurrence of competitive microbial growth. Of all the above features, temperature and water activity play an important role in supporting the mould growth leading to spoilage and decline in shelf life (Moss, 1999). For instance, growth of *A. ochraceus* requires 15.2% moisture and 0.76 a_w and OTA production is found to occur between 4-32°C, with optimum at 25°C. The OTA production usually occurs between 7-14 days of fungal contamination and the toxin level is detectable.

Mycotoxins are the toxic secondary metabolites produced by various fungi. Mould contaminated seeds usually consist of mycotoxins, which results in deleterious effects to humans and livestock (D'Mello and MacDonald, 1997; Peraica *et al*., 1999). The main concern of Mycotoxin in coffee are the aflatoxins and ochratoxin A. Studies have shown the toxic, mutagenic and teratogenic effects of aflatoxins in human hepatic and extra hepatic carcinogenesis (Massey *et al*., 1995). The OTA is known for nephrotoxic, teratogenic, carcinogenic and immunotoxic effects (Bayman *et al*., 2002; Höhler, 1998; IARC, 1993) and the provisional daily intake of OTA in foods is extremely low (1-16 ng/kg body wt) (Codex, 1998). Some of the common fungi producing OTA are *Aspergillus carbonarius*, *A. niger*, *A. ochraceus* and *Penicillium* spp. (Moss, 1996; Joosten *et al.,* 2001). *Aspergillus ochraceus* is thought to be the most important OTA producing mould in coffee beans (Le-Bars and Le-Bars, 2000; Pitt, 2000). However, about 80% of *A. carbonarius* strains collected from coffee beans have also been shown to produce OTA (Fungaro *et al*., 2004). Figure 3 shows the colonization of a typical Arabica coffee bean by *A. ochraceous*. Like all other mycotoxins, ochratoxins are also the product of polyketide pathway of fungi that are produced after its active growth has stopped. Even though OTA is heat stable, reports are available on reduction of Ochratoxin A during the process of roasting (Van-der Stegen, 2001; El-Sayed and Shalaby, 1996). Roasting results in two-third reduction of OTA, but this limitation for OTA is not uniform. Each importing nation has its own quarantine strategies about OTA. The presence of aflatoxins in infected coffee beans (infected mainly by *A. flavus*) have been reported by Soliman (2002), wherein the infected samples showed 4.28 and 2.85μg/kg to be present in green coffee beans and in ground roasted coffee beans, respectively. He has also reported the highest yields of aflatoxins in the decaffeinated green coffee beans (24.29 μg/kg) and roasted coffee beans (16.00 μg/kg). However, roasting lowered the concentration of aflatoxins in green coffee beans, wherein it reduced by about 42.2-55.9% indicating aflatoxin degradation on roasting.

Table 1 provides maximum permissible limits imposed for OTA in coffee products by various importing countries. Studies in Europe have indicated the

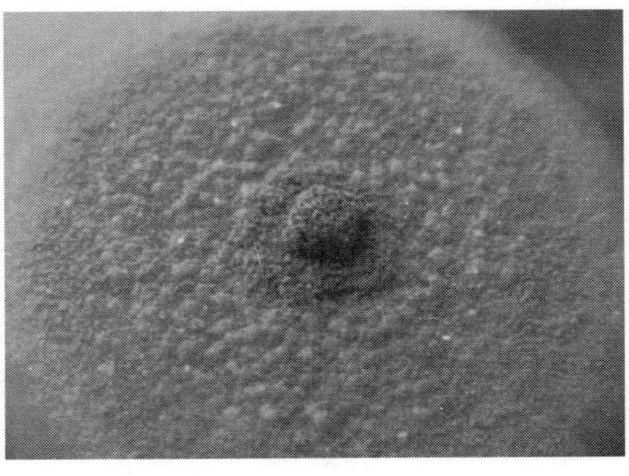

Fig. 3. *Aspergillus ochraceus* on coffee beans.

occurrence of OTA in various commercially available raw, roasted and instant coffee samples originated from different countries. Natural occurrence of OTA in green coffee was first reported by Levi *et al.* (1974) followed by extensive studies (Micco *et al.*, 1989; Studer-Rohr *et al.*, 1994; Nakajima *et al.*, 1997; Patel *et al.*, 1997, Van der Stegen, 1997). The consumption of coffee contaminated with *A. ochraceus* followed by OTA production has been reported to cause health hazards (Boorman, 1989). There are no comprehensive studies on the exact stage of infection/contamination during processing and OTA production in coffee. However, Romani *et al.* (2000) attributed the occurrence of OTA in coffee beans to both environmental (climate, length of storage and transportation) and the processing conditions (wet, mechanical or dry processes). Similarly, OTA has been detected in coffee beans prior to storage, indicating the possibility that harvesting and post-harvest handling of coffee cherries could be the critical stage leading to fungal contamination (Bucheli *et al.*, 2000).

Table1. Maximum permissible limits imposed for Ochratoxin A (OTA) in coffee products by various importing countries (*No limits are prescribed yet)

Countries	Limits fixed for OTA in coffee products (ppb)		
	Green coffee	Roasted coffee	Instant coffee
Czech Republic	10	10	10
Finland	05	05	05
Germany	*	03	06
Greece	20	*	*
Hungary	15	10	10
Italy	08	04	04
The Netherlands	*	10	10
Portugal	08	04	04
Spain	08	04	04
Switzerland	05	05	05
Japan	05	*	*

Processing and Harvesting

Quality has been defined as 'the totality of features and characteristics of a product or service that bears on its ability to satisfy the implied needs of the customer' (ISO 8402-1986). In the liberalized market scenario, quality assumes a paramount importance among the export-oriented coffee. Visual and organoleptic quality of coffee is a cumulative contribution of various parameters. The visual assessment of various parameters in a typical coffee bean contributing to coffee quality is shown in Figure 4. The intrinsic quality of coffee is established at the estate level apart from the internal and external factors such as altitude, chemical composition, bean size and density, age of plants, farm management and climatic conditions. Processing techniques adopted at the estate level are the most important operation, which determines the quality of coffee beverage. The importance of Good Agricultural Practices (GAP), Good

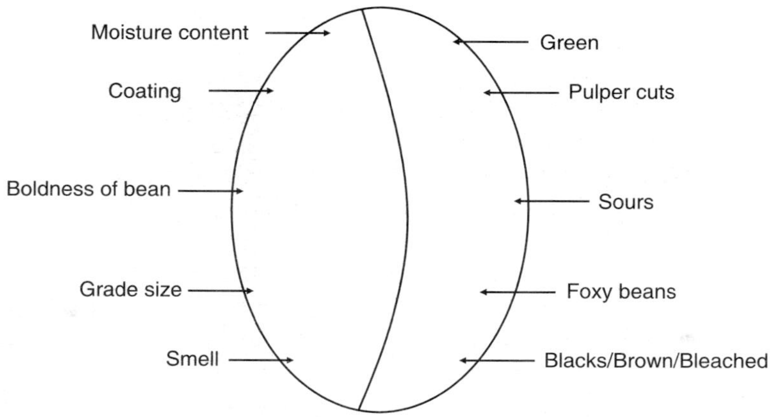

Fig. 4. Parts of a coffee bean determines various quality criteria.

Manufacturing Practices (GMP), Good Hygienic Practices (GHP) and Sanitary and Phyto-sanitary measures (SPS) assume utmost significance to improve the overall quality of coffee at the on-farm level. Processing techniques adopted at the estate levels are the most important operation that determines the quality of the coffee beverage. While careful handling of coffee fruits and appropriate processing techniques can enhance the inherent quality of coffee, faulty and unclean processing will cause an 'off taste' or 'off flavour' in the cup. There are many qualities that decide the processing steps in coffee, among them the fermentation of coffee is the most important one.

Generally, during wet processing, two to three rounds of selective picking are practiced followed by the main harvest and final striping, while in cherry preparation it is uneconomical for selective harvesting and hence stripping is followed. However, the harvested coffee berry has to be sorted out according to the stages of maturity and with the presence of at least greens and ripes for quality improvement. Coffee should never be allowed to dry while on the tree as it encourages mould growth that leads to an inferior quality. It is always advisable to use harvest mats to avoid direct fruit contact with soil and to reduce the quantum of gleanings (gleanings are the fallen fruits obtained during hand picking and will be generally collected after completion of harvest) generated. Time lag between harvest and processing should be minimum (<4 hr) for once the fruits are plucked from the plant it is susceptible for any microbial attack. The harvested fruits (berry) should always be stored in perforated bags under shade to avoid fruit fermentation within bags due to elevated temperature. Tree dried cherries contain 30-45% moisture and over ripe berries are highly susceptible to mould attack compared to ripe berries. Thus, timely harvesting is of utmost important to avoid quality deterioration.

Based on our experience, higher mould incidence occurs in gleanings, tree-dried and over-ripe fruits than to ripe fruits. Figure 5 shows various categories

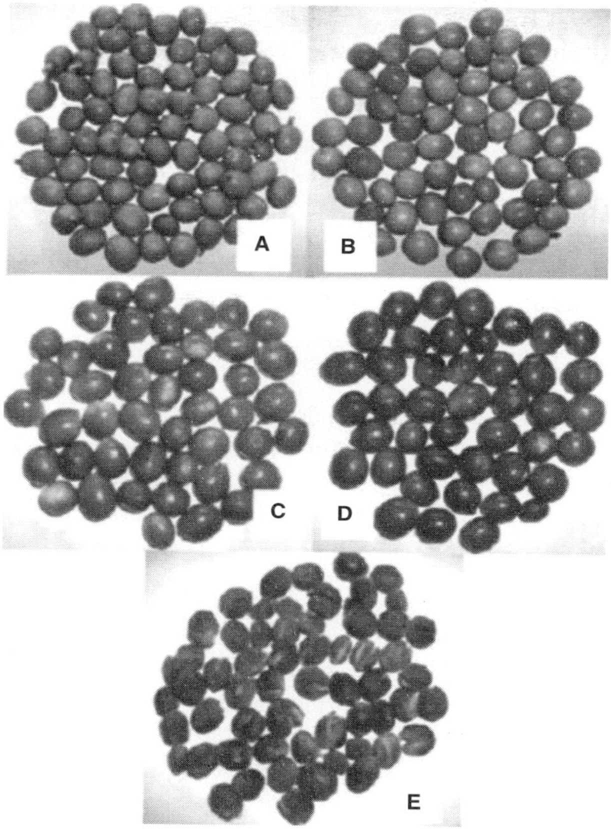

Fig. 5. Categories of coffee berries: greens (a), half-ripe (b), ripe (c), over-ripe (d) and tree dried/splits/diseased (e).

of harvested coffee fruits/berry, while Figure 6 depicts the colonization of coffee fruit and husk by various moulds. Although greens possess lesser fungal contamination, the cup quality was found to be reduced (raw/grassy taste). If the fruit lot consists of different mature categories, the drying gets affected due

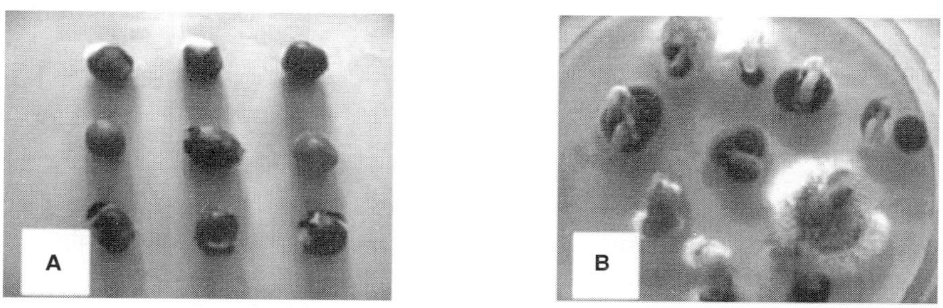

Fig. 6. Moulds grown on coffee berries (a) and husk (b)

to moisture differences in different categories. While evaluating and comparing the effects of the frequency of the mixing of coffee cherries put out for sun drying (on cement concrete) on fungal growth, it revealed that more frequently the coffee cherries were mixed (at least 4-5 times a day) the faster they dry leading to a reduction of fungal growth and proliferation. A delay in processing (time lag between harvest and processing) was found to increase mould incidence, spoiling quality and increasing mycotoxin production. The over-ripe beans, rotten beans or beans that have started fermenting due to delay between picking and processing need to be avoided as they are more susceptible for mould attacks.

Sorting, Grading and Floatation

It is difficult to pick only the ripe fruits during the main harvest, but the harvest lot can be segregated into different fruit maturity categories apart from the removal of leaves and twigs from the coffee lot. Segregation of fruits prior to pulping aids in easy and efficient pulping. Fruits may be graded into lots of two sizes using a suitable sieve to minimize the pulper cuts. Grading of fruits before pulping has been found to reduce the pulper cuts as evident in Arabica.

One of the most important constituents of wet processing is the removal of floats including tree dried fruits in a floatation tank before pulping, which is not achievable in cherry preparation. According to the trials undertaken by us, it was found that pre-washing of coffee cherries before pulping/drying was found to reduce the quantum of mould on the surface in both Arabica as well as Robusta varieties. Similarly, spoiled and rotten (due to mould contamination) cherries float on the surface, while ripe cherries sink to the bottom.

Pulping, Demucilisation, Fermentation and Soaking

Pulped coffee fruits generally do not have the same size and result in higher pulper cuts in beans and more of skinned beans due to more green fruits. Thus, it is advised to adjust the pulper discs and components of pulper to suit the coffee varieties. During fermentation, the pulp/skin in the pulped beans contributes for dark colour (tannins/polyphenols) to the washed parchment apart from inviting unwanted fungal community. In case of direct washing, the fruit skin is found to harbor more harmful fungi as a result of high moisture and a_w during the course of drying. Frank (2001) has indicated that depulping will significantly reduce the risk of OTA contamination during subsequent fermentation and drying steps, but studies on moulds of these processes are necessary. The water used for pulping should be of good quality without silt particles. Maintenance of cleanliness and hygiene is a must in pulping section. We observed that there was high fruit injury in machine washed coffee, which leads to mould contamination. Similarly, pulp contamination in fermenting beans (even at 1%) was known to be prone for mould contamination leading to deterioration in the quality. Residues left unattended in pulper section (e.g., uncleaned fermentation vats, siphon tanks, carry baskets, fruit skins and pulp water) are found to harbor toxigenic moulds that can re-infect the fresh coffee lots if conditions are conducive. Water recycled for more than two days leads to bad odor thereby lowering the cup quality. Sometimes, the water used for cleaning might also harbor toxigenic mould propagules. Hence, recycling of water at pulping section should be restricted to one day and at any point of time, the recycled water should not be used over two days as it deteriorates the coffee quality (leads to ferment taste/foxy beans) apart from microbial contamination in the finished beans.

The demucilisation of coffee is performed by several methods (e.g., natural fermentation, alkali wash, enzyme treatment, and direct machine wash). Among these, naturally fermented coffee is considered to be the washed coffee in the international market, though the difference between fermented and machine washed coffee is not so wide. Fermentation in coffee simply means the process, which is concerned with the degradation/removal of mucilage (a slippery layer of 0.5-2 mm thick around the seed) by combined action of naturally occurring enzymes in coffee fruits and the growth of microorganisms. Our study reveals that semi washed (removal of coffee fruit skin and dried without washing) coffee produced cup quality in between washed and unwashed coffee (cherry coffee). The advantage of semi washed coffee is to avoid pollution due to pulping/washing, but the problem of delivery of pulped fruits to the drying yard and subsequent drying becomes difficult. Apart from this, semi-washed coffee invites more dust, which harbor fungal spores.

Fermentation of coffee, apart from producing good cup quality, favors multiplication of wide variety of bacterial and yeast flora (*Lactobacillus* spp., *Bacillus* spp., *Pseudomonas* spp., *Saccharomyces* spp., *Shizosaccharomyces* sp., *Torula* spp.), which aids in flavor and aroma development during fermentation, apart from inhibiting harmful fungal population (e.g., *Aspergillus niger, A. nidulans*, *A. terreus* and *A. tamari*, *Cladosporium* sp., *Fusarium* sp., *Mucor* sp., *Penicillium* sp. and *Rhizopus* sp.). Our studies revealed that fermentation improved the overall quality of coffee compared to machine washed and fermented coffee and lesser mould incidence compared to non-fermented coffee.

After fermentation/direct machine wash, coffee beans are soaked under freshwater to improve colour and appearance. During soaking, the excess colour and bitterness- producing compounds like phenols and tannins are removed to give a soft cup. Soaking of beans apart from removing unwanted colour, removes residual mucilage left after washing giving a proper appearance to the final dried beans. However, at this stage, care should be taken to minimize the risk of fungal contaminants and the use of freshwater should be encouraged.

Positive Implication of Coffee Moulds

Even though colonized moulds in coffee pose a major risk for the economy, studies showed the occurrence of some of the non-toxigenic moulds in coffee beans and produce industrially useful secondary metabolites (e.g., enzymes and coloring pigments). It is evident that some of the moulds are known to induce flavour (e.g., *Penicllium* sp. in cheese). This is evidenced in 'monsooned coffee' commonly referred to as 'moulded coffee'. Studies need to be initiated to explore the possibilities by employing non-toxigenic fungi to induce flavor in coffee beans. The main constraint involved in the use and popularity of coffee is the presence of caffeine, an alkaloid present in appreciable quantity known to stimulate the central nervous system. Some studies oriented towards decaffeination of coffee beans using chemical methods (e.g., organic solvents or water or supercritical carbon dioxide) (Ramalakshmi and Raghavan, 1999). Some reports are available on the efficiency of microbes in decaffeination of green coffee (Hakil *et al*., 1998; Tagliari *et al.,* 2003). The role of natural non-toxigenic fungi associated with berries and beans of coffee might be of immense value in decaffeination.

Drying

Drying is one of the most important criteria that determine the shelf life of coffee beans. Appropriate drying at adequate moisture and a_w assumes importance to prevent the colonization of storage moulds. Different drying methods have been followed for early drying of the processed coffee beans. An appropriate drying yard is a prerequisite for the production of high quality coffee beans. The drying yard has to be always located in an elevated place of the farm, where maximum sunshine is expected and without any risk of flooding and contamination by animals. A better drying yard should be with a concrete floor, tiled, brick stones, granite or any other hard surfaces. The processed coffee beans should never be dried on bare soil or soil smeared with cow dung slurry since these surfaces produce inferior coffee quality and increases the risk of microbial contamination. In coffee plantations, those who do not have an appropriate drying yard facility can dry their coffee on plastic or tarpaulin sheets (picking mats), provided the coffee is turned frequently to evaporate the condensed moisture in the bottom layer of plastic sheets. The drying yard should be smooth with no crevices and cracks to avoid microbial propagules to harbor and flourish. Our observation revealed that parchment and cherry coffee dried on soil and cow dung smeared floors consist of a high incidence of fungi and mycotoxin contamination apart from increasing the duration required for drying. Such practice lowered the cup quality when compared to the usage of the standard drying surfaces (concrete, tiles and tarpaulin).

Tray drying, thickness and stirring

Once the coffee is washed, it is advised to go for tray drying (surface drying in sunlight) to remove the excess moisture on the surface of coffee beans. We observed that the drying of coffee in a tray (2 m × 1 m × 5 cm) hastens the drying process by removing excess moisture accompanied with a reduction of splitting of parchment. Tray dried coffee beans were found to attract lesser dust and fungal contaminants compared to direct dried beans.

The duration required for drying parchment and cherry varies based on the drying yard used, thickness of the layer, number of stirring/day, the initial water content of beans, and the average sunshine hours at the site, yard temperature as well as the relative humidity. The maintenance of correct layer thickness aids in faster and uniform drying and avoids splitting of parchment and casing of cherry without any mould formation. Coffee should be raked/stirred at least 4-6 times a day for faster removal of moisture to prevent mould growth. The raking devices used should not be hard as it splits the parchment cover and should be provided with rubber beadings. Fresh cherries should not be mixed with partially dried ones. Our experience reveals that beans dried above 5cm thickness in parchment and above 8cm in cherry was found to lower cup quality, increase days of drying and are prone to more fungal and mycotoxin contamination. Stirring and raking 4-6 times /day was found to improve the quality, reduce the drying days and mould contamination in both cherry and parchment coffee.

Prevention of rewetting

During parchment preparation, although covering of beans is a regular practice, it is not followed in cherry due to practical difficulties. Parchment coffee with high moisture should not be covered until it reaches the desired level of moisture (~35%) and requires daily covering after moisture is reached. Before covering, the coffee lots should be heaped, and allowed to attain equilibrium and then covered. If the lot is covered during the hot sun, sweating of beans will lead to rewetting. Our experience revealed that rewetting of coffee lot (particularly during later stage of drying records more fungal incidence compared to initial stage of drying. Rewetting beans were also found to increase duration of drying and lead to bleaching of coffee during storage.

Floats, Gleanings and Defects in Beans

During wet processing, the floats are usually removed by floatation. However, in dry coffee preparation, the floatation method is not used. Instead, the unsorted lot is dried resulting in high mould incidence during drying as well as in the final dried beans. Float coffee is generally weak coffee and are prone to higher colonization by moulds during processing stage rather than while on the plant. Hence, it is necessary to remove floats before processing. The best method to remove floats in cherry preparation is prewashing of fruits before drying. Prewashing of floats not only removes surface mould contamination but also reduces the mould load during drying courses. Gleanings on the other hand are potential sources of mould and mycotoxin contamination in coffee as the number of days for coffee contact with soil increases. Since soil is the reservoir of mould propagules, a long contact of the fruit with soil leads to high contamination and loss of quality. Toxigenic moulds will displace the initial beneficial yeast population on fruit surface as the days of contact with the soil increases. Gleanings should be collected in a short period and must be dried and processed as a separate lot. It should never be mixed with a good lot. Some of the common defects that are encountered in a coffee bean include cut beans, blacks, browns, faded and bleached beans, pulper cuts and insect infested beans. We examined these beans and found highly contaminated with various fungi species (e.g., *Aspergillus flavus, A. niger, A. tamarii, Cladosporium* sp., *Eurotium* sp.) and dominated by *A. niger* like unripe seeds. Hence, as these beans adversely affect the overall quality and acceptability of coffee in trading, care should be taken to minimize or eliminate such defected beans. Other possible means of contamination is the packaging material used (e.g., gunny bags). Bags used for packing chemical fertilizers leads to off flavour and tastes to the coffee beans. This suggests the necessity of employing suitable and safe packing material. Use of toxic fumigants (e.g., ethylene-di-bromide, methyl bromide) to prevent mould contamination in beans during storage will be highly dangerous and result in adverse effect on human health.

Heaping and Winnowing

Heaping of coffee is generally followed in small and marginal plantations, wherein there is inadequate drying yard facility, less fruit quantity for spreading and based on traditional knowledge, heaping hastens drying. Our research finding reveals that heaping fruits for even one day causes deterioration of coffee quality with one hundred percent incidence of the presence of moulds and increase in mycotoxin (mainly OTA). Temperature in a heaped lot was found to rise steadily based on the height of the heap and the moisture in fruits, which favor mould development and finally the loss of quality. As experienced by the farmers, heaping of coffee fruits is found to reduce drying time due to the oozing out of the internal contents of fruits (i.e., loss of moisture), but this always generates poor cup quality and high mycotoxin incidence.

Winnowing is a general practice followed in cherry preparation, where once the coffee reaches the desired moisture content, it will be winnowed before bagging. Winnowing, in addition to improve the visual quality of coffee, removes the inactive and dry mould spores and dust/soil material from the lot. Dust contamination is found to be a potential source of mould contamination in coffee, which cannot be avoided by any means during drying. Hence, winnowing before bagging is the only option left to reduce the dust material in the final stored coffee. If the coffee is not winnowed and stored, there is a higher chance of reinfection by mould species as they get sufficient moisture to grow under inadequate storage conditions. However, air sampling during coffee winnowing recorded higher mould counts apart from recording toxigenic moulds in dust (e.g., *A. ochraceus, Cladosporium* sp. *Fusarium* sp., *Wallemia* sp.). Similarly, dust samples of gleanings and floats showed higher mould population than good coffee lot.

Storage, Curing and Shipment

A clear-cut demarcation between different lots of green coffee beans (processed/ dried at various stages) has to be maintained during storage to retain the identity of individual lots in respect to quantity and quality. Care should also be taken to check that the coffee stacks are not leaning against the walls of the godowns (underground storage buildings), which sometimes might be harboring contaminating moulds. The stacks of raw coffee should be provided with dunnage-made of parchment husk filled in gunny bags, while the clean coffee stacks should be provided with wooden dunnage. Coffee godowns should be rodent-proof and volatile materials (e.g., fertilizers, pesticides) should not be stored inside or near coffee godowns. It is always better to transport the coffee for curing before the onset of the monsoon, as the warm humid conditions favor fast growth of moulds.

In curing works, dust accumulation is found to be a potential source of mould contamination in coffee. From a study carried at curing works during the peak curing activity for dust settling rate, it was found that a higher quantum of dust was generated by huller followed by polisher among curing machineries and the mould spores were found to be more on polisher followed by destoner and huller. The historic dust (dust accumulated over a period of time) was also found to harbour high percentage of toxigenic moulds. Hence, proper facility should be provided in curing works to remove/avoid the dust particle settling on the curing machineries to prevent contamination by moulds and mycotoxins.

Coffee beans being hygroscopic absorb atmospheric moisture readily and this poses a problem during shipment to long distances. Thus, during shipment, care should be exercised on the technicality (such as air-tight containers, containers free of holes and free from corrosion on the roof or sides, intact door locks, rubber and sealing devices). Bags should be appropriate and without leaks, slack or torn, wet and stained. The saddle stow is the best way to stow bags in a container as it minimizes air circulation between the bags and reduces transport of moisture to cold spots.

Waste management

There are two major waste products of coffee processing that can be considered: firstly, solid (coffee pulp/skin) and secondly, liquid waste (coffee waste water). Generally, the solid wastes produced during coffee processing are not considered to be a waste or a major pollutant due to its use as compost in coffee farms. However, we found that coffee wastes, if not completely composted, harbours various harmful moulds (particularly *A. ochraceus*) that can survive on the partially composted material in the vicinity. This may find a way to enter coffee habitat and contaminate the system. However, compared to solid wastes, the liquid waste accumulated during coffee processing poses serious risks to the environment. The main problem in wastewater is the removal of solids, which are found to degrade faster upon non-removal by the action of microbes resulting in lowering the pH and elevates BOD, COD and total solids. Hence, a proper neutralization (lime, 1g and 1.5 g/liter of Arabica and Robusta effluent, respectively) of coffee effluent is a necessity. We observed that the initial ochraceous load in fruits came down from 74.0 to 11.5% in dried beans after passing through fermentation, washing and drying during wet processing. The total reduction of *A. ochraceus* was found to be 84.5%. This clearly indicates that even if coffee gets infected with ochraceous during the initial stages, its load could be brought down through wet processing. Experiments were also carried out on the status of OTA in coffee seeds that are artificially contaminated by *A. ochraceus* and accordingly the data has been presented in Table 2. It was found that washed parchment is highly susceptible for mould infection compared to fruit and pulped beans during wet processing. So also, the data on the status and survival of inoculated *A. ochraceus* in fruits to dried beans have been depicted in Figure 7, wherein higher survival rates were recorded in freshly harvested fruits.

Table 2. Ochratoxin A contamination in artificially infected coffee samples (Source: Panneerselvam *et al.* 2004)

Infection	OTA (ppb)
Artificial infection in fruits	0.40
Artificial infection in pulped beans	0.48
Artificial infection in washed parchment	1.25
Control (no infection)	Not detectable

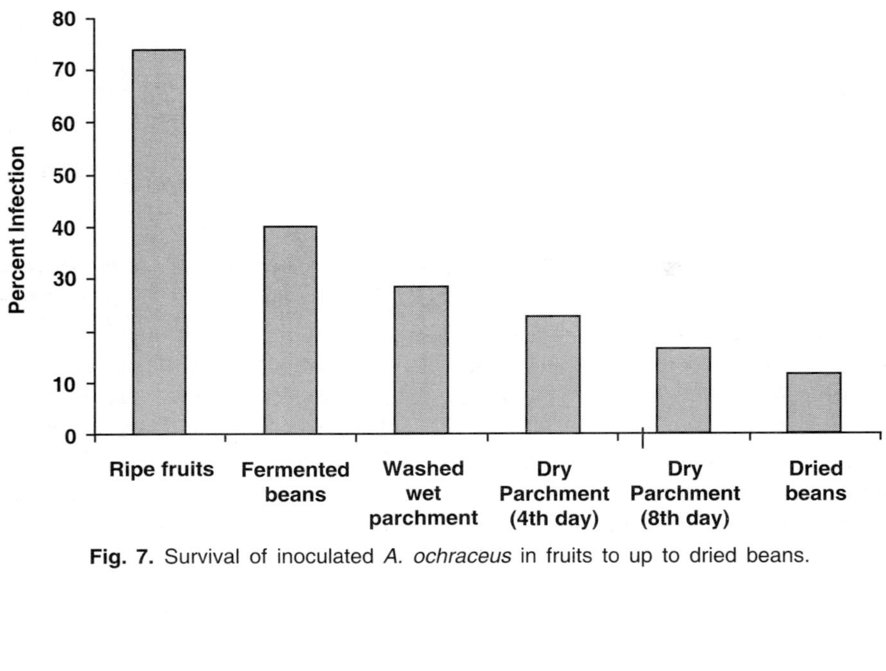

Fig. 7. Survival of inoculated *A. ochraceus* in fruits to up to dried beans.

Conclusions and Outlook

It is evident that coffee beans harbor a wide range of moulds and toxins, which cause major economic loss to the producing country. Hygienic and good manufacturing practices need to be followed at each step of processing for the production of quality coffee to fetch a good price. As coffee seeds are highly hygroscopic, they tend to absorb moisture from the surrounding environment particularly during storage and are highly susceptible for storage moulds. Future investigations have to be aimed at employing alternate methods of preservation (e.g., modified atmospheric packaging, natural biopreservatives, ionizing radiations) along with the recent advances of hurdle technology (stress reactions, metabolic exhaustion of microorganisms, homeostasis and multi target prevention) with Hazard Analysis and Critical Control Point (HACCP) approach for the maintenance of safety and quality of coffee beans.

Acknowledgements

The authors are grateful to Dr. R. Naidu, Y. Raghuramulu, T.N. Gopinadhan and P. Paneerselvam, Coffee Board, Government of India for the technical suggestions and timely help.

References

Bayman, P., Baker, J.L. and Mahoney, N.E. 2002. *Aspergillus* on tree nuts: incidence and associations. *Mycopathologia 155*, 161-169.

Bhushan, B., Bhat, R. and Sharma, A. 2003. Status of free radicals in Indian monsooned coffee beans g- irradiated for disinfestation. *J. Agric. Food Chem. 51*, 4960-4964.

Boorman, G.A. 1989. Toxicology and carcinogenesis studies of Ochratoxin A in F344/N rats. In: *NTP Technical Report # 23*. National Toxicology Program, Research Triangle Park, NC.

Boutrif, E. 1999. Recent international developments in the field of mycotoxins. In: *18th ASIC Coffee Conference*, Helsinki, Finland, 239-247.

Boutrif, E. and Bessy, C. 2001. Global significance of mycotoxins and phycotoxins. In: *Mycotoxins and Phycotoxins in Perspective at the Turn of the Millennium* (Ed. De Koe, W.J., Samson, R.A., Van Egmond, H.P., Gilbert, E. J. and Sabino, M.), 3-16.

Bucheli, P., Kanchanomai, C., Meyer, I. and Pittet, A. 2000. Development of ochratoxin A during Robusta (*Coffea canefora*) coffee cherry drying. *J. Agric. Food Chem. 48*, 1358-1362.

CODEX 1998 (Codex Alimentarius Commission). Position paper on ochratoxin A. Food and Agricultural Organization of the United Nations, Rome, Italy.

D'Mello, J.P.F. and MacDonald, A.M.C. 1997. Mycotoxins. *Animal Sci. Technol. 69*, 155-166.

El-Sayed, A.M.A. and Shalaby, A.R. 1996. Incidence of green coffee beans contamination with ochratoxin-A in Egypt and its decomposition during coffee roasting. *J. Agric. Sci. 27*, 137-142.

Frank, J.M. 2001. On the activity of fungi in coffee in relation to ochratoxin A production. In: *19th ASIC Coffee Conference*, May 14-18, 2001, Trieste, Italy.

Fungaro, M.H., Vissotto P.C, Sartori D., Vilas-Boas, L.A., Furlaneto, M.C. and Taniwaki M.H. 2004. A molecular method for detection of *Aspergillus carbonarius* in coffee beans. *Curr. Microbiol. 49*, 123-127.

Hakil, M., Denis, S., Viniegra-Gonzalez, G. and Augur, C. 1998. Degradation and product analysis of caffeine and related dimethylxanthines by filamentous fungi. *Enzyme Microb. Technol. 22*, 355-359.

Höhler, D. 1998. Ochratoxin A in food and feed: Occurrence, legislation and mode of action. *Zeitschrift für Ernährungswissenschaft 37*, 2-12.

IARC 1993 (International Agency for Research on Cancer). Ochratoxin A - Monographs on the Evaluation of Carcinogenic Risks to Humans: Some Naturally Occurring Substances, Food Items and Constituents. In: *Heterocyclic Amines and Mycotoxins*, IARC, Lyon, 489-521.

Joosten, H.M.L.J., Goetz, J., Pittet, A., Schellenberg, M. and Bucheli, P. 2001. Production of ochratoxin A by *Aspergillus carbonarius* on coffee cherries. *Int. J. Food Microbiol. 65*, 39-44.

Le-Bars, J. and Le-Bars, P. 2000. Mycotoxigenesis in Grains- application to mycotoxic prevention in coffee. In: *Coffee Biotechnology and Quality*. Kluwer Academic Publishers, Sera, Netherlands, 355-368.

Levi, C., Trenk, H.L. and Mohr, H.K. 1974. Study of the occurrence of Ochratoxin A in green coffee beans. *J. AOAC. 57*, 866-870.

Massey, T.E., Stewart, R.K., Daniels, J.M. and Ling, L. 1995. Biochemical and molecular aspects of mammalian susceptibility to aflatoxin B_1 carcinogenicity. *Proc. Soc. Experimental Biol. Med. 208*, 213-227.

Micco, C., Grossi, M. Miraglia, M., Brera, C. 1989. A study of the contamination of Ochratoxin A of green and roasted coffee beans. *Food Add. Contam. 6*, 933-939.

Moss, M.O. 1996. Mode of formation of ochratoxin A. *Food Add. Contam. 13*: 5-9.

Moss, M.O. 1999. Problems caused by fungi. In: *Encyclopedia of Food Microbiology* (Ed. Robinson, R.K., Batt, C.A. and Patel, P.D.). Academic Press, London, 2056-2062.

Muralidhara, H.R, Bhat, R. and Velmourougane, K. 2004. Biodiversity in coffee plantations - Need of conservation. *Planter's Chronicle*, March 2004, 11-16.

Nakajima, M., Tsubouchi, H., Miyabe, M. and Ueno, Y. 1997. Survey of aflatoxin B1 and ochratoxin A in commercial green coffee beans by high-performance liquid chromatography linked with immunoaffinity chromatography. *Food Agric. Immunol. 9*, 77-83.

Panneerselvam, P., Velmourougane, K., Gopinandhan, T.N., Raghuramulu, Y. and Naidu, R. 2004. Fate of artificially inoculated *Aspergillus ochraceus* and OTA production during wet processing of coffee. *Proceedings of 20th ASIC*. Bangalore, India, October, 2004.

Patel, S., Hazel C.M., Winterton, A.G.M. and Gleadle, A.E. 1997. Survey of ochratoxin A in UK retail coffees. *Food Add. Contam. 14*, 217-222.

Peraica, M., Radiæ, B., Luciæ, A. and Pavloviæ, M. 1999. Toxic effects of mycotoxins in humans. In: *Bulletin of WHO # 77*, 754-766.

Pitt, J.I. 2000. Toxigenic fungi: which are important? *Medical Mycol. 38*, 17-22.

Ramalakshmi, K. and Raghavan, B. 1999. Caffeine in coffee: its removal. Why and how? *Crit. Rev. Food Sci. Nutr. 39*, 441-456.

Romani, S., Sacchetti, G., Lo´pez, C.C., Pinnavia, G.G. and Rosa, M.D. 2000. Screening on the occurrence of ochratoxin A in green coffee beans of different origins and types. *J. Agric. Food Chem. 48*, 3616-3619.

Soliman, K.M. 2002. Incidence, level, and behavior of aflatoxins during coffee bean roasting and decaffeination. *J. Agric. Food Chem. 50*, 7477-1481.

Studer-Rohr, L.D.R., Schlatter, D.J. and Schlatter, C. 1994. The occurrence of Ochratoxin A in Coffee. *Food Chem. Toxicol. 33*, 341-355.

Tagliari, C.V., Sanson, R.K., Zanette A., Franco, T. T. and Soccol, C. R. 2003. Caffeine degradation by *Rhizopus delemar* in packed bed column bioreactor using coffee husk as substrate. *Braz. J. Microbiol.* 34, 102-104.

Ueno, Y. 2000. Risk of multiexposure to natural toxins. *Mycotoxins 50*, 13-22.

Van der Stegen, G., Jörissen, U., Pittet, A., Saccon, M., Steiner, W., Vincenzi, M., Winkler, M., Zapp, J. and Schlatter, C. 1997. Screening of European coffee final products for occurrence of ochratoxin A (OTA). *Food Addit. Contam. 14*, 211-216.

Van der Stegen, G.H., Essens, P.J. and Van der Lijn. J. 2001. Effect of roasting conditions on reduction of ochratoxin A in coffee. 2001. *J. Agric. Food-Chem. 49*, 4713-4715.

Variyar, P.S., Ahmad, R., Bhat, R., Niyas, Z. and Sharma, A. 2003. Flavoring components of raw monsooned arabica coffee and their changes during radiation processing. *J. Agric. Food Chem. 51*, 7945-7950.

Frontiers in Fungal Ecology, Diversity and Metabolites (edited by K.R. Sridhar), I.K. International Publishing House Pvt. Ltd., New Delhi, 2009.

16

Fungal Infections in Humans

Jerome Pinto[1], M. Manjunath Shenoy[1*], B.S. Girisha[1] and M. Suchitra Shenoy[2]

[1]*Department of Skin and Venereal Diseases, K.S. Hedge Medical Academy, Deralakatte 574 160, Mangalore, Karnataka, India*

[2]*Department of Microbiology, Kasturba Medical College, Mangalore 575 001, Karnataka, India*

*(*E-mail: manjunath576117@yahoo.co.in)*

Introduction

Medical mycology is the study of fungal epidemiology, ecology, pathogenesis, diagnosis, and treatment. About 200 species of fungi are known to cause or associated with human disease (Malcolm and David, 2003). In healthy individuals, fungal infections are usually benign, but a few of them cause life-threatening illness. The incidence of fungal infections in the human has increased in the last decade. This may be due to immuno-compromised status such as use of antibacterials, immunosuppressive agents, anticancer agents, transplantations and diseases like AIDS (Chander, 2002; Lortholary and Dupont, 1997). Most of the fungal infections are acquired from an exogenous source in the environment. The mode of transmission could be through inhalation, ingestion or traumatic implantation.

The pathogenic fungi can exist as yeasts or as moulds. Yeasts are unicellular organisms that reproduce by budding, while moulds are filamentous or mycelial and are septate or non-septate. Some fungi occur in both unicellular and mycelial forms. Fungi can be classified as endogenous or exogenous depending on the route of infection (Fridkin and Jarvis, 1996). Endogenous infections originate from normal flora when the host immune system is suppressed. Fungi may be primary pathogens or opportunistic pathogens depending on their interaction with the host's immune response. Primary pathogens cause disease even when the host immunity is intact, while opportunistic pathogens cause disease in the immuno-compromised host (Musial *et al.*, 1988).

Clinically, fungal infection can be classified according to the type and degree of tissue involved. The fungal infections may therefore be classified as superficial fungal infections, subcutaneous fungal infections, systemic fungal infections, and opportunistic fungal infections. This Chapter gives a brief account on the common fungal diseases of human beings. The causative fungus, clinical disease, pathogenesis, diagnosis and the line of treatment in each case of infection have been described briefly to emphasize the importance of the increased rate of fungal infection in human beings.

Pathogenesis, Immunity and Diagnosis

Fungal infection depends on the inoculum size and the resistance of host. The severity of the infection depends on the immunological status of the host. Some of the virulence factors of fungi responsible for human infection include thermo-tolerance, enzyme/toxin production and the ability to evade or suppress host defense mechanisms. The thermo-tolerance helps the fungus to survive and replicate at 37°C. The enzymes/toxins may lead to tissue necrosis (Hogan *et al.*, 1996). The surface components like cell wall and capsule help fungus to adhere to the host tissues. Most fungi can resist the effects of the active oxygen radicals released during the respiratory burst of phagocytes (Chander, 2002). The fungi spread through the blood stream (hematogenous) to distant sites in the body via the phagocytes.

The host's defense to fungal infections involves both cell-mediated and humoral immunity. Cell-mediated immunity (CMI) appears to be more important because patients with defects in CMI usually suffer more severe fungal infections than those with depressed humoral immunity. In fungal infections, non-specific barriers play a significant role as the primary defense. The intact skin and mucous membrane, the fatty acids in sebum (secretions of sebaceous glands) and the antifungal substances secreted by the surfaces contribute as major primary barriers to fungal infections (Wagner and Sohnle, 1995). The normal bacterial flora of humans and the epithelial turnover rate prevent colonization by fungi. Neutrophils are the first line of defense. The activated Neutrophils ingest and kill the fungi. They also release hydrogen peroxide and other cellular products, which are toxic to fungi. The monocyte-macrophage system appears to be the most efficient defense against systemic fungal infections.

The clinical symptoms and the clinical history may provide the diagnosis of infection with fungal etiology. Depending on the type of infection, specimens are collected and tests are performed. Microscopic examination of the skin scrapings, nail fragments and hair might show the presence of dermatophytes (Fig. 1). Vaginal discharge or respiratory secretions may be taken in case of candidiasis. In tissue biopsy, two fragments will be taken; one is used for histological examination using special stains, while the other for culture on simple media like Sabouraud's agar with antibiotics (gentamicin or chloramphenicol) (Fig. 2). Most of the fungi grow at optimum temperature

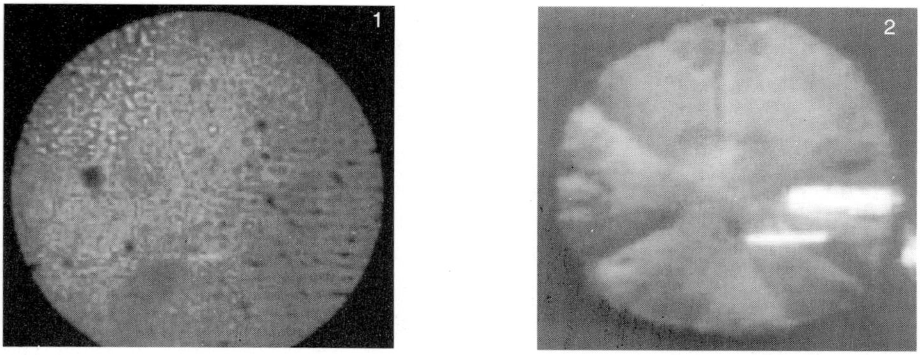

Fig. 1–2. (1) A KOH mount of skin scrapings showing septate hyphae of a fungus; (2) A culture on Sabouraud's agar showing growth of *Trichophyton violaceum.*

(25-30ºC), while the dimorphic fungi, the parasitic phase can be seen on incubating the cultures at 37ºC in an enriched medium (e.g., brain heart infusion agar). In some fungal diseases, serology may be used to detect fungal antigens or host antibodies (Fisher and Cook, 1998). Most of the individuals are exposed to the fungus in the environment and will form antibodies spontaneously, which makes the interpretation of serology results quite difficult. However, the detection of fungal antigens in the lesions is very useful. The recent molecular biological techniques have the ability to detect and identify specific fungal pathogens in this era of increasing fungal infections (Yeo and Wong, 2002).

Treatment and Prevention

There are fewer antifungal agents compared to antibacterial agents. Often fungi infect less vascularized tissues. Thus, the penetration of the drugs is less and the dose is high, which may be toxic to the human tissue (Chander, 2002). It is difficult to treat fungal diseases as both human and fungal cells are eukaryotic and therefore, the drugs become toxic to human tissues. The antifungal drugs act on the sterol in the fungal cell membrane (ergosterol), while in humans it acts on cholesterol. Azoles, allylamine antifungal drugs inhibit the ergosterol synthesis and result in death of the fungus (White *et al.*, 1998). Polyenes combine with fungal cell membrane and interfere with vital cellular processes (e.g., respiration and glucose utilization). Other antifungal agents inhibit DNA and RNA synthesis (e.g., 5-fluorocytosine) or mitosis (e.g., griseofulvin) (Ghannoum and Rice, 1999; Mukherjee *et al.*, 2005). In patients with underlying disease such as diabetes, the management of the disease will reduce the incidence of the fungal infections. In patients who are under cancer chemotherapy, suffering from AIDS, or have undergone organ transplantation, antifungal agents are given as prophylaxis to prevent any fungal infections (Lortholary and Dupont, 1997).

Thus, personal hygiene is important and prevention of fungal disease is always better than treatment.

Superficial Mycoses

Superficial mycoses refer to the infection of the outer layers of the stratum corneum of skin or the extra-follicular part of the hair shaft (Table 1). As the infection is confined to the dead keratin tissue, there will be no direct immune response and minimal physical discomfort to the host. In such cases, the patients visit the doctor more for cosmetic reasons.

Table 1. Causative agents and clinical presentations of superficial mycoses.

Disease	Causative agent	Clinical presentation
Dermatophytosis	*Epidermophyton* (*E. floccosum*) *Microsporum* (*M. audouinii*) and *Trichophyton* (*T. tonsurans*)	Ring shaped/arcuate/polycyclic lesions with active margins
Piedra	*Trichosporon beigelii* (white piedra)	Light-colored, soft and loosely attached nodules occur on the beard hairs
	Piedraia hortai (black piedra)	Dark brown or black, gritty and firmly-attached nodules occur on the scalp hairs
Tinea nigra	*Exophiala werneckii*	Flat, brown or black patches with irregular pigmentation localized to palms of hands and soles of feet.
Tinea/Pityriasis versicolor	*Malassezia furfur*	Areas of hypo- or hyper-pigmentation of the skin on the back and trunk or limbs.

Dermatophytosis

The term, dermatophytosis denotes infection of the skin, hair, and nail by a group of filamentous fungi called dermatophytes (Wagner and Sohnle, 1995; Weitzmani and Summerbell, 1995). These fungi belong to three genera, namely *Epidermophyton, Microsporum* and *Trichophyton*. Out of about 40 species, 10 are commonly associated with human infection (Malcolm and David, 2003). They are geophilic, zoophilic or anthropophilic in nature. They are keratinophilic and infect the keratinized layer of the skin (stratum corneum), hair on the scalp and the nails. Clinically, they cause infection of various body regions named depending on the type of infection, *Tinea capitis*: dermatophyte infections of skin and hair of scalp region; *Tinea faciale*: infections of non hairy face; *Tinea barbae*: infections of the hairy face; *Tinea corporis*: infections of the trunk (Fig. 3) and limbs excluding the hands, feet and groin; *Tinea cruris*: infections of the groins, pubic region and perianal region; *Tinea manuum*: infections of the hands; *Tinea pedis*: infections of the feet; *Tinea unguium*: infections of nails (Fig. 4).

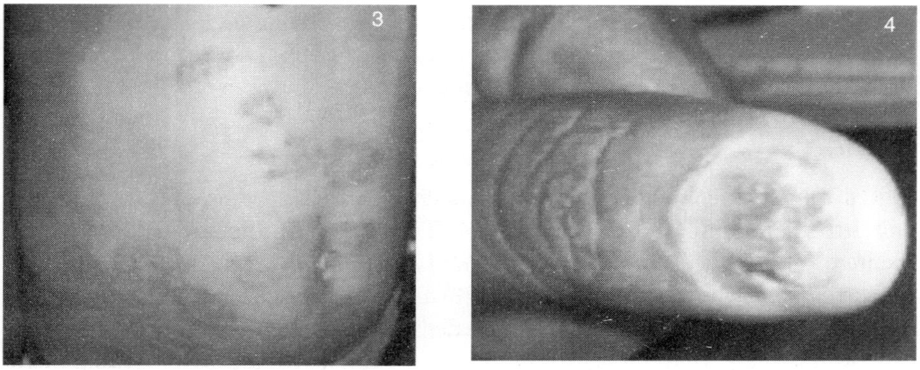

Fig. 3–4. (3) *Tinea corporis* involving abdominal skin; (4) *Tinea unguium* involving the finger nail.

Appearance of lesions is variable but in a typical case they are characterized by well-circumscribed reddish, scaly and dry ring-like lesions ('ring worm'). As it extends peripherally, fresh ring-like lesions appear around the margins and may fuse to give a gyrated appearance. The immuno-suppressed people like HIV-infected patients often show less obvious but extensive lesions. Scalp infections vary from a mild scaling to an extensive area of hair loss, to a highly inflammatory lesion. The latter is called 'Kerion' and is caused by zoophilic fungi. Favus is a special type of scalp infection that is seen with inflamed and crusted lesions of the scalp eventually showing through patches of permanent hair loss. *Trichophyton schoenleinii* is the sole etiologic agent of favus. Foot infections commonly present with maceration of toe clefts with itching, peeling fissuring and pain. Other presentations with thickening of sole of foot (hyperkeratotic variety) and blistering of feet (vesicular variety) can also occur. Nails involved with dermatophytosis show thickening, friability, discoloration and onycholysis (separation of nail plate from the nail bed).

Skin diseases such as dandruff, psoriasis, candidiasis and eczemas can resemble the dermatophyte infections and may have to be distinguished by suitable laboratory tests. Mount of the skin scrapings in 10% KOH may demonstrate septate hyphae and spores (Fig. 1). Skin scrapings are cultured on Sabouraud's (SAB) agar incubated for 2-4 weeks and causative fungus can be identified by growth characteristics and biochemical reactions (Fig. 2). Dermatophyte infections are generally treated with local application of antifungal agents: imidazole compounds (miconazole, clotrimazole, ketaconazole, oxiconazole, bifonazole) and allylamine compounds (terbinafine) are the commonly used topical antifungals. They are used for a period of 2-4 weeks and all compounds give a similar high cure rates. When they do not respond adequately to topical therapy and the infection is extensive, oral treatment with antifungal drugs is needed for the scalp and nails. Griseofulvin, azoles (ketaconazole, itraconazole and fluconazole) and allylamines (terbinafine) are the various compounds that are given orally for a variable period of time and they are effective in managing these infections.

Piedra

It is a fungal infection of the hair shaft characterized by the presence of firm and irregular nodules (Chander, 2002). Two types of Piedra includes, white piedra: caused by *Trichosporon beigelii* (*T. cutaneum*), affects the beard and mustache; black piedra: caused by *Piedraia hortai* affects the scalp hair. The hair can be plucked and a KOH mount demonstrates the presence of the nodule around the hair shaft. The fungi may be seen as a mass of hyphae with ascospores (*Piedraia hortai)* or arthrospores (*Trichosporon beigelii)*. Culture of the hair on SAB agar or mycosel can also be used for the laboratory diagnosis. Usually, removal of hairs gives a complete relief from the disease.

Pityriasis versicolor (Tinea versicolor)

It is a superficial fungal infection of the horny layer of the epidermis, which is characterized by irregular patches of white or black pigmentary changes usually present on the trunk but occasionally on arms, legs, neck or face as well (Wagner and Sohnle, 1995; Marconi and Powell, 1992). It is caused by *Malassezia furfur (Pityrosporum orbiculare, P. ovale, P. furfur)*. This lipophilic yeast is part of the normal flora of humans (>95%). Heat and moisture may be contributing factors for infection and it presents as patches or areas of hypo- or hyper-pigmentation localized to back, trunk, limbs and rarely on the face. Lesions are covered with dust like scales and are mostly asymptomatic. Clinically, observation under the Wood's lamp may show a pale yellow fluorescence and 10% KOH mount of the skin scrapings may demonstrate round, budding yeasts with short hyphae commonly known as 'spaghetti' and 'meatball' appearance. Skin scrapings are cultured on SAB agar with an olive oil overlay. Application of 20% sodium hyposulfite or selenium sulfide gives good results. Azole antifungals given as local application or as oral tablets are also effective, but recurrences are common.

Tinea nigra

It is a superficial fungal infection of the palms caused by *Exophiala werneckii*, which presents as brown to black patches (Chander, 2002). The KOH mount of the skin scrapings show brown septate hyphae. Culture on SAB agar and microscopic examinations will confirm the diagnosis. For the treatment, local application of keratinolytic agents is effective.

Subcutaneous Mycoses

These are infections caused by a heterogenous group of fungi characterized by the development of lesions at the site of traumatic inoculation into the skin. Fungi causing subcutaneous mycoses can cause disease with high morbidity and mortality if there is disruption of the skin barrier or underlying systemic conditions (e.g., diabetes mellitus, AIDS, immunosuppressive therapy following organ transplantation). Usually, infection involves the dermis, subcutaneous tissue, fascia and sometimes bone or other tissues at the site of inoculation. However, some of them may lead to pulmonary, gastrointestinal and renal, or disseminated disease via the lymphatics or hematogenously (via blood). The causative fungal agents are commonly saprophytes. Some of these fungi have a worldwide distribution, while some are restricted to tropical or subtropical regions (Table 2). Routes of infection include, direct inoculation into deeper tissues (e.g., thorn prick, stepping on a stick, burn, motor vehicle accident), inhalation of air-borne fungal cells and implantation into the paranasal sinuses and lungs, and iatrogenic (e.g., intravenous catheter, contaminated dressings). The subcutaneous mycoses are difficult to treat and medical or surgical (excision or amputation intervention) modalities may have to be combined.

Table 2. Causative agents and diagnostic feature of subcutaneous mycoses.

Disease	Causative agent	Diagnostic feature
Chromoblastomycosis	*Fonseceae compacta, F. Pedrosoi, Phialophora verrucosa*	Verrucous plaques and microscopy shows 'copper pennies'
Lobomycosis	*Loboa loboi*	Painless warty nodules and ulcers, and microscopy demonstrates the intra cellular yeasts
Mycetoma	*Acremonium* spp., *Madurella grisea, M. mycetomatis*	Sinuses with discharge of grains of different colours
Phaeohyphomycosis	*Alternaria* spp., *Bipolaris* spp., *Cladosporium bantianum, Exophiala jeanselmei*	Subcutaneous nodules and cysts
Sporotrichosis	*Sporothrix schenckii*	Ulcerating nodules in linear fashion and microscopy shows 'asteroid bodies'
Zygomycosis	Mucorales: *Absidia, Apophysomyces, Cokeromyces, Cunninghamella, Mortierella, Mucor,Rhizomucor, Rhizopus, Saksenaea* Entomophthorales: *Basidiobolus* and *Conidiobolus*	Involve paranasal sinuses, orbit, face, nose and brain

Chromoblastomycosis (Chromomycosis)

It is a localized, chronic fungal infection of skin and subcutaneous tissues characterized by the development of warty or tumor-like lesions usually confined to the lower legs and feet (Chander, 2002; Revankar, 2007). The causative agents are soil inhabiting fungi, *Fonseceae compacta, F. pedrosoi* and

Phialophora verrucosa, which are common in tropics. The disease is characterized by raised, crusted, and warty lesions. The lesions are localized to the site of fungal infection and get secondary infection by bacteria, which complicates the situation. The involvement of bone and muscle is very rare.

Microscopic examination of crusts in KOH shows the presence of irregular, 10-20 μm large, brown-walled elements with transverse septa, 'sclerotic cells' or 'copper pennies' (Fig. 5). A histological examination of the biopsy may show the sclerotic bodies and the granulomatous reaction of the host. Culture on Sabouraud agar with antibiotics may help to isolate the fungus. The slow growing dematiaceous fungi are difficult to speciate. Chromomycosis is extremely resistant to antifungal treatment. Itraconazole (200-400 mg/day) and terbinafine (500 mg/day) if given for 6-12 months has resulted in a high degree of cure rate. Local heat application and cryotherapy (freezing) for lesions with limited extent has given good results. New azoles such as variconazole and posaconazole have high response *in vitro*, but further clinical data are necessary to use them routinely. For smaller lesions, a surgical excision with grafting can also be performed.

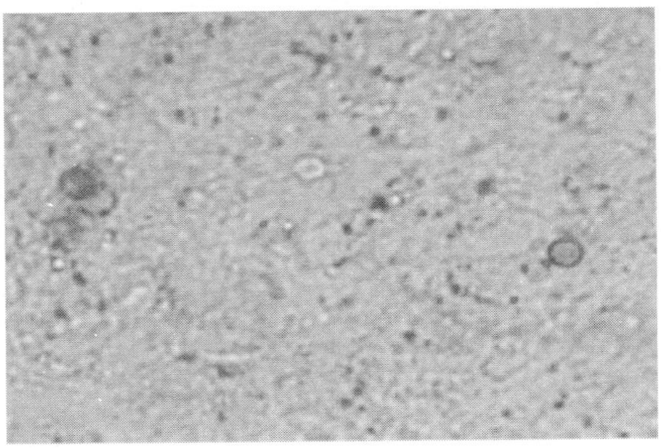

Fig. 5. A KOH mount showing 'copper pennies' in chromoblastomycosis.

Lobomycosis

It is a chronic progressive nodular and ulcerative fungal infection of the skin and subcutaneous tissue (Chander, 2002). The causative agent, *Loboa loboi* can be isolated from the soil in rural areas of South America, Central America and Amazon region. Lobomycosis affects the exposed areas like the limbs and the face. It often develops at sites of minor trauma as small papules or pustules, which are itchy. The lesions become verrucoid or lobulated and have crusty plaques. They are well defined and painless. They do not adhere to the deeper tissues so they are freely mobile. There can be lymphatic spread to proximal sites from the extremities. There are no systemic symptoms, but the histopathological examination of the skin biopsy demonstrates the presence of spherical intracellular yeast cells. These cells possess a thick cell wall and are present singly or in chains within the macrophages. The yeast cells may be linked by tubules. Grocott silver-methenamine or haematoxylin and eosin stains can be used to stain the organism. Asteroid bodies in giant cells may be seen in the biopsies. The culture of the organism has not yet been obtained. Only surgical treatment with total resection of the small lesions is feasible. Repeated cryotherapy is also useful, however, there is no medical line of treatments.

Mycetoma (Madura foot/Maduromycosis)

This is a chronic suppurative lesion characterized by localized, swollen lesions of skin and subcutaneous tissues with the later involvement of muscle and bone (Lichon and Khachemoune, 2006). It has a triad of symptoms: tumefaction (swelling), draining sinuses and discharge of grains. Mycetomas are caused both by fungal (eumycotic mycetoma, eumycetoma) and also actinomycetes (actinomycotic mycetoma, actinomycetoma). The fungi causing eumycetoma are all soil saprophytes. They are found worldwide but are mostly confined to the subtropical areas (Table 2). Various agents causing eumycotic mycetoma include: *Madurella mycetomatis, M. grisea, Leptospheria senegalensis, Acremonium* species. Actinomycotic mycetomas are caused by *Actinomadura pelleteri, A. madurae, Nocardia brasiliencis, N. asteroides*, and *Streptomyces somaliensis*. Usually, it affects the hand or foot following traumatic implantation. Infection is characterized by an initial nodule at the site of the inoculation, which ulcerates and forms draining sinuses that discharge purulent fluid containing 'grains', the compact masses of the etiologic agent. The colour of the grain depends on the etiology: black granules are always of fungal origin (e.g., *Madurella mycetomatis*); small red granules are specific for the actinomycotic (e.g., *Streptomyces pelletieri*); small white grains are specific for nocardial (*Nocordia*); whitish-yellow granules are fungal or actinomycotic in nature. Lesion typically enlarges and involves underlying tissues and leads to severe bone damage (Fig.6).

The differential diagnosis between fungal and actinomycotic mycetomas is

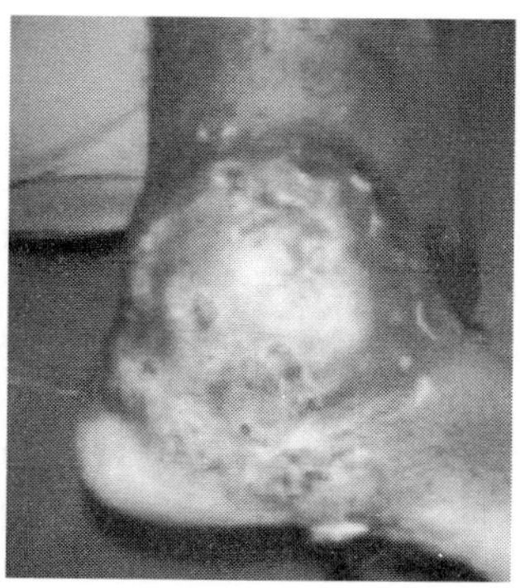

Fig. 6. Mycetoma involving left foot and ankle.

based on the examination of the granules and/or culture. In the direct examination of a crushed granule in KOH mount, the presence of true hyphal elements can be demonstrated. A histological examination of the tissue biopsy may be helpful to demonstrate the presence of the fungus and the tissue reaction. Culturing the grains on SAB agar with or without antibiotics and incubation at 30ºC will help to isolate the causative fungal or bacterial agent of mycetoma. Itraconazole or ketoconazole has been used and they yield the best results for the treatment of fungal mycetomas. For actinomycetoma, the first choice treatment is a combination treatment of two drugs, such as streptomycin with dapsone or cotrimoxazole. Amikacin and rifampicin can also be used as part of a combination treatment. In resistant cases, the surgical removal of the whole affected area was found to be successful. In advanced cases, amputation is required.

Phaeohyphomycoses

It is a subcutaneous deep fungal infections caused by dematiaceous fungi (possess brown pigment, melanin: e.g., *Alternaria* spp., *Bipolaris* spp., *Cladosporium bantianum* and *Exophiala jeanselmei*) (Chander, 2002; Marconi and Powell, 1992). These fungi are common in the soil or plants and are acquired by traumatic implantation or by inhalation. They usually cause infection in the patients with immunosuppression, neutropenia and diabetes mellitus. Phaeohyphomycoses may infect any subcutaneous or deep tissue depending on the site and mode of infection. Usually, it is present as skin nodules and cysts, sinusitis, keratitis and invasive or disseminated infection (e.g., cerebral infection). Based on the site of infection, the patient may have a nasal obstruction, corneal scarring or headache. The KOH mount of specimen may demonstrate the presence of fungal hyphae and histopathologic examinations show the presence of brown septate hyphae. By preparing a culture on the SAB agar and on incubation, the fungus may grow within a week. Culture is a must to identify the causative fungus. Treatment is usually surgical excision, although relapses can occur especially in immuno-compromised patients.

Rhinosporidiosis

Rhinosporidiosis is a chronic granulomatous disease that is usually present as vascular friable polypoidal growths arising from the nasal mucosa or external structures of the eye. Cutaneous lesions can occur and rarely disseminated forms have also been reported. The organism causing rhinosporidiosis is *Rhinosporidium seeberi,* which is not a classical fungus and its taxonomy is not clearly understood. The presumed mode of infection is through the traumatized epithelium, most commonly the nasal mucosa. Diagnosis is based on identifying the presence of sporangia and spores on cytology and histopathological investigations. Treatment is surgical excision coupled with dapsone therapy. However, recurrence rates reported are very high.

Sporotrichosis (Rose gardeners' disease)

It is a chronic infection, characterized by nodular lesions of the skin or subcutaneous tissues and the adjacent lymphatics (Morris-Jones, 2002). The causative agent *Sporothrix schenckii* is commonly found on decaying vegetation. The disease has worldwide distribution and is the most common infection in Mexico. Following traumatic implantation of the fungus into the hand or forearm, the lesion begins as a small, hard, painless nodule, which enlarges over a period to form a fluctuant mass and may eventually ulcerate. More nodules may develop along the lymphatics draining the initial site.

Lympho-cutaneous sporotrichosis is the most common type of manifestation in which the lesions do not spread beyond the regional lymphatics, draining the initial lesion. Pulmonary sporotrichosis results from the inhalation of spores and not a subcutaneous lesion. The disease may resemble tuberculosis or the other respiratory mycoses or a secondary colonizer of cavitations resulting from another lung disease. Varieties like fixed cutaneous and extracutaneous disseminated sporotrichosis are other clinical forms and the latter being particularly troublesome in AIDS patients.

Direct examination of the tissue in KOH mount is not very reliable due to the paucity of organisms. The fungus may be recognized by the presence of 'cigar bodies'. Histopathologic identification is not very reliable or sensitive, since the organism is generally not present in abundance, but sometimes 'asteroid bodies' (yeasts surrounded by radiating eosinophilic material) are seen. Surface crusting can be cultured on the brain heart infusion agar. The dimorphic mould, *S. schenckii* at 37°C, the yeast phase is obtained with oblong cells and the mycelial phase grows at 25ºC. The Cutaneous form is treated with saturated solution of potassium iodide given orally with water or milk with gradual increasing dosage for a long time. Terbinafine, itraconazole and local heat therapy can also be effective. All other forms are treated with itraconazole 100-200 mg/day and in systemic form amphotericin B is employed.

Zygomycoses

The term, zygomycosis describes in the broadest sense any infection caused by a member of Zygomycetes. These are primitive, fast growing, terrestrial, largely saprophytic fungi with a cosmopolitan distribution (Spellberg *et al.*, 2005). It is a progressive infection of the subcutaneous tissue and paranasal sinuses, which may be fatal. The disease usually involves the head and neck region, lungs, gastrointestinal tract, skin and less commonly other organs. It is often associated with diabetes (ketoacidosis), starvation, major trauma, severe burns, intravenous drug abuse, hematologic malignancies and immunosuppressive therapy or the use of corticosteroids. The fungus can invade blood vessels of the arterial system causing embolism and subsequent necrosis of tissues. Medically important orders and genera include: Mucorales, causes skin, soft tissue, and internal organ infections (Mucormycosis) (*Absidia, Apophysomyces, Cokeromyces,*

Cunninghamella, Mortierella, Mucor, Rhizomucor, Rhizopus and *Saksenaea*); Entomophthorales, causes subcutaneous tissue infections (Entomophthoromycosis) (*Basidiobolus* and *Conidiobolus*). There are rhinocerebral (paranasal sinuses, orbit, face, nose and brain), pulmonary (lungs), gastrointestinal (alimentary canal), cutaneous (skin) and disseminated varieties of zygomycoses. Head injuries and intravenous drug abuse may cause central nervous system infections. Entomophthoromycosis are generally restricted to skin and subcutaneous tissue in healthy individuals and is characterized by slowly progressive infections with severe inflammatory response and lack of vascular invasion. Histopathology of the biopsy tissue demonstrates fungal elements, granulomata and eosinophils. A culture can be established on SAB agar with antibiotics and incubated at 30ºC and the growth will confirm the diagnosis. Infections caused by Zygomycetes require an early diagnosis, control of underlying disease (e.g., diabetes mellitus), surgical debridement and systemic antifungal therapy, which may have to be continued until all infected necrotic tissue is removed. Potassium iodide solution, azole antifungals, amphotericin B and trimethoprim-sulfamethoxazole have been employed with various success rates.

Systemic Mycosis

These are fungal infections that involve all the internal organs and the skin at any stage of the disease (Table 3). Salient Features of the systemic mycosis include: dimorphic fungi, existing as one form in the environment and the other in the host. The morphology is thermally regulated (i.e., hyphae at room temperature, yeast at 37ºC) (Chakrabarti and Shivaprakash, 2005). Infection occurs via the respiratory route through inhalation of infectious mould particles followed by conversion to the pathogenic and host-adapted form. Direct infection is also possible in laboratory workers. The primary site of infection is the lung and the fungus may remain localized or disseminate to other sites. Thus, these fungi are called 'systemic' or 'deep' fungal pathogens. Patients infected with any of the four fungi (*Blastomyces dermatitidis, Coccidioides immitis, Histoplasma capsulatum* and *Paracoccidioides braziliensis*) can display skin manifestations as an important sign of dissemination. These fungi infect and cause diseases in apparently immunocompetent individuals ('true pathogens'). However, all fungi may cause severe infection in immuno-compromised individuals (Tappero *et al.*, 1995).

Table 3. Causative agents and distribution of systemic mycosis

Disease	Causative agent	Region
Blastomycosis	*Blastomyces dermatitidis*	North America
Coccidioidomycosis	*Coccidioides immitis*	Southwest US
Histoplasmosis	*Histoplasma capsulatum*	South America and Central US
Paracoccidioidomycosis	*Paracoccidioides braziliensis*	South America

Blastomycosis (Chicago disease/Gilchrist's disease/North American blastomycosis)

It is a chronic granulomatous and suppurative disease having an initial pulmonary stage that is often followed by dissemination to other body sites (Wheat, 1995). This disease can occur in the humans as well as dogs. The causative organism, *Blastomyces dermatitidis* can be isolated from the soil enriched with animal excreta and decaying organic material and the infection occurs via inhalation of spores. Spores grow as yeasts, which cause the lung disease and sometimes spread haematogeneously to other organs (including skin, bone or uro-genital system). The incidence of the disease is associated with areas near large rivers in North America and sporadically in other parts of the world (Africa, Middle East, India and Mexico). Many pulmonary infections are asymptomatic and resolution can occur spontaneously. In the progressive disease, signs and symptoms resemble tuberculosis. The patient may have a mild respiratory infection, dry cough, hoarseness and low-grade fever. Some of the cases may progress to acute pneumonia, which may spread rapidly to other organs haematogeneously. The skin involvement resembles cancerous lesions and genital involvement such as chronic epididymitis can also occur. Bones, such as ribs and vertebrae are frequently affected. Untreated disseminated blastomycosis is usually progressive and can be fatal.

Radiological studies usually reveal pulmonary infiltrates and enlarged hilar lymph nodes. Progressive pulmonary blastomycosis resembles tuberculosis or a neoplasm. The organism is found in clinical specimens as a thick, double-walled cell, 5-20 μm in diameter, sometimes even reaching 30 μm. Some yeast cells have a single bud. Gomori's methenamine silver stain and PAS staining are useful for biopsies. The mother cell is connected to the daughter cell through a wide connection. Culture can be established on brain heart infusion agar with blood and incubated at 25°C and 37°C. The mycelial form can be demonstrated in culture at 25°C, while yeast at 37°C. Serology is not very useful, while itraconazole (200-400 mg/day) is used as the first-choice treatment and is effective in most cases on administration for a long time. In recalcitrant cases, 800 mg itraconazole per day or a switch to IV amphotericin B will be effective. In order to identify a relapse a follow up should be continued for several years. Untreated disseminated blastomycosis is usually progressive and can also be fatal.

Coccidioidomycosis (San joaquin Valley fever)

It is a benign and unapparent systemic mycosis usually present with a mild upper respiratory tract infection, which usually resolves and results in life-long immunity (Wheat, 1995; Cox and Magee, 2004; Pappagiants and Zimmer, 1990). In the immuno-compromised patient, it can be present as an acute or disseminating fatal mycosis. The causative agent, *Coccidioides immitis* is extensively isolated from soil in endemic regions of Southwest US. In primary coccidioidomycosis, 60% of those infected exhibit no symptoms or show only minor respiratory disorders. Approximately 40% will develop lower respiratory tract infections after 1-3 weeks, sometimes with skin lesions (erythema nodosum or erythema multiforme). About 5% of patients retain cavities and nodules in their lungs, while symptomatic disease patients present with fever, pain, cough, respiratory distress and anorexia. In secondary coccidioidomycosis, very rarely normal people progress to systemic disease. Less than 0.5% of cases may proceed to chronic lung disease with no dissemination. The disease may stabilize, but dissemination could occur at any time whenever there is immune disturbance (e.g., old age, debilitation, malnutrition). The meninges as well as the bones will also be target sites for this disease.

During direct examination of the tissue biopsy, no yeasts can be detected. Instead spherules, large, spherical elements of 20-80 μm diameter can be seen. These spherules contain numerous spores (Tappero *et al.*, 1995). Culture can be established on brain heart infusion agar with blood on incubation at 25°C and 37°C. The mycelial form of fungus can be demonstrated in culture at 25°C. The mycelial phase may show the presence of barrel shaped arthroconidia with alternating empty cells. The culture should be performed in a level 3 laboratory to prevent any accidental infection to the laboratory worker. Coccidioidin skin test and other serological tests (latex agglutination, immunodiffusion, complement fixation, ELISA) can be employed for the detection of antibodies. In the primary disease, supportive treatment will help the healthy individuals. However, in case of secondary infection or in immunosupressed individuals, antifungal treatment needs to be given. Oral fluconazole or itraconazole are found to be effective in the management of the disease. If there is no response, intravenous amphotericin B can be administered.

Histoplasmosis (Darling's disease)

It is a relatively common systemic granulomatous disease of South and Central America. It is known as 'the syphilis (venereal disease caused by *Treponema pallidum*, a great imitator of many diseases) of the fungus world', meaning, histoplasmosis is one of the 'great imitators' (Wheat, 1995; Eissenberg and Goldman, 1991). The infection is acquired by the inhalation of spores and results in a variety of clinical manifestations, primarily affecting the reticuloendothelial system (RES). The causative agent, *Histoplasma capsulatum* is commonly isolated from soil with high nitrogen content in endemic areas of major river valleys of Central US and South America. Predominantly, this infection presents as lung disease. African histoplasmosis is geographically confined to Central Africa caused by *H. capsulatum* var. *duboisii* . Patients with this chronic disease exhibit skin, bone and lymph node involvement, and ultimately nonspecific systemic disease.

In most cases, contact with a low dose results in an asymptomatic infection (>90%) results. In larger inoculations, a lung disease (acute pulmonary histoplasmosis) can manifest. In pulmonary disease, 95% of all infections are asymptomatic and is present as a mild flu-like disease. Epidemic histoplasmosis occurs when a large number of spores are inhaled and leads to high fever, cough, weakness and chest pain. In case of chronic pulmonary histoplasmosis, the primary disease does not resolve and progresses to a systemic disease. It manifests as a productive cough, progressive weakness and low-grade fever. By disseminated disease, the infection spreads from the lungs to other organs like skin, bone and lymph node especially RES and is often fatal, if untreated.

In chest X-ray, the infiltrations seen are indistinguishable from tuberculosis. An examination of the sputum, blood or bone marrow shows the presence of intracellular yeasts within the macrophages (Tappero *et al.*, 1995). The skin test using histoplasma antigen becomes positive within two weeks of exposure but is of not much significance (Eissenberg and Goldman, 1991). The serological tests are based on the detection of specific antibodies and recently through the detection of specific antigens. A microscopic examination of specimen is not of much use. A culture can be established in brain heart infusion agar with blood and incubated at 25°C and 37°C. As it is a dimorphic fungus, the mycelial form with tuberculate macroconidia and teardrop microconidia can be demonstrated in a culture at 25°C, while the diagnostic phase of yeast at 37°C. In case of healthy patients, the disease resolves spontaneously uneventfully. They may require some supportive therapy to relieve the pulmonary discomfort. However, once the disease becomes systemic, it requires therapy. Amphotericin B and ketoconazole are effective drugs, and the recovery is usually rapid.

Paracoccidioidomycosis (South American blastomycosis)

It is a chronic, progressive granulomatous infection presenting with an initial pulmonary stage followed by dissemination to other body sites especially the mucocutaneous region of the nose and mouth (Wheat, 1995; Brummer *et al.*, 1993). The causative agent, *Paracoccidioides braziliensis* is commonly isolated from the soil of South America and occasionally from Central America and Mexico. Pulmonary involvement occasionally leads to mild symptoms, including cough and sputum production although the radiograph of the chest can show dramatic involvement of the lungs. There may be a dry cough and infections tend to be very chronic but not fatal. From the pulmonary focus, there may be progressive mycosis with lesions of the skin, mucous membranes (especially in the mouth, lips and nose). The infection spreads to the lymphatics and hematogenously to the internal organs especially of the gastrointestinal tract due to the systemic involvement through the draining lymph nodes.

Demonstration of the yeasts in tissue/secretions is possible by the Gomori staining of biopsies. Yeasts are usually abundantly present in the edges of a slowly expanding ulcer. In tissue specimens, the yeasts are rather large (15 μm) with multiple buds, thereby resembling the steering wheel of a boat. The yeast cells are thick walled and the daughter cells are connected to the mother cell by a narrow connection. A culture can be established on brain heart infusion agar with blood and have to be incubated at 25ºC and 37ºC. As it is a dimorphic fungus, the mycelial form can be demonstrated in culture at 25ºC, while the diagnostic phase of yeast at 37ºC. Serology by immunodiffusion is positive in about 98% of cases. Azoles drugs, especially itraconazole is usually effective in the treatment. Amphotericin B is used for recalcitrant disease. Sulfonamides can also be used, but shows moderate activity and should be given for long periods. Prognosis depends on duration and dissemination of the disease. Lesions heal by fibrosis but persisting fungi can always cause relapse after cessation of therapy.

Opportunistic Mycosis

The fungi causing opportunistic mycoses generally do not cause disease in normal individuals (Table 4) but cause disease in individuals with immune defect (e.g., AIDS) or internal disease (e.g., diabetes mellitus). These illnesses have gained more significance at the beginning of the AIDS epidemic in the later half of the twentieth century.

Table 4. Causative agents and salient features of opportunistic mycosis.

Disease	Causative agent	Salient features
Aspergillosis	*Aspergillus flavus, A. fumigatus, A. niger* and *A. terreus*	Mimic pulmonary tuberculosis and bronchial asthma
Candidiasis	*Candida albicans, C. glabrata, C. guillermondii, C. krusei, C. parapsilosis,C. pseudotropicalis* and *C. tropicalis*	Normal inhabitant of alimentary tract and the vagina; oropharyngeal and genital candidiasis are the common presentation
Cryptococcosis	*Cryptococcus neoformans*	Causes meningitis. Demonstration of capsular material is of diagnostic
Penicillosis	*Pencillium marneffei*	Presence of oval or elliptical yeast cells within the macrophages

Aspergillosis

It is an opportunistic fungal infection caused by *Aspergillus* spp. and the disease may be localized or systemic and fatal (Latge, 1999; Kurup and Kumar, 1991). *Aspergillus* is a wide spread saprophyte found in the soil and dead/decaying matter. Among 300 species of *Aspergillus* known, about eight species have been implicated in disease etiology (the most common include: *A. flavus, A. fumigatus, A. niger* and *A.terreus*). *Aspergillus* spp. cause diseases of a wide spectrum such as allergy/hypersensitivity, pneumonitis/allergic bronchopulmonary aspergillosis (ABPA). On inhalation, the spores of the *Aspergillus* spp. give rise to allergic manifestations in some individuals. Patient may have mild symptoms to full-blown asthma, bronchial constriction and mucus plugs resulting in severe and debilitating respiratory distress, which may progressively worsen over time. The disease is due to the host's response to the fungus. In pulmonary aspergillosis, the fungus may cause a localized infection in the nasal sinuses or may colonize any cavity present in the lungs due to past infection e.g., tuberculosis. The 'aspergilloma' will be obvious in the chest X-ray within the cavity. The patient has hemoptysis due to erosion of the fungal ball into some blood vessel. In some instances with neutropenia or immunosuppression, the disease may spread and become systemic. In mycotoxicosis, certain species like *A. flavus* can produce aflatoxins, which may be found in raw or processed seeds or nuts. They are carcinogenic and toxic to liver, and infection with viable fungi is not necessary as the fungal toxin alone is responsible for this.

The material obtained by sinus can be subjected to direct or histological examination in sinusitis due to *Aspergillus* (*Aspergillus* 'heads' and branched hyphae are found). Identification of the pathogen is based on culture. In aspergilloma, direct examination (presence of hyphal fragments) and sputum culture are not necessarily always positive. Specific precipitins to *A. flavus* and *A. fumigatus* are found in about 90% of cases, which can be detected by serologic tests. The same precipitins are also found in the case of allergy. The diagnosis of invasive pulmonary aspergillosis and disseminated aspergillosis is based on clinical observation and laboratory tests, particularly the determination of the patient's immune status and recurrent isolation of the fungus. In patients with impaired resistance, precipitins are not produced or produced only at a late stage. The best currently available serological technique is the antigen detection. When systemic aspergillosis is associated with skin lesions, the diagnosis can be performed by histological examinations. It shows the presence of dichotomously branching hyphae within the tissue. In non-invasive sinusitis, rinsing with azole or polyene antifungals is useful. Co-administration of itraconazole is recommended in the event of the presence of predisposing factors or in the case of invasive sinusitis. Aspergillomas are surgically removed. The severity of haemoptysis will determine whether or not to operate. Alternatively, intracavitary instillations of itraconazole or oral treatment with itraconazole are also useful. In all forms of invasive aspergillosis, intervention should be as rapid and effective as possible. The treatment consists of IV amphotericin B or oral itraconazole for the less serious cases.

Candidiasis (Candidosis/Moniliasis/Thrush)

Candidiasis refers to infections caused by opportunistic fungal pathogens belonged to genera, *Candida* and characterized clinically by a minor infection of the skin, nail and mucosa in healthy individuals to a florid invasive systemic infection in the immuno-compromized patients. *Candida albicans* is the most common species causing infection. Other species, which cause candiadiasis, include: *C. glabrata, C. guillermondii, C. krusei, C. parapsilosis, C. pseudotropicalis* and *C. tropicalis* (Fridkin and Jarvis, 1996). The *C. albicans* is a normal inhabitant of the alimentary tract (20-40%) and the vagina. The infection sets in the presence of any predisposing factors like debilitated conditions, prolonged use of antibiotics, physiological changes like pregnancy or steroid use. Most commonly, candidiasis is present as cutaneous, mucosal (oropharyngeal, vaginal and penile infections) and nail disease (Wagner and Sohnle, 1995). There is an entity called the chronic mucocutaneous candidiasis, an uncommon inherited clinical syndrome characterized by recalcitrant cutaneous, nail and mucosal infections by *C. albicans.* Several clinical forms of chronic mucocutaneous candidiasis are recognized, which are coexisting with immunological and endocrinal disorders. Cutaneous form of the disease is characterized by the involvement of body folds such as finger and toe clefts, groins and gluteal folds ('intertrigo'). In the finger and toe webs, it is present as whitish fissures with surrounding redness and maceration. In the groins and intergluteal folds, it is characterized by papules and pustules that become confluent with margins showing satellite pustules. Soreness and itching are always present. Oropharyngeal candidiasis is often asymptomatic and characterized by the presence of white, raised ('curd-like') lesions on the tongue, palate, tonsil and buccal mucosa (acute pseudomembranous candidiasis). If it involves the throat, swallowing becomes painful. The vaginal infections are associated with itching and pain with or without discharge. Direct examination by KOH mount or Gram stain shows the presence of Gram- positive yeasts with or without pseudohyphae. The presence of pus cells is significant to confirm the presence of infection (Jones, 1990). Culture can be established on SAB agar with antibiotics and incubated at 37°C. Growth is identified by Gram's stain, germ tube formation, chlamydoconidia production (Staib and Morschha, 2006) and assimilation and fermentation sugar tests. *Candida* isolated from sterile sites like cerebro spinal fluid and blood even once is significant. Most often removal of the underlying predisposing factor will help to heal the infection spontaneously. For instance, removal of sharp tooth may help to cure the oral ulcer by *Candida*. Amphotericin B and fluconazole are administered for systemic infections, while for local application, nystatin or azole antifungals are effective.

Cryptococcosis (European blastomycosis)

It is a subacute or chronic pulmonary, systemic or meningitic infection that has worldwide distribution caused by *Cryptococcus neoformans* (Mitchell and Perfect, 1995; Musial *et al.*, 1988). There are two varieties of *Cryptococcus*: *C. neoformans* var. *neoformans* and *C. neoformans* var. *gattii*, which can be distinguished on the basis of serological, morphological, physiological and epidemiological characteristics. The pigeon is the maintenance host in nature and pathogens have been isolated from pigeon droppings. The *C. neoformans* var. *gattii* is exclusively tropical and has never been isolated from bird droppings. The portal of entry is the inhalation of the yeast from the environment like pigeon droppings or from the eucalyptus tree. The organism spreads to other organs similar to meninges. In normal individuals, there may be mild infection, which may go unnoticed. But often the infection is widespread in individuals with compromised cell-mediated immunity as in AIDS patients. Pure pulmonary infection is rare and in most cases is asymptomatic. Usually, pulmonary manifestations are in systemic infections. The patient may have a cough, malaise, chest pain and low-grade fever. The CNS invasion has a sequence of pulmonary infection. The patient complains of a headache, fever and stiff neck, which is often a fatal infection. Cutaneous infection is usually a manifestation of systemic disease but primary cutaneous disease is not rare.

In meningitis, CSF India ink preparation can be performed to demonstrate the presence of the capsule of *C. neoformans* and budding yeast cells. Gram stain can be used to demonstrate the presence of pus cells and yeast cells. In histological specimens, mucicarmine stains the capsule and silver impregnation stain (Gomori-Grocott) can be used to stain the yeast wall. The presence of capsular antigen in the CSF can be detected using latex agglutination tests. The specimens are cultured on the SAB agar without actidione. The growth is identified by Gram stain, and the demonstration of capsule and biochemical tests. On special media, Niger seed agar, *C. neoformans* forms brown colonies. Amphotericin B and flucytosine is given in the case of systemic cryptococcosis. Maintenance therapy must be given in case of AIDS patients with fluconazole to prevent any relapse. Cutaneous infection can be treated with itraconazole.

Penicilliosis

It is an opportunistic infection caused by *Penicillium* species. *Penicillium marneffei* is the most important of about 150 species known (Vanittanakom *et al.*, 2006). It causes infection in the individuals with suppressed T cell immunity especially with AIDS. The *P. marneffei* has been isolated from soil in the regions of South East Asia and China, where the disease is prevalent. It is a thermally dimorphic fungus, which exists as mould in nature and as yeast-like cells in the tissues. Infection is acquired by inhalation of the spores present in the air. Usually, the infection is asymptomatic until there is any T cell-mediated immunological defect. The patient has pulmonary symptoms like a fever, cough and weakness, but the disease soon spreads throughout the various organ systems. The patient then has skin lesions, hepatosplenomegaly and lymphadenitis.

Smears from bone marrow, or skin and lymph node biopsies can be stained with Wright's stain to demonstrate the presence of oval or elliptical yeast cells within the macrophages. These yeast cells do not bud but possess cross walls. The fungus can be isolated on SAB agar after incubating at 25ºC for 1-3 weeks. The mould colonies grow with the production of red pigment, which diffuses into the medium. Antigen of *P. marneffei* can be detected in the tissue of the HIV-infected patients using ELISA or immunodiffusion methods. The detection of antibodies against the fungus is helpful in HIV-negative individuals possessing good humoral immunity. The drugs, amphotericin B and iatroconazole are the best choice in the treatment of disseminated disease by *P. marneffei*.

Conclusions and Outlook

For nearly 30 years, amphotericin B (which causes significant nephro-toxicity) was the sole drug available to control serious fungal infections (White *et al.*, 1998). With the advent of imidazoles and triazoles, the treatment of fungal infections has become safe and effective. However, extensive use of these drugs increases the resistance (Ghannoum and Rice, 1999). The development of resistance can be reduced to a great extent by appropriate antifungal in adequate doses or in combinations. In prophylaxis measures, the antifungals should be started at specific time. For instance, antifungal prophylaxis should be initiated at the time of chemotherapy as to be active when the neutrophil count drops and once the counts recover, the prophylaxis can be discontinued (Uzun and Anaissie, 1995). Primary prophylaxis in AIDS patients must consider many factors like the prevalence and severity of the infections, efficacy and safety of antifungal drugs, the risk of emergence of resistance and the cost of antifungal treatment (Lortholary and Dupont, 1997). In the recent past, there have been tremendous increases in fungal infections, especially the opportunistic fungi. With the advent of new potent antifungal drugs, it is the responsibility of clinical microbiologists to provide an accurate identification of the pathogen. The conventional methods of diagnosis are time consuming due to cultural studies of the clinical specimen and identification of the fungus. For rapid diagnosis of infections, molecular methods like PCR and use of DNA probes specific for fungi have been developed (Chakrabarti *et al.*, 2000). The PCR can detect very small quantities of fungal DNA, thus enabling diagnosis during early stages of the infection (Iyer *et al.*, 2002) and facilitate appropriate treatment to reduce morbidity and mortality.

References

Brummer, E. Castaneda, E. and Restrepo, A. 1993. Paracoccidioidomycosis: An update. *Clinical Microbiol. Rev. 6*, 89-117.

Chakrabarti, A. and Shivaprakash, M.R. 2005. Microbiology of systemic fungal infections. *Indian J. Med. Microbiol. 51*, 16-20.

Chakrabarti, A., Kaur, R. and Das, S. 2000. Molecular methods in diagnosis and epidemiology of fungal infections. *Indian J. Med. Microbiol. 18*, 146-152.

Chander J. 2002. *Text Book of Medical Mycology*. Second edition, Mehta Publishers, New Delhi, India.

Cox, R.A. and Magee, D.M. 2004. Coccidioidomycosis: Host response and vaccine development. *Clinical Microbiol. Rev. 17*, 804-839.

Eissenberg, L.G. and Goldman, W.E. 1991. Histoplasma variation and adaptive strategies for parasitism: New perspectives on histoplasmosis. *Clinical Microbiol. Rev. 4*, 411-421.

Fisher, F. and Cook, N.B. 1998. *Fundamentals of Diagnostic Mycology*. W.B. Saunders Company, Phildelphia.

Fridkin, S.K. and Jarvis, W.R. 1996. Epidemiology of nosocomial fungal infections. *Clinical Micoribiol. Rev. 9*, 499-511

Ghannoum, M.A. and Rice, L.B. 1999. Antifungal agents: Mode of action, mechanisms of resistance and correlation of these mechanisms with bacterial resistance. *Clinical Microbiol. Rev. 12*, 501-517.

Hogan, L.H., Klein, B.S. and Levitz, S.M. 1996. Virulence factors of medically important fungi. *Clinical Microbiol. Rev. 9*, 469-488.

Iyer, R.S., Pal, R.B., Patel, R.Y. and Banker, D.D. 2002. Polymerase chain reaction based diagnosis of systemic fungal infections and sensitivity testing of the fungal isolates. *Indian J. Med. Microbiol. 20*, 132-136.

Jones, J.M. 1990. Laboratory diagnosis of invasive candidiasis. *Clinical Microbiol. Rev. 3*, 32-45.

Kurup, V.P. and Kumar, A. 1991. Immunodiagnosis of aspergillosis. *Clinical Microbiol. Rev. 4*, 439-456.

Latge, J.-P. 1999. *Aspergillus fumigatus* and aspergillosis. *Clinical Microbiol. Rev. 12*, 310-350.

Lichon, V. and Khachemoune, A. 2006. Mycetoma: A review. *Am. J. Clinical Dermatol. 7*, 315-321.

Lortholary, O. and Dupont, B. 1997. Antifungal Prophylaxis during Neutropenia and Immunodeficiency. *Clinical Microbiol. Rev. 10*, 47-504.

Malcolm, D.R. and David W.W. 2003. *Fungal Infection – Diagnosis and Management.* Third edition, Blackwell Publishing, Massachusetts.

Marconi, M.J. and Powell, D.A. 1992. Human infections due to *Malassezia* spp. *Clinical Microbiol. Rev. 5*, 101-119.

Mitchell, T.G. and Perfect, J.R. 1995. Cryptococcosis in the era of AIDS - 100 years after the discovery of *Cryptococcus neoformans Clinical Microbiol. Rev. 8*, 515-548.

Morris-Jones, R. 2002. Sporotrichosis. *Clin. Exp. Dermatol. 27*, 427-431.

Mukherjee, P.K., Sheehan, D.J., Hitchcock, C.A. and Ghannoum, M.A. 2005. Combination treatment of invasive fungal infections. *Clinical Microbiol. Rev. 18*, 163-194.

Musial, C.E., Cockerill III, F.R. and Roberts,G.D. 1988. Fungal infections of the immunocompromised host: Clinical and laboratory aspects. *Clinical Microbiol. Rev. 1*, 349-364.

Pappagiants, D. and Zimmer, B.L. 1990. Serology of coccidioidomycosis. *Clinical Microbiol. Rev. 3*, 247-268.

Revankar, S.G. 2007. Dematiaceous fungi. *Mycoses 50*, 91-101.

Spellberg, B., Edwards, J. and Ibrahim1, A.Jr. 2005. Novel perspectives on mucormycosis: Pathophysiology, presentation, and management. *Clinical Microbiol. Rev. 18*, 556-569.

Staib, P. and Morschha, J. 2006. Chlamydospore formation in *Candida albicans* and *Candida dubliniensis* – An enigmatic developmental. *Mycoses 50*, 1-12.

Tappero, J.W., Perkins, B.A., Wenger, J.D. and Berger, T.G. 1995. Cutaneous manifestations of opportunistic infections in patients infected with human immunodeficiency virus. *Clinical Microbiol. Rev. 8*, 440-450.

Uzun, O. and Anaissie, E. J. 1995. Antifungal prophylaxis in patients with hematologic malignancies: A reappraisal. *Blood 86*, 2063-2072.

Vanittanakom, N., Coopoer, Jr. C.R., Fisher, M.C. and Sirisanthana, T. 2006. *Penicillium marneffei* infection and recent advances in the epidemiology and molecular biology aspects. *Clinical Microbiol. Rev. 19*, 95 -110.

Wagner, D.K. and Sohnle, P.G. 1995. Cutaneous defenses against dermatophytes and yeasts. *Clinical Microbiol. Rev. 8*, 317-335.

Weitzmani, I. And Summerbell, R.C. 1995. The dermatophytes. *Clinical Microbiol. Rev. 8*, 240-259.

Wheat, J. 1995. Endemic mycoses in AIDS: A clinical review. *Clinical Microbiol. Rev. 8*, 146-159.

White, T.C., Marr, K.A. and Bowden,A.A. 1998. Clinical, cellular and molecular factors that contribute to antifungal drug resistance. *Clinical Microbiol. Rev. 11*, 382-402.

Yeo, S.F. and Wong, B. 2002. Current status of nonculture methods for diagnosis of invasive fungal infections. *Clinical Microbiol. Rev. 15*, 465-484.

Frontiers in Fungal Ecology, Diversity and Metabolites (edited by K.R. Sridhar), I.K. International Publishing House Pvt. Ltd., New Delhi, 2009.

17

Plasma Membrane Transport of Monocarboxylates in Yeasts

F. Cássio[1*], C. Leão[2] and F. Baltazar[2]
[1]*Centre of Biology, Department of Biology, University of Minho, 4710-057 Braga, Portugal*
[2]*Life and Health Sciences Research Institute (ICVS), School of Health Sciences, University of Minho, 4710-057 Braga, Portugal*
*(*E-mail: fcassio@bio.uminho.pt)*

Introduction

Organic acids are widely spread in nature and occur in many foodstuffs either as natural compounds (e.g., in fresh fruits) or as a result of microbial activity during the production of alcoholic beverages (wine and beer), fermented vegetables (sauerkraut and pickles), and fermented milks (yogurt and kefyr) (Wood, 1998). Organic acids, including short-chain carboxylic acids, are weak acids that in solution, they exist as both dissociated and undissociated forms. The predominance of each form depends on the dissociation constant(s) of the acid and pH of the medium. The undissociated form of the acids can permeate microbial plasma membranes by simple diffusion, which makes them useful food preservatives (Jay, 1992). The presence of organic acids greatly influences the microbiological stability as well as the organoleptic characteristics of a wide variety of food products.

About one half of the total listed yeast species contain strains that are able to use carboxylic acids as a carbon and an energy source (Barnett *et al.*, 2000). Thus, they can potentially constitute a serious threat in the food industry. Yeasts can be the dominant contaminants of foods and beverages whereas the competition from other microorganisms, particularly bacteria and moulds, is restricted mainly by low pH and the presence of preservatives (Tudor and Board, 1993). On the other hand, the ability of yeasts to grow on organic acids allows their exploitation as sources of high nutritional value (proteins, enzymes and vitamins) with applications in the health and food industry. Therefore, knowledge on carboxylic acid transport across the yeast plasma membrane and its regulation is critical to efficiently explore the potential of these microorganisms in different scenarios.

The purpose of this chapter is to overview the most relevant developments on the biology of monocarboxylate transporters in yeasts. Firstly, we summarize the physiological data of the transporters obtained by using either whole cells or cell-free systems. Secondly, we present the studies that were aiming towards identification of the genesor proteins associated with monocarboxylate transport. Thirdly, we include the currently available structural details on plasma membrane carboxylate permeases, since they give unequivocal contribution to clarify the permease transport mechanism. Finally, we compare the data of mammalian monocarboxylate transporters with those from yeast cells, because yeast cells share basic structural and functional organization with higher eukaryotes and are easier to manipulate, thereby making them potential models to study important mechanisms for mammalian cells.

Monocarboxylic Acid Transport Systems in Whole Cells and Cell-free Systems

The first step of the utilization of carboxylic acids by yeasts implies that they are able to cross plasma membrane of cells, either as dissociated or undissociated forms. Although the undissociated form of the acids can permeate the yeast plasma membrane by simple diffusion, the uptake of the ionized form(s) implies the presence of membrane proteins. From the early 1980s, different transport systems for monocarboxylic acids were characterized in several yeast species (Table 1). Those studies were aimed at the characterization of transport systems mainly from the kinetics and energetic point of view. They comprise the measurement of the initial rates of uptake of radiolabeled monocarboxylic acids and the search for the movements of proton, since protons could putatively be co-transported by those transporters. In addition, the energetic dependence of the substrate transport was investigated by examining the effect of protonophores, such as m-chlorophenylhydrazone (CCCP) or p (trifluoromethoxy) phenylhydrazone (FCCP), on the substrate accumulation by yeast cells. In most cases, the uptake of monocarboxylates follows a Michaelis-Menten kinetic, proton movements are associated with the uptake of substrates, and a rapid efflux of accumulated monocarboxylates occurs found when protonophores are added (see references in Table 1). These findings point mainly to secondary transport systems, like proton co-transporters, as the mechanisms involved on the transport of monocarboxylates in yeasts. Specifically, proton symports were found to be involved in the uptake of lactate and other monocarboxylates in *Saccharomyces cerevisiae*, *Candida utilis*, *C. albicans*, *Dekkera anomala*, *Torulaspora delbrueckii* and *Zygosaccharomyces bailii*, while a uniport system for lactate was identified in *Kluyveromyces marxianus*.

The monocarboxylate proton symports that are characterized in yeasts exhibit different substrate specificities (Table 1). Lactate proton symports characterized in cells grown in lactic acid are commonly shared by pyruvate, acetate and propionate, being the case of *T. delbrueckii*, *S. cerevisiae*, and *Candida* species. On the other hand, acetate proton symports, which are not able to accept lactate, are induced in acetic acid grown cells of *S. cerevisiae T. delbrueckii*, *D. anomala* and *Z. bailii.* Therefore, in the former yeast species, two acetate proton symports co-exist in lactic acid grown cells, which, besides exhibiting different specificities, also show different kinetic parameters (Table 1). In addition, the data shown in Table 1 suggest that, under certain environmental conditions, more than one transport mechanism may co-exist for a specific substrate in yeasts. Indeed, in *C. utilis* cells grown in lactic acid, the anionic form of lactic acid (lactate) is transported by a proton symport (high-affinity system), while the undissociated form of the acid is transported by facilitate diffusion (low-affinity system). The latter system shows a broad specificity, being able to accept mono-, di-, tri-carboxylic acids and some amino acids.

In terms of regulation, most of the transport systems described above are inducible by certain carboxylic acids (exceptions are the acetate proton symport of *S. cerevisiae* and *Z. bailii* characterized in acetic acid- and glucose-grown cells, respectively) (Tablc 1). In addition, all transport systems are repressed by glucose (with the exception of the proton symport of *Z. bailii* specific for acetate). The studies mentioned herein were performed in intact yeast cells and, in spite of their unequivocal contribution for the knowledge of transport of monocarboxylic acids across plasma membranes, there are some limitations to such approaches, mainly due to the metabolism of acids and/or cell compartmentation. To overcome these limitations, a vesicular system was prepared by fusing plasma membranes of lactic acid-grown cells of *C. utilis* with liposomes containing cytochrome *c* oxidase, as a proton motive force generating system. In this cell-free system, lactate transport showed substrate affinity and specificity pattern similar to the one characterized in whole cells of *C. utilis* (Gerós *et al.*, 1996). In addition, with this vesicle system, a stoichiometry of 1:1 for the lactate proton symport, independent of the extracellular pH (pH 5.5-7.0) was established. Using a theoretical model, this stoichiometric value was previously predicted at pH 5.0, but it seemed to change with the extracellular pH (Cássio *et al.*, 1993).

Identification of A 43-kDa Polypeptide Associated with Lactate Transport in *Candida utilis*

Despite increasing knowledge on genes or proteins involved in monocarboxylate uptake in yeasts, there are no reports of studies using purified permeases, which would have greatly contributed to the better understanding of the mechanisms of monocarboxylate transport. A first attempt to identify the lactate transport protein in yeasts was based on the fact that, under certain conditions, L-[U-^{14}C] lactic acid is able to tightly bind to the mammalian lactate permease, which

Table 1. Characterization of monocarboxylate transport systems in yeasts

Yeast species	Carbon and energy source[a]	Transport mechanism	Kinetic parameters	Specificity	Regulation	Reference
Candida albicans	Lactic acid	Mediated, probably a proton symport	pH 5.0: K_m 0.339 ± 0.09 mM; V_{max} 0.859 ± 0.06 nmol s^{-1} mg^{-1} dry wt	D-, L-lactate, pyruvate and propionate	Inducible and glucose repressed	Soares-Silva *et al.*, 2004
Candida utilis	Lactic acid	Proton symport	pH 5.5: K_m 0.06 mM; V_{max} 1.40 nmol s^{-1} mg^{-1} dry wt	D-, L-lactate, pyruvate, acetate, propionate and fumarate	Inducible and glucose repressed	Eddy and Hopkins, 1985; Leão and van Uden, 1986
	Lactic acid	Facilitate diffusion	pH 3.5[b]: K_m 14.9 mM; V_{max} 9.9 nmol s^{-1} mg^{-1} dry wt	Mono-, di- and tricarboxylic acids, and glycine and glutamic acid	Inducible and glucose repressed	Cássio and Leão, 1991, 1993
Dekkera anomala	Acetic acid	Proton symport	pH 5.0: K_m 0.12 ± 0.04 mM; V_{max} 0.10 ± 0.01 nmol s^{-1} mg^{-1} dry wt	Propionate, Formate, Sorbate	Inducible and glucose repressed	Gerós *et al.*, 2000a
Kluyveromyces marxianus	Lactic acid	Uniport	pH 5.4: V_{max} 1.00 ± 0.13 mmol h^{-1} g^{-1} dry wt; Ks = 0.42 ± 0.08 mM	D-, L-lactate, pyruvate, acetate, bromoacetate, propionate and glicolate	Partially inducible and partially glucose repressed	Fonseca *et al.*, 1991
Saccharomyces cerevisiae	Lactic acid	Proton symport	pH 5.5: K_m 0.13 mM; V_{max} 0.34 nmol s^{-1} mg^{-1} dry wt	D-, L-lactate, pyruvate, acetate and propionate	Inducible and glucose repressed	Cássio *et al.*, 1987
	Acetic acid	Proton symport	pH 5.0: K_m 0.34 ± 0.058 mM; V_{max} 1.32 nmol s^{-1} mg^{-1} dry wt	Acetate, propionate and formate	Constitutive and glucose repressed	Casal *et al.*, 1996
Torulaspora delbrueckii	Lactic acid	Proton symport	pH 5.0: K_m 0.05 mM; V_{max} 0.38 nmol s^{-1} dry wt;	Lactate, acetate, pyruvate, propionate and formate	Inducible and glucose repressed	Casal and Leão, 1995
	Acetic acid	Proton symport	pH 5.0: K_m 0.55 mM; V_{max} 2.93 nmol s^{-1} mg dry wt^{-1};	Acetate, propionate and formate	Inducible and glucose repressed	Casal and Leão, 1995
Zygosaccharomyces bailii	Acetic acid	Proton symport	pH 5.0: K_m 2.6 ± 0.4 mM; V_{max} 4.3 ± 0.4 nmol s^{-1} mg^{-1} dry wt;	Acetate, propionate and formate	Inducible and glucose repressed	Sousa *et al.*, 1996
	Glucose	Proton symport	pH 5.0: K_m 45 ± 8.0 mM; V_{max} 4.8 ± 1.3 nmol s^{-1} mg^{-1} dry wt	Acetate	Constitutive	Sousa *et al.*, 1996

[a], substrate used as carbon and energy to characterize the transport system

[b], above pH 5.0 the facilitate diffusion system is not measurable.

allowed its identification as a radiolabeled polypeptide, after plasma membrane protein separation by SDS-PAGE (Poole and Halestrap, 1993). Using this methodology, a 43-kDa polypeptide associated with lactate transport was identified in plasma membranes of *C. utilis* (Gerós *et al.*, 2000b). Subsequently, this lactate transporter was partially purified by DEAE-Sepharose chromatography and the activity of the lactate carrier was further confirmed by its reconstitution into liposomes (Baltazar *et al.*, 2006). This allowed a more positive identification of the protein involved in the lactate transport of *C. utilis*.

Genes involved in the Transport of Monocarboxylates

In the last decade, with the widespread use of gene sequencing technology, data on the biology of transmembrane transport proteins have grown exponentially. In view of the importance of these proteins to living cells and aiming to promote a comprehensive understanding of their basic functions, Saier (1998) proposed a Transporter Classification (TC) system, based on both function and phylogeny of transport proteins. This system has the potential to include all known permeases for which sequence data are available and has the flexibility to accommodate new sequences of permeases that are likely to be revealed by genome sequencing and the biochemical analysis of plasma membrane transporters in other organisms. The TC system has been continuously updated (Busch and Saier, 2002; Saier, 2000; Saier *et al.*, 2006), and is now freely available in the Transporter Classification Database (TCDB; http://www.tcdb.org).

While tracing the evolution of membrane transporters, De Hertogh *et al.* (2002) constructed the Yeast Transporter Information (YETI database) comprising all membrane proteins from *S. cerevisiae*. Currently, YETI also includes data from complete genome sequences of other species of the phylum Hemiascomycete: *Candida glabrata*, *Debaryomyces hansenii*, *Kluyveromyces lactis* and *Yarrowia lipolytica.* Among the 2480 Hemiascomycete transporters, the secondary transporters were the most widespread, followed by transporters energized by ATP. A comparative analyses of 97 small phylogenetic transporter subfamilies, comprising a total of 355 transporters, allowed the identification of four evolutionary patterns, namely, "species-specific" transporters, "ubiquitous" transporters (that are conserved throughout the entire phylum), "phylum gained or lost" transporters, and "homoplasic" transporters (De Hertogh *et al.*, 2006).

A free access to data from the whole genomic analysis of *S. cerevisiae* transporter proteins that also includes the TC classification is the Yeast Transport Protein database (YTPdb) (http://rsat.scmbb.ulb.ac.be/ytpdb). While searching YTPdb for plasma membrane transporters of short-chain monocarboxylates, only four different genes are displayed: *JEN1* (ORF YKL217w), *ADY2* (ORF YCR010c), *PDR12* (ORF YPL058c) and *AZR1* (ORF YGR224w). The first two genes appear to code for permeases responsible for the uptake of monocarboxylates by yeast cells, while the others are associated with permeases involved in the extrusion of carboxylates, including monocarboxylates.

JEN1 and ADY2

Both lactic acid and acetic acid are non-fermentable carbon substrates metabolized to acetyl-CoA, which enters the TCA cycle and needs glyoxylate cycle to fulfill needs of cell's energy and biosynthetic metabolites (Flores *et al.*, 2000). *JEN1* was the first gene associated with the transport of monocarboxylates across plasma membrane of fungi, which is required for the uptake of lactate and pyruvate in *S. cerevisiae* (Casal *et al.*, 1999). According to the TC classification, *JEN1* belongs to the Major Facilitator Superfamily (MFS) and to the sialate:proton symporter family (Paulsen *et al.*, 1998; Saier 2000; Saier *et al.*, 1999). The functionality of Jen1p of *S. cerevisiae* was confirmed in *P. pastoris* cells expressing *JEN1*, by measuring lactate uptake either in whole cells or hybrid vesicles, obtained by fusion of *P. pastoris* plasma membranes with liposomes able to maintain a proton motive force (Soares-Silva *et al.*, 2003). *JEN1* expression in *S. cerevisiae* is repressed by glucose and induced by non-fermentable substrates, such as lactate and ethanol (Lodi *et al.,* 2002). Isogenic strains that are deleted in gene encoding enzymes of the lactic acid metabolism do not express *JEN1,* indicating an association between the intracellular metabolism of lactic acid and the transcription of the gene permease (Andrade and Casal, 2001). *JEN1* expression is repressed by Mig1p and Mig2p and requires the transcription factor Cat8p for full de-repression (Bojunga and Entian, 1999; Lodi *et al.,* 2002) and the protein kinase Snf1p, which is reported to de-repress the glucose-repressed genes in *S. cerevisiae* (Celenza and Carlson 1986). Snf1p also inhibits Migp1 repression and regulates the transcriptional activator Cat8p, stimulating transcription of *JEN1* (Lodi *et al.,* 2002).

In addition to these regulators, the transcriptional factor Hap2/3/4/5 complex, important for de-repression of genes encoding mitochondrial functions that are related to shifts from fermentative to oxidative metabolism, also play a role in *JEN1* regulation. Lodi *et al.* (2002) found that Hap2/3/4/5 complex interacts with the CAAT-box element in the *JEN1* promoter to de-repress its transcription. Similarly to many genes encoding proteins involved in the catabolism of non-fermented carbon sources, the expression of *JEN1* requires oxygen but do not seem to require haem synthesis (Lodi *et al.,* 2002). Further, Andrade *et al.* (2005) showed that *JEN1* encodes three different mRNA transcripts, starting at positions -51, +391 or +972, and proposed that the transcript *JEN1*(+391) may function as a sensor, regulating the glucose-triggered degradation of *JEN1*(-51) mRNA coding for Jen1p. Using the *S. cerevisiae JEN1* sequence to search for similar monocarboxylate permeases, Soares-Silva *et al.* (2004) identified Ca*JEN1* in *C. albicans*, the disruption of which resulted in the loss of all measurable lactate transport activity. Additionally, by using the green fluorescent protein-fused Ca*JEN1,* the authors showed that CaJen1p is localized at the yeast plasma membrane. Consistent with the data of *JEN1* regulation in *S. cerevisiae*, the expression of Ca*JEN1* was not found in glucose-repressed cells of *C. albicans* and CaCat8p was important for induction of Ca*JEN1* (Soares-Silva *et al.*, 2004).

In *K. lactis*, two genes (Kl*JEN1* and Kl*JEN2*) displaying high similarity to *JEN1* were identified (Lodi *et al.*, 2004). The physiological characterization of null mutants showed that Kl*JEN1* is a functional homologue of *JEN1*, whereas Kl*JEN2* encodes a transporter of di-carboxylic acids. This is consistent with an earlier report describing a transport system for succinate, fumarate and malate in *K. lactis*, which was absent in glucose-grown cells (Zmijewski and MacQuillan, 1975). Similar to *JEN1* and Ca*JEN1*, both Kl*JEN1* and Kl*JEN2* are under glucose repression. In *K. lactis*, glucose control is mediated by Fog1p and Fog2p, which are homologous to *S. cerevisiae* Gal83p and Snf1p, respectively (Celenza and Carlson, 1986; Goffrini *et al.*, 1996), and by KlCat8p. Kl*JEN1* and Kl*JEN2* expression displays different response to de-repressing conditions and induction by non-fermentable carbon sources (Kl*JEN1* is specifically induced by lactate, while Kl*JEN2* is induced by ethanol, acetate and succinate, but not by lactate).

In addition to *JEN1* homologues, the *ADY2* gene was reported to play a key role in acetate transport across the plasma membrane of *S. cerevisiae*, because its disruption abolished the active transport of acetate in this yeast species (Paiva *et al.*, 2004). These data are in accordance to previous reports indicating the presence of two acetate transporters in cells of *S. cerevisiae* (Table 1) that are distinguishable by their specificity (Jen1p carrying acetate, propionate and pyruvate, while Ady2p transports acetate, propionate and formate, but not lactate) (Casal *et al.*, 1996; Paiva *et al.*, 2004). However, *ADY2* was also associated with the ammonia transport by *S. cerevisiae* (Palkova *et al.*, 2002); therefore, more studies are essential to clarify the actual function of the Ady2p.

In *S. cerevisiae*, the shift from glucose to acetic acid showed a remarkable increase in the expression of genes involved in gluconeogenesis (*PCK1* and *FBP1*), in acetate conversion to acetyl-CoA (*ACS1* and *ACHI*), in genes encoding enzymes of the glyoxylate and TCA cycles as well as in the peroxysomal production of acetyl-CoA (Paiva *et al.*, 2004). Respiratory metabolism was also shown by the over expression of genes encoding components of the oxydative phosphorylation chain and haem biosynthesis. Moreover, regulators of the catabolite de-repression, such as the transcription factors *CAT8* and *SIP4* , which are associated with the gluconeogenesis and glyoxylate cycles, were activated (Haurie *et al.*, 2001). *ADY2* exhibits homology to the *Yarrowia lipolytica GPR1* gene, which is implicated in acetic acid sensitivity of this yeast (Tzschoppe *et al.*, 1999). In contrast to *ADY2* gene, deletion of *GPR1* did not affect growth in the presence of acetic acid (Augstein *et al.*, 2003). In addition, the expression of *GPR1* is only moderately repressed by glucose, whereas *ADY2* is under glucose repression and does not seem to be involved in the acetate adaptation of *S. cerevisiae* as demonstrated by transport and micro-array analyses (Paiva *et al.*, 2004).

Other Genes

Assigned to the plasma membrane ATP-Binding Cassette (ABC) superfamily

(TC classification) is the *PDR12* gene. In *S. cerevisiae*, expression of this gene is strongly induced by sorbate, benzoate and other moderately lipophilic short-chain carboxylates, such as lactate (Kawahata *et al.*, 2006), but not by high levels of acetate (Hatzixanthis *et al.*, 2003). The Pdr12p (mediating active efflux of carboxylic acid anions from cells) is probably associated with the maintenance of low intracellular levels of these compounds. Indeed, benzoate efflux was severely impaired in mutants of *S. cerevisiae* lacking *PDR12*, which were hypersensitive to the food preservatives (sorbic, benzoic, propionic and acetic acids) at low pH (Piper *et al.*, 1998). In addition, Schüller *et al.* (2004) while comparing micro-array data with those from phenotypic screening showed that among more than 100 genes induced by weak acid stress, PDR12 was the only one that was required for weak acid resistance.

The *AZR1* is another gene associated with multidrug resistance. This is a putative proton antiporter from the MFS superfamily. The *AZR1* was implicated in the resistance of *S. cerevisiae* to acetic acid and azole compounds (ketoconazole and fluconazole) (Tenreiro *et al.*, 2000), but incontrast to Pdr12p (Piper *et al.*, 1998; Holyoak *et al.*, 1999) and Azr1p is not directly involved in the active extrusion of acetate (Tenreiro *et al.*, 2000). Further studies from Bauer *et al.* (2003) reported that acetate resistance of *S. cerevisiae* was restricted to aromatic amino acids auxotrophic mutants (*trp1*), and therefore they concluded that the Azr1p or Pdr12p permeases do not confer resistance of *S. cerevisiae* to acetate. Conversely, they also concluded that, in prototrophs, Azr1p and Pdr12p are detrimental for acetate resistance. Thus, more studies are needed to explicitly understand the spoilage by *S. cerevisiae* of certain foodstuffs preserved by acetic acid.

Structure of Monocarboxylate Transport Proteins

The knowledge on structural details of membrane permeases is of utmost importance in biotechnology and biomedicine since most drug targets are integral membrane proteins. An analysis of the three-dimensional (3-D) structure of these proteins can reveal more information (for instance, identification of the substrate-binding site and residues that are involved in substrate recognition), which could ultimately contribute to unravel the permease transport mechanism. The first high-resolution structure of a membrane protein, the subunits in the photosynthetic reaction centre of *Rhodopseudomonas viridis*, became available in 1985 (Deisenhofer *et al.*, 1985). However, only seventeen years later, Hirai *et al.* (2002) revealed the 3-D structure of the first MFS protein involved in the exchange of oxalate for formate in *Oxolobacter formigenes.* About one hundred protein structures currently available are mostly of prokaryotic origin (www.mpibp-frankfurt.mpg.de/michel/public/memprotstruct.html).

Despite recent efforts, in many cases membrane proteins are under-represented in structural databanks due to the difficulties related to sample preparation (expression and crystallization) (Hildebrand *et al.*, 2006; Klammt *et al.*, 2006). Nevertheless, with the widespread use of gene sequencing technology, the primary structures of many eukaryotic plasma membrane permeases are becoming available and sequence-based prediction methods can be of value to elucidate the physiological processes mediated by these proteins (Park and Helms, 2006). Tertiary structure prediction methods are now being used aimed at gaining information on the 3-D structure of membrane proteins not yet crystallized (Hildebrand *et al.*, 2006). Both computational and structural methods have been used to study membrane protein folding and function by establishing the mechanisms of the transmembrane helice organization (Liu *et al.*, 2004). From the structure of 14 transmembrane proteins known so far, six are multi-drug efflux transporters from bacteria, three are *Escherichia coli* MFS transporters, and five are ABC transporters. MFS is the largest superfamily of secondary permeases that are currently recognized, in which high-resolution structural data are available for three of its members: the lactose proton symporter (LacY), the a-glycerophosphate phosphate antiporter (GlpT) and the oxalate formate antiporter (OxlT). Although these three proteins are very divergent in sequence, they essentially exhibit the same 3-D structures (Abramson *et al.*, 2003; Hirai *et al.*, 2003; Huang *et al.*, 2003). Almost all MFS families exhibit a uniform topology with 12 Transmembrane Segments (TMSs) and both N- and C-termini in the cytoplasm. A member with just six TMSs, several members with 14 TMSs, and one member with 24 TMSs have been identified (Pao *et al.*, 1998; Saier, 2003).

The prediction of membrane spanning regions for the lactate permease of *C. albicans*, CaJen1p, indicated 12 transmembrane segments. This protein showed 41% of identity with the *S. cerevisiae* Jen1p (Soares-Silva *et al.*, 2004). Similarly, the lactate permease of *K. lactis*, KlJen1p, showed twelve transmembrane helices associated with all topological characteristics that are typical of solute-transporting proteins (André, 1995; Lodi *et al.*, 2004). Protein structure prediction of KlJen2p indicated that this protein also displays a typical permease structure, with the presence of multiple transmembrane helices, but the exact number of helices was not clarified (Lodi *et al.*, 2004). In addition, the alignment of deduced amino acid sequence of Kl*JEN2* and Kl*JEN1* showed that the two proteins are 39% identical. Alignment of the KlJen1p and KlJen2p amino acid sequences with amino acid sequences deduced from other genome sequences allowed the identification of two genes homologous in *Kluyveromyces marxianus*, six genes in *Yarrowia lypolitica* and four genes in *Debaryomyces hansenii*. These findings are consistent with the ability of these yeast species to grow on succinate (Barnett *et al.*, 2000). The predicted topology of Pdr12p includes 12 putative transmembrane-spanning a-helices and two nucleotide-binding domains (Piper *et al.*, 1998), consistent with the common architecture of ABC proteins comprising two cytosolic nucleotide-binding domains (NBD) and two transmembrane domains (TMD). TMDs are mainly involved in substrate recognition and transport, while NBDs are responsible for the ATP binding/ hydrolysis and energizing substrate transport (Ramaen *et al.*, 2006).

Emerging Issues on Monocarboxylate Transporters

Transport of monocarboxylates across the plasma membrane of mammalian cells is mediated by a group of membrane proteins known as MonoCarboxylate Transporters (MCTs). Based on sequence homologies, 14 members of the MCT family were identified (although only a few have been functionally characterized). *MCT1-MCT4* are proton symporters, which have lactate as a common substrate, and each isoform exhibits different biochemical properties, such as substrate specificity and affinity as well as inhibitor sensitivity (Halestrap and Meredith, 2004). Substrate affinity has been associated with the specific distribution of each transporter in mammalian tissues, depending on their metabolic activity. *MCT2* is present in tissues that use lactate as substrate (e.g., liver) (Pierre and Pellerin, 2005), while *MCT4* is present in highly glycolytic tissues (e.g., slow-twitch muscle) (Dimmer *et al.*, 2000). *MCT1* has a ubiquitous distribution as it is present in both lactate consuming and lactate producing tissues (Pierre and Pellerin, 2005), whereas *MCT3* is confined to the epithelia of the basal membrane of the retinal pigment and of the choroid plexus in the brain (Halestrap and Meredith, 2004). MCTs have been recognized to play an important role in pathologies involving low cellular oxygen availability. For example, solid tumors become highly dependent on anaerobic glycolysis, resulting in an increased production of acids. Despite this, the intracellular pH of tissues in these tumors is either normal or higher than normal tissues, and only the interstitial pH is low (Helmlinger *et al.*, 2002). It is believed that an accumulation of lactic acid is the main cause of tumor acidity and that MCTs are the most important transporters that are involved in the maintenance of tumor intracellular pH, by promoting the efflux of the accumulating acid. Using *in vitro* and *ex vivo* models, Wahl *et al.* (2002) demonstrated that MCTs are major players in the regulation of pH in melanoma cells. There are also evidences for the up-regulation of MCTs in other tumors, such as high-grade glial neoplasms (Mathupala *et al.*, 2004), colorectal carcinomas (Koukourakis *et al.*, 2006) and alveolar sarcoma of soft tissues (Ladanyi *et al.*, 2002). Taking these facts into account, MCTs may constitute an attractive target for cancer therapy (Enerson and Drewes, 2003). Differences in the expression of MCT isoform and the activity between the tumor and the normal cells may allow selective inhibition, and thereby inducing tumor intracellular acidification and possibly cell death.

The search of *S. cerevisiae* genome for similarities to MCTs revealed five putative monocarboxylate permease homologues (Mch1-5p), that display transporter like topological features (André, 1995; Nelissen *et al.*, 1997). Makuc *et al.* (2001) investigated the role of these *S. cerevisiae* MCH genes (ORFs YKL221w, YNL125c/ESBP6, YOR306c, YOL119c and YDL054c), and they found that none of them appeared to be involved in the uptake of monocarboxylates across yeast plasma membrane. The biological functions of Mch1-4 are presently unknown and Mch5p (ORF YOR306c) was recently described to catalyze riboflavin (vitamin B2) uptake in *S. cerevisiae* (Reihl and

Stolz, 2005). The closest human homolog of Mch5p, *MCT8*, was found to catalyze the uptake of the thyroid hormones, which are negatively charged monocarboxylates (Friesema *et al.*, 2003). Despite MCT related genes are present in the genome of *S. cerevisiae*, they do not appear to be involved in the uptake of monocarboxylates, such as lactate. The functional characterization of the lactate transporter of the yeast *C. utilis* showed several similarities with the transport systems of mammalian cells. *C. utilis* has a proton symport, similar to MCT 1 - MCT 4, that accepts lactate and pyruvate (Leão and van Uden, 1986) and is inhibited by known mammalian MCT inhibitors, such as CHC (á-cyanohydroxycinamic acid) and DIDS (4,4' di-isothiocyano-stilbene-2,2'-disulfonic acid) (Gerós *et al.*, 2000b). In addition, the polypeptide of approximately 40 kDa associated with *C. utilis* lactate transport (Baltazar *et al.*, 2006; Gerós *et al.*, 2000b) is within the range of molecular weights identified for mammalian lactate transporters (Allen and Brooks, 1994; Poole and Halestrap, 1992, 1994). Since MCTs are recognized as potential targets for cancer therapy, the development of adequate and simple cellular models to study the mechanisms of monocarboxylate transport inhibition will be of great value. In this perspective, yeasts are promising because they share the basic structural and functional organization with higher eukaryotes, and have the advantage of being easier to manipulate.

Conclusions and Outlook

The first step for the utilization of monocarboxylic acids by yeasts is their transport across the plasma membrane, which requires the activity of permeases in the transportation of dissociated forms of acids. In the last two decades, monocarboxylate transport systems of various yeast species were characterized from the kinetics and energetic point of view. Results of such studies indicated that the transport of monocarboxylates in yeasts is mainly mediated by secondary transporters, like proton symporters. In addition, monocarboxylate transporters show different substrate affinity and specificity (some are rather specific, while others appear to transport several monocarboxylates). Gene sequencing technology allowed the identification of several genes associated with the transport of monocarboxylates. In *Saccharomyces cerevisiae*, the genes *JEN1* and *ADY2* appear to encode plasma membrane permeases for the uptake of lactate-pyruvate and acetate respectively, while the genes *PDR12* and *AZR1* seem to code for permeases implicated in the control of intracellular concentration of carboxylic acid anions, including lactate and acetate. However, other studies reported that acetate resistance of *S. cerevisiae* that is associated with Azr1p or Pdr12p is restricted to aromatic amino acids auxotrophic mutants, and the loss of these permeases increased acetate resistance of prototroph strains. Further, Ady2p was also implicated in the transport of ammonia. Therefore, there is still some missing data linking the expression of these genes to their actual function in yeast cells. The use of micro-arrays in conditions of high permease gene expression is expected to contribute to a rapid progress in elucidating permease regulation. These studies are also likely to explain the physiological significance of the apparent redundancy of many monocarboxylate transport activities. In addition, an analysis of high-resolution structures of yeast monocarboxylate permeases along with advances in bioinformatics will contribute to unravel the molecular transport mechanism and may allow the exploitation of these proteins as drug targets. Mammalian proton-linked MCTs show altered expression in several tumors and, therefore, they are pointed out as potential targets for cancer therapy. However, adequate models are needed to study monocarboxylate transport inhibition and possibly screen for new inhibitors. Although *S. cerevisiae* MCT homologues do not appear to transport lactate, we cannot discard the possibility that other yeast species have MCT homologues involved in monocarboxylate transport. Further, yeasts may be used for heterologous expression of mammalian MCTs and explored for the identification of novel drug MCT targeted therapies.

References

Abramson, J., Smirnova, I., Kasho, V., Verner, G., Kaback, H.R. and Iwata, S. 2003. Structure and mechanism of the lactose permease of *Escherichia coli*. *Science, 301*, 610-615.

Allen, P.J. and Brooks, G.A. 1994. Partial purification and reconstitution of the sarcolemmal L-lactate carrier from rat skeletal muscle. *Biochem. J. 303*, 207-212.

Andrade, R.P. and Casal, M. 2001. Expression of the lactate permease gene *JEN1* from the yeast *Saccharomyces cerevisiae*. *Fungal Genet. Biol. 32*, 105-111.

Andrade, R.P., Kötter, P., Entian, K.-D. and Casal, M. 2005. Multiple transcripts regulate glucose-triggered mRNA decay of the lactate transporter *JEN1* from *Saccharomyces cerevisiae*. *Biochem. Biophys. Res. Commun. 332*, 254-262.

André, B. 1995. An overview of membrane transport proteins in *Saccharomyces cerevisiae*. *Yeast 11*, 1575-1611.

Augstein, A., Barth, K., Gentsch, M., Kohlwein S.D. and Barth, G. 2003. Characterization, localization and functional analysis of Gpr1p, a protein affecting sensitivity to acetic acid in the yeast *Yarrowia lipolytica*. *Microbiology 149*, 589-600.

Baltazar, F., Cássio, F. and Leão, C. 2006. Functional purification of the monocarboxylate transporter of the yeast *Candida utilis*. *Biotechnol. Lett. 28*, 1221-1226.

Barnett, J.A., Payne, R.W. and Yarrow D. 2000. *Yeasts: Characteristics and Identification*. 3rd Edition, Cambridge University Press, Cambridge, UK.

Bauer, B.E., Rossington, D., Mollapour, M., Mamnun, Y., Kuchler. K. and Piper, P.W. 2003. Weak organic acid stress inhibits aromatic amino acid uptake by yeast, causing a strong influence of amino acid auxotrophies on the phenotypes of membrane transporter mutants. *Eur. J. Biochem. 270*, 3189-3195.

Bojunga, N. and Entian, K.D. 1999. Cat8p, the activator of gluconeogenic genes in *Saccharomyces cerevisiae*, regulates carbon source-dependent expression of NADP-dependent cytosolic isocitrate dehydrogenase (Idp2p) and lactate permease (Jen1p). *Mol. Gen. Genet 262*, 869-875.

Busch, W. and Saier, M.H., Jr. 2002. The transporter classification (TC) system. *Crit. Rev. Biochem. Mol. Biol. 37*, 287-337.

Casal, M. and Leão, C. 1995. Utilization of short-chain monocarboxylic acids by the yeast *Torulaspora delbrueckii*: specificity of the transport systems and their regulation. *Biochim. Biophys. Acta 1267*, 122-130.

Casal, M., Cardoso, H. and Leão, C. 1996. Mechanisms regulating the transport of acetic acid in *Saccharomyces cerevisiae*. *Microbiology 142*, 1385-1390.

Casal, M., Paiva, S., Andrade, R.P., Gancedo C. and Leão, C. 1999. The lactate-proton symport of *Saccharomyces cerevisiae* is encoded by *JEN1*. *J. Bacteriol. 181*, 2620-2623.

Cássio, F., Corte-Real, M. and Leão, C. 1993. Quantitative analysis of proton movements associated with the uptake of weak carboxylic acids. The yeast *Candida utilis* as a model. *Biochim. Biophys. Acta 1153*, 59-66.

Cássio, F. and Leão, C. 1991. Low- and high-affinity transport systems for citric acid in the yeast *Candida utilis. Appl. Environ. Microbiol. 57*, 3623-3628.

Cássio, F. and Leão, C. 1993. A comparative study on the transport of L(-)malic acid and other short-chain carboxylic acids in the yeast *Candida utilis*: evidence for a general organic acid permease. *Yeast 9*, 743-752.

Cássio, F., Leão, C. and van Uden, N. 1987. Transport of lactate and other short-chain monocarboxylates in the yeast *Saccharomyces cerevisiae. Appl. Environ. Microbiol. 53*, 509-513.

Celenza, J.L. and Carlson, M. 1986. A yeast gene that is essential for release from glucose repression encodes a protein kinase. *Science 233*, 1175-1180.

De Hertogh, B., Carvajal, E., Talla, E., Dujon, B., Baret, P. and Goffeau, A. 2002. Phylogenetic classification of transporters and other membrane proteins from *Saccharomyces cerevisiae. Funct. Integr. Genomics 2*, 154-170.

De Hertogh, B., Hancy, F., Goffeau, A. and Baret, P.V. 2006. Emergence of species-specific transporters during evolution of the Hemiascomycete Phylum. *Genetics 172*, 771-781.

Deisenhofer, J., Epp, O., Miki, K., Huber, R. and Michel, H. 1985. Structure of the protein subunits in the photosynthetic reaction centre of *Rhodopseudomonas viridis* at 3Å resolution. *Nature 318*, 618-624.

Dimmer, K.S., Friedrich, B., Lang, F., Deitmer, J.W. and Broer, S. 2000. The low-affinity monocarboxylate transporter MCT4 is adapted to the export of lactate in highly glycolytic cells. *Biochem. J. 1*, 219-227.

Eddy, A.A. and Hopkins, P.G. 1985. The putative electrogenic nitrate-proton symport of the yeast *Candida utilis*. Comparison with the systems absorbing glucose or lactate. *Biochem. J. 231*, 291-297.

Enerson, B.E. and Drewes, L.R. 2003. Molecular features, regulation, and function of monocarboxylate transporters: implications for drug delivery. *J. Pharm. Sci. 92*, 1531-1544.

Flores, C.L., Rodriguez, C., Petit, T. and Gancedo, C. 2000. Carbohydrate and energy-yielding metabolism in non-conventional yeasts. *FEMS Microbiol. Rev. 24*, 507-529.

Fonseca, A., Spencer-Martins, I. and van Uden, N. 1991. Transport of lactic acid in *Kluyveromyces marxianus*: evidence for a monocarboxylate uniport. *Yeast 7*, 775-780.

Friesema, E.C., Ganguly, S., Abdalla, A., Manning Fox, J.E. and Halestrap, A.P. and Visser, T.J. 2003. Identification of monocarboxylate transporter 8 as a specific thyroid hormone transporter. *J. Biol. Chem. 278*, 40128-40135.

Gerós, H., Cássio, F. and Leão, C. 1996. Reconstitution of lactate proton symport activity in plasma membrane vesicles from the yeast *Candida utilis. Yeast 12*, 1263-1272.

Gerós, H., Cássio, F. and Leão, C. 2000a. Utilization and transport of acetic acid in *Dekkera anomala* and their implications on the survival of the yeast in acidic environments. *J. Food Prot. 63*, 96-101.

Gerós, H., Baltazar, F., Cássio, F. and Leão, C. 2000b. L-(U-^{14}C]Lactate binding to a 43 kDa protein in plasma membranes of *Candida utilis. Microbiology 146*, 695-699.

Goffrini, P., Ficarelli, A., Donnini, C., Lodi, T., Puglisi, P.P. and Ferrero, I. 1996. FOG1 and FOG2 genes, required for the transcriptional activation of glucose-repressible genes of *Kluyveromyces lactis*, are homologous to GAL83 and SNF1 of *Saccharomyces cerevisiae. Curr. Genet. 29*, 316-326.

Halestrap, A.P. and Meredith, D. 2004. The SLC16 gene family-from monocarboxylate transporters (MCTs) to aromatic amino acid transporters and beyond. *Pflugers Arch. 447*, 619-628.

Hatzixanthis, K., Mollapour, M., Seymour, I., Bauer, B.E., Krapf, G., Schüller, C., Kuchler, K. and Piper, P.W. 2003. Moderately lipophilic carboxylate compounds are the selective inducers of the *Saccharomyces cerevisiae* Pdr12p ATP-binding cassette transporter. *Yeast 20*, 575-585.

Haurie, V., Perrot, M., Mini, T., Jenö, P., Sagliocco, F. and Boucherie, H. 2001. The transcriptional activator Cat8p provides a major contribution to the reprogramming of carbon metabolism during the diauxic shift in *Saccharomyces cerevisiae. J. Biol. Chem. 276*, 76-85.

Helmlinger, G., Sckell, A., Dellian, M., Forbes, N.S., and Jain, R.K. 2002. Acid production in glycolysis-impaired tumors provides new insights into tumor metabolism. *Clin. Cancer Res. 8*, 1284-1291.

Hildebrand, P.W., Lorenzen, S., Goede, A. and Preissner, R. 2006. Analysis and prediction of helix-helix interactions in membrane channels and transporters. *Proteins 64*, 253-262.

Hirai, T., Heymann, J.A.W., Maloney, P.C. and Subramaniam, S. 2003. A structural model for 12-helix transporters belonging to the major facilitator superfamily. *J. Bacteriol. 185*, 1712-1718.

Hirai, T., Heymann, J.A.W., Shi, D., Sarker, R., Maloney, P.C. and Subramaniam, S. 2002. Three-dimensional structure of a bacterial oxalate transporter. *Nat. Struct. Biol.* 9, 597-600.

Holyoak, C.D., Bracey, D. Piper, P.W., Kuchler, K. and Coote, P.J. 1999. The *Saccharomyces cerevisiae* weak acid-inducible ABC transporter Pdr12 transports fluorescein and preservative anions from the cytosol by an energy-dependent mechanism. *J. Bacteriol. 181*, 4644-4652.

Huang, Y., Lemieux, M.J., Song, J., Auer, M. and Wang, D.-N. 2003. Structure and mechanism of the glycerol-3-phosphate transporter from *Escherichia coli. Science 301*, 616-620.

Jay, J.M. 1992. *Modern Food Microbiology.* 4th Edition, Chapman and Hall, New York.

Kawahata, M., Masaki, K., Fujii, T. and Iefuji, H. 2006. Yeast genes involved in response to lactic acid and acetic acid: acidic conditions caused by the organic acids in *Saccharomyces cerevisiae* cultures induce expression of intracellular metal metabolism genes regulated by Aft1p. *FEMS Yeast Res. 6*, 924-936.

Klammt, C., Schwarz, D., Lohr F., Schneider, B., Dotsch, V. and Bernhard, F. 2006. Cell-free expression as an emerging technique for the large scale production of integral membrane protein. *FEBS J. 273*, 4141-4153.

Koukourakis, M.I., Giatromanolaki, A., Harris, A.L. and Sivridis, E. 2006. Comparison of metabolic pathways between cancer cells and stromal cells in colorectal carcinomas: a metabolic survival role for tumor-associated stroma. *Cancer Res. 66*, 632-637.

Ladanyi, M., Antonescu, C.R., Drobnjak, M., Baren, A., Lui, M.Y., Golde, D.W. and Cordon-Cardo, C. 2002. The precrystalline cytoplasmic granules of alveolar soft part sarcoma contain monocarboxylate transporter 1 and CD147. *Am. J. Pathol. 160*, 1215-1221.

Leão, C. and van Uden, N. 1986. Transport of lactate and other short-chain monocarboxylates in the yeast *Candida utilis. Appl. Microbiol. Biotechnol. 23*, 389-393.

Liu, W., Eilers, M., Patel, A.B. and Smith, S.O. 2004. Helix packing moments reveal diversity and conservation in membrane protein structure. *J. Mol. Biol. 337*, 713-729.

Lodi, T., Fontanesi, F. and Guiard, B. 2002. Co-ordinate regulation of lactate metabolism genes in yeast: the role of the lactate permease gene JEN1. *Mol. Genet. Genomics 266*, 838-847.

Lodi, T., Fontanesi, F., Ferrero, I. and Donnini, C. 2004. Carboxylic acids permeases in yeast: two genes in *Kluyveromyces lactis. Gene 339*, 111-119.

Makuc, J., Paiva, S., Schauen, M., Krämer, R., André, B., Casal, M., Leão, C. and Boles, E. 2001. The putative monocarboxylate permeases of the yeast *Saccharomyces cerevisiae* do not transport monocarboxylic acids across the plasma membrane. *Yeast 18*, 1131-1143

Mathupala, S.P., Parajuli, P. and Sloan, A.E. 2004. Silencing of monocarboxylate transporters via small interfering ribonucleic acid inhibits glycolysis and induces cell death in malignant glioma: an *in vitro* study. *Neurosurgery 55*, 1410-1419.

Nelissen, B., De Wachter, R. and Goffeau, A. 1997. Classification of all putative permeases and other membrane plurispanners of the major facilitator super family encoded by the complete genome of *Saccharomyces cerevisiae. FEMS Microbiol Rev. 21*, 113-134.

Paiva, S., Devaux, F., Barbosa, S., Jacq, C. and Casal, M. 2004. Ady2p is essential for the acetate permease activity in the yeast *Saccharomyces cerevisiae. Yeast 21*, 201-210.

Palková, Z., Devaux, F., Ricicová, M. Mináriková, L., Le Crom, S. and Jacq, C. 2002. Ammonia pulses and metabolic oscillations yeast colony development. *Mol. Biol. Cell. 13*, 3901-3914.

Pao, S.S., Paulsen, I.T., Saier, M.H., Jr. 1998. The major facilitator super family. *Microbiol. Mol. Biol. Rev. 62*, 1-32.

Park, Y. and Helms, V. 2006. How strongly do sequence conservation patterns and empirical scales correlate with exposure patterns of transmembrane helices of membrane proteins? *Biopolymers 83*, 389-399.

Paulsen, I.T., Sliwinski, M.K., Nelissen, B., Goffeau, A. and Saier, M.H., Jr. 1998. Unified inventory of established and putative transporters encoded within the complete genome of *Saccharomyces cerevisiae*. *FEBS Lett. 430*, 116-125.

Pierre, K. and Pellerin, L. 2005. Monocarboxylate transporters in the central nervous system: distribution, regulation and function. *J. Neurochem. 94*, 1-14.

Piper P., Mahé, Y., Thompson, S., Pandjaitan, R., Holyoak, C., Egner, R., Mühlbauer, M., Coote, P. and Kuchler, K. 1998. The pdr12 ABC transporter is required for the development of weak organic acid resistance in yeast. *EMBO J. 17*, 4257-4265.

Poole, R.C. and Halestrap, A.P. 1992. Identification and partial purification of the erythrocyte L-lactate transporter. *Biochem. J. 283*, 855-862.

Poole, R.C. and Halestrap, A.P. 1993. Transport of lactate and other monocarboxylates across mammalian plasma membranes. *Am. J. Physiol. 264*, C761-C782.

Poole, R.C. and Halestrap, A.P. 1994. N-terminal sequence analysis of the rabbit erythrocyte lactate transporter suggests identity with the cloned monocarboxylate transporter protein MCT1. *Biochem J. 303*, 755-759.

Ramaen, O., Leulliot, N., Sizun, C., Ulryck, N., Pamlard, O., Lallemand, J.-Y., van Tilbeurgh, H. and Jacquet, E. 2006. Structure of the human multidrug resistance protein 1 nucleotide binding domain 1 bound to Mg^{2+}/ATP reveals a non-productive catalytic site. *J. Mol. Biol. 359*, 940-949.

Reihl, P. and Stolz, J. 2005. The monocarboxylate transporter homolog Mch5p catalyzes riboflavin (vitamin B2) uptake in *Saccharomyces cerevisiae*. *J. Biol. Chem. 280*, 39809-39817.

Saier, M.H., Jr. 1998. Molecular phylogeny as a basis for the classification of transport proteins from bacteria, archaea and eukarya. *Adv. Microb. Physiol. 40*, 81-136.

Saier, M.H., Jr. 2000. A functional-phylogenetic classification system for trans-membrane solute transporters. *Microbiol. Mol. Biol. Rev. 64*, 354-411.

Saier, M.H., Jr., 2003. Tracing pathways of transport protein evolution. *Mol. Microbiol., 48*, 1145-1156.

Saier, M.H., Jr., Beatty, J.T., Goffeau, A., Harley, K.T., Heijne, W.H.M., Huang, S.-C., Jack, D.L., Jahn, P.S., Lew, K., Liu, J., Pao, S.S., Paulsen, I.T., Tseng, T.-T. and Virk, P.S. 1999. The major facilitator superfamily. *J. Mol. Microbiol. Biotechnol. 1*, 257-279.

Saier, M.H., Jr., Tran, C.V., Barabote, R.D. 2006. TCDB: the Transporter Classification Database for membrane transport protein analyses and information. *Nucleic Acids Res. 3*, D181-D186.

Schüller, C., Mamnun, Y.M., Mollapour, M., Krapf, G., Schuster, M., Bauer, B.E., Piper, P.W. and Kuchler, K. 2004. Global phenotypic analysis and transcriptional profiling defines the weak acid stress response regulon in *Saccharomyces cerevisiae*. *Mol. Biol. Cell. 15*, 706-720.

Soares-Silva, I., Paiva, S., Kotter, P., Entian, K.D. and Casal, M. 2004. The disruption of JEN1 from *Candida albicans* impairs the transport of lactate. *Mol. Membr. Biol. 21*, 403-411.

Soares-Silva, I., Schüller, D., Andrade, R.P., Baltazar, F., Cássio, F. and Casal, M. 2003. Functional expression of the lactate permease Jen1p of *Saccharomyces cerevisiae* in *Pichia pastoris*. *Biochem. J. 376*, 781-787.

Sousa, M.J., Miranda, L., Côrte-Real, M. and Leão, C. 1996. Transport of acetic acid in *Zygosaccharomyces bailii*: effects of ethanol and their implications on the resistance of the yeast to acidic environments. *Appl. Environ. Microbial. 62*, 3152-3157.

Tenreiro, S., Rosa, P.C., Viegas, C.A. and Sá-Correia, I. 2000. Expression of the *AZR1* gene (ORF YGR224w), encoding a plasma membrane transporter of the major facilitator superfamily, is required for adaptation to acetic acid and resistance to azoles in *Saccharomyces cerevisiae*. *Yeast 16*, 1469-1481.

Tudor, E.A. and Board, R.G. 1993. Food-spoilage yeasts. In: *The Yeasts*. Volume 5, 2nd Edition (Ed. Rose, A.H. and Harrison J.S.). Academic Press, London, 435-516.

Tzschoppe, K., Augstein, A., Bauer, R., Kohlwein, S.D. and Barth, G. 1999. *Trans*-dominant mutations in the *GPR1* gene cause high sensitivity to acetic acid and ethanol in the yeast *Yarrowia lipolytica*. *Yeast 15*, 1645-1656.

Wahl, M.L., Owen, J.A., Burd, R., Herlands, R.A., Nogami, S.S., Rodeck, U., Berd, D., Leeper, D.B. and Owen, C.S. 2002. Regulation of intracellular pH in human melanoma: potential therapeutic implications. *Mol. Cancer Ther. 1*, 617-628.

Wood, B.J.B. 1998. *Microbiology of Fermented Foods*. Volume 1, 2nd Edition, Blackie Academic and Professional, London.

Zmijewski, M.J., Jr. and MacQuillan, A.M. 1975. Dual effects of glucose on dicarboxylic acid transport in *Kluyveromyces lactis*. *Can. J. Microbiol. 21*, 470-480.

Frontiers in Fungal Ecology, Diversity and Metabolites (edited by K.R. Sridhar), I.K. International Publishing House Pvt. Ltd., New Delhi, 2009.

18

Cell Permeabilization–An Approach to Increase the Biocatalysis in Yeasts

Rajendra Upadhya
Department of Pathology, Albert Einstein College of Medicine, Bronx, New York 10461, USA
(E-mail: rupadhya@aecom.yu.edu)

Introduction

Enzymes are a large and ubiquitous group of protein molecules which act as biocatalysts in catalyzing various reactions that constitute the cellular metabolism. They are more remarkable than other conventional chemical catalysts for their ability to efficiently catalyze reactions under mild conditions and for maintaining strict stereo- and regio-specificity. Enzyme catalyzed reactions are always found to be superior due to their high degree of conversion, good product uniformity and less side reactions. The biocatalyst of choice may be a single enzyme, a couple of enzymes or a cell containing enzyme(s). Enzymes can be isolated from plant, animal and microbial sources. Although they are produced by all living organisms, enzymes from microbial sources are most suitable for industrial purposes due to rapid production, cost effectiveness and possibility of mass production. Recently, enzymes have been modified to increase their efficiency to catalyze reactions without the loss of reaction specificity at high temperature, extreme pH, and in non-aqueous systems.

Microbial enzymes can be classified into two major categories: (i) extracellular enzymes (secreted by the organisms into the culture medium) and (ii) intracellular enzymes (retained within the cell). Most of the industrially biocatalyzed processes utilize inexpensive extracellular enzymes such as proteolytic, lipolytic and amylolytic enzymes. On the other hand, the applications involving isolated intracellular enzymes are greatly hampered by the laborious expensive isolation procedures and the relative enzyme instability (Kierstan and Coughlan, 1991). Thus, as an alternate to isolated enzymes, the whole cells are gaining importance in biocatalysts. In a whole cell, the catalytic activities of a number of enzymes including a few labile ones are preserved and these cells can be useful in catalyzing sequential reactions involving several enzymes. Moreover, whole cells also possess the capacity to regenerate and retain co-enzymes, which are required for certain catalytic functions. Microbial cells are characterized by high catalytic efficiency and specificity and they catalyze many reactions of contemporary value in industries and medicine.

Many microbial cells possess enzymes that are inducible. These properties of microbial cells along with factors such as an enhanced bioconversion and very little unwanted side reactions make the use of microbial cells a pragmatic and profitable possibility. However, the main limitation in the use of whole cells for catalysis is the permeability barrier of the cell wall or a membrane to the entry and exit of substrate(s) and product(s), respectively. Another demerit of whole cell catalysis is the occurrence of unwanted side reactions, which may take place due to the presence of other enzymes within the cells. These hurdles can be solved by subjecting the whole microbial cells into a process called permeabilization. This chapter discusses the methods of cell permeabilization, detection, and applications of permeabilization in yeasts.

Permeabilization

Permeabilization is defined as a process, which alleviates the permeability barrier of the cell membrane or wall permitting the free mobility of substances across the cell wall or membrane. This includes substrate(s) and product(s) (Fig. 1). Cells can be permeabilized without lysis or the destruction of their inner organization by different methods. Based on several factors such as cell type, nature of the enzyme, efficiency of biotransformation, the process of permeabilization can be made either partial or total. The choice of a particular

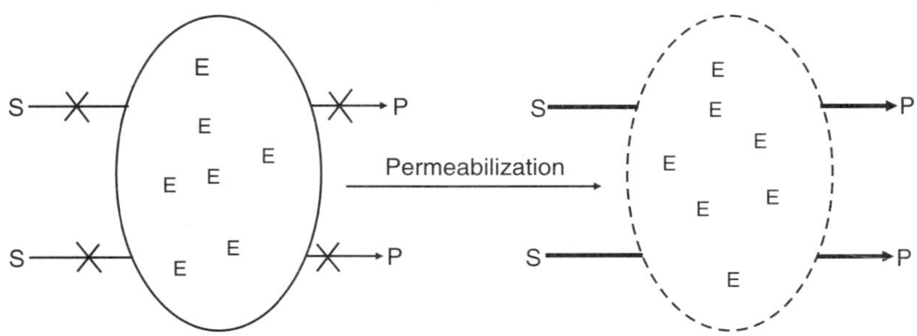

Fig. 1. Schematic representation of the microbial cells before and after permeabilization. In the schematic sketch "E" is the symbol used for the enzyme, "S" and "P" symbolizes substrate and product, respectively. In the whole cells, due to the cell membrane permeability barrier the passage of the substrates and products across the cell wall/ membrane is restricted. Once the cells are permeabilized, there is a free flow of substrates into the inside environment of the cell, where it is converted into a respective product by the action of the intracellular enzyme and the product leaves to the outside of the cell.

method of permeabilization depends on the composition of the cell wall or cell membrane and the nature of the bioconversion reaction. The extent of permeabilization also depends on the concentration and the time of incubation of the permeabilizing agent with the cell (Felix, 1982). Permeabilization of the cell membrane is a delicate task, as a severe treatment may lead to the leakage of proteins and other macromolecules from the cell. Hence, an ideal method of permeabilization is one that opens up the membrane just enough to facilitate the passage of substrate(s) or product(s) across the membrane without either allowing the enzymes or other necessary biomolecules to leave from within the cell.

Methods of Permeabilization

Physical methods

Physical methods of permeabilization involve either drying of yeast cells on a filter paper at room temperature (Rodriguez *et al.*, 1992) or with the addition of salts. Both have been reported to cause permeabilization by rendering the cells permeable to certain substrates, which could be attributed to osmotic shock. Freeze thawing, a temperature-mediated permeabilization can also lead to the release of cellular components including proteins and phospholipids. Electroporation is a widely accepted concept to describe the phenomenon of cell membrane discharge (reversible eletroporation) and cell membrane breakdown (irreversible electroporation) under the application of short electric field pulses (Golzio *et al.*, 2004). The study of the electroporation on biomembranes has become one of the most exciting topics in biophysical and biotechnological areas. The extent of electropermeabilization depends on various factors such as membrane properties, external medium, and the protocols of electroporators.

Chemical methods

Organic Solvents

A single organic solvent or a combination of two solvents at appropriate concentrations has been used to permeabilize cells. Permeabilization of growing cells of *Escherichia coli* with toluene has been studied for the release of protein and RNA (Jackson and Demoss, 1965). In addition, solvents such as ether, dimethyl sulfoxide (DMSO) and acetone have also been reported to increase the intracellular enzyme activities in various organisms (Goyal and Chand, 1996). Ethanol caused the permeabilization of cells of *Kluyveromyces* and was reported to be advantageous for the biocatalysts to be used in the food industry (Flores *et al.*, 1994). Recombinant cells of *Pichia pastoris* over-expresssing selective enzymes have been used for catalysis after permeablization with isopropyl alcohol (Liu *et al.*, 2000).

Detergents

Detergents are reported to cause both reversible and irreversible permeabilization of cells. Lysolecithin brings reversible permeabilization of animal cells. Detergents such as Triton X-100 (Galabova *et al.*, 1996), sodium dodecyl sulphate, sodium deoxycholate, Oxgall, Brij-58 and the copolymer surfactant Pluronic F-68 have been reported to be effective permeabilizing agents (Somkuhi and Steinberg, 1994). Cetyltrimethylammonium bromide (CTAB), a cationic detergent has been widely used for permeabilizing *E. coli, Kluyveromyces fragilis, Saccharomyces cerevisiae* and *Zymomonas mobilis* cells to exploit a number of intracellular enzymes (Joshi *et al.*, 1987). Digitonin, a plant glycoside has been used for the permeabilization of yeast cells. The whole cell fumarase activity in *S. cerevisiae* has been greatly increased after treating the cells with surfactants such as deoxycholate and cetylpyridinium chloride (CPC) (Neufeld *et al.*, 1991). Several factors such as time and temperature of incubation were found to influence the permeabilization of yeast cells with CTAB (Upadhya *et al.*, 1999). In addition, it also depends on the concentration of the detergent and the detergent to yeast cell ratio. Therefore, the optimum permeabilization for obtaining maximum enzyme activity depends on the cell type and the biotransformation reaction under investigation. The exact mechanism by which CTAB causes the permeabilization of cells is not clearly understood. However, earlier reports indicated that protamine sulphate and chitosan positively charged molecules interact with the negative charges localized on the yeast cell membrane rendering the cells permeable. It is quite possible that the detergent CTAB might act in a similar manner. Moreover, it is likely that the detergent may remove certain proteinaceous components leading to the perforation of the membrane.

Detection of Permeabilization

The event of permeabilization can be monitored by observing the release of endogenous small molecular weight compounds either mainly nucleotides or radiolabeled elements, which are previously loaded into the cells via $Na^{+}K^{+}$ exchange pump. The leakage of cytosolic enzymes such as lactate dehydrogenase proved to be a good marker for monitoring cell permeability. The process of permeabilization can also be monitored by following the influx of an impermeant molecule like ethidium bromide, which forms a fluorescent complex with DNA in the cells. Detection of the activity of several intracellular enzymes such as β–galactosidase, alcohol dehydrogenase, phenylalanine ammonia lyase and *catalase* are some of the widely employed methods for monitoring the process of permeabilization (Joshi *et al.*, 1989b; Srinivasan *et al.*, 1994).

Side Reactions

A cell is a complex system with a broad spectrum of metabolic potential and hence there is a strong possibility of interference from other cellular enzyme activities. There are several methods to either eliminate or minimize these side reactions. One of them is the process of permeabilization itself, which causes the removal of a number of small molecular weight compounds (e.g., cofactors and essential ions). Bile extract treatment of *Brevibacterium ammonigenes* for the production of L-malic acid was reported to suppress the formation of unwanted *succinic* acid (Yamamoto *et al.*, 1976). Permeabilization by acetone and detergents has also been used to suppress the side reactions. *Archromobacter liquidium*, which was employed for the production of urocanic acid was incubated at 70°C for 30 min to eliminate urocanase activity. The esterase activity of *Rhodotorula toruloides* was inhibited specifically by heat treatment to exploit intracellular D-amino acid oxidase (DAAO) activity (Lee *et al.*, 1994). The whole cells of *Trigonopsis variabilis* were used for the production of 7-Amino cephalosporanic acid, after specifically inactivating the catalase activity by alkaline treatment of whole cells (Vicenzi and Hansen, 1993). Hahn-Hagerdal and Mattiasson (1982) reported the production of ethanol under aerobic conditions by the addition of azide to the fermentation medium. In the production of glycerol from glucose using *S. cerevisiae*, sodium sulphite has been employed as a metabolic inhibitor (Bisping and Rehm, 1982).

Enzyme Stability

The main advantage of the use of whole or permeabilized cell system compared to purified enzyme is the ability of using whole cell catalyst system repeatedly for the bioconversion reaction. However, the repeated use of permeabilized cells either in free or immobilized state may result in the gradual leakage of enzymes from the permeabilized cells (Joshi *et al.*, 1989a; Upadhya *et al.*, 2000). Thus, enzymes have to be restricted within the permeabilized cells.

Enzymes can be confined within permeabilized cells by various membrane sealing agents, such as Ca^{2+}, Mg^{2+} and glycerol. Enzymes and cells can be immobilized by cross-linking, using several bi-functional and multifunctional agents (Peter and Richards, 1977). Some of the most frequently used cross-linking agents are glutaraldehyde, malonaldehyde, cyanuric chloride, isocyanates, isothiocyanates, aryl halides and imidates. Argeon *et al.* (1980) prepared intracellular enzymes in erythrocytes for *in situ* studies by treating the cells with bi-functional cross-linking agents. Glutaraldehyde (GA), a bifunctional cross-linking agent has been widely employed as a convenient cross-linking agent because of its ready availability, inexpensive and generally regarded as safe (GRAS) status (Upadhya *et al.*, 2000). It is stable in aqueous solution and reacts mainly with amino groups and also with thiol, phenol and imidazole groups of amino acids and proteins, thereby stabilizing the enzyme within the cells, thus, preventing their leakage (Fig. 2). A heat-stable β–galactosidase of *Bacillus* sp. cross-linked with GA, is one of the commercial preparations marketed by Novozyme. The GA-stabilized penicillin acylase activity in CTAB permeabilized cells of *E. coli* has been found to be very efficient for the production of 6-aminopenicillanic acid (Prabhune *et al.*, 1992). If the enzyme

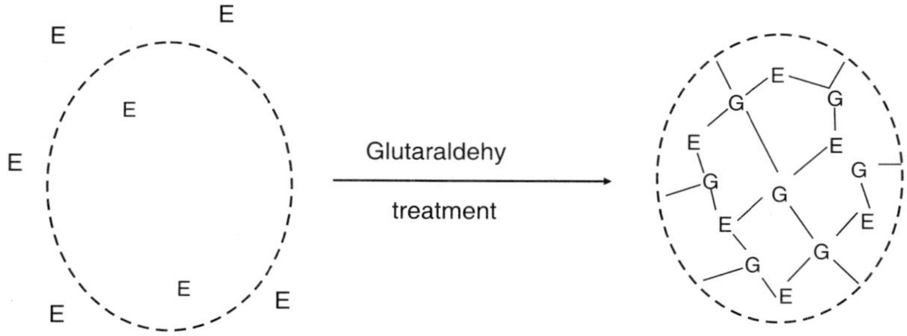

Fig. 2. Schematic representation of the cross-linking of the enzymes within the permeabilized cells by bi-functional cross-linking agent *glutaraldehyde* (GA). In the permeabilized cells, there is a gradual leakage of the intracellular enzymes (E) to the surrounding medium, which may bring about the inactivation of the enzyme. The treatment of either permeabilized cells with GA or the permeabilization of GA treated cells results in the cross-linking of the intracellular enzymes inside the cell, preventing their leakage and thereby, significantly increasing the half-life of the whole cell enzyme activity.

of interest is very unstable, GA treatment may help to prevent the inactivation of the enzyme. The GA treatment of the cells can be carried out either before or after the process of permeabilization. A dramatic stabilization of DAAO and catalase was found in CTAB permeabilized cells of *Rhodotorula gracilis* after GA treatment. The half-life of DAAO and catalase within the GA treated CTAB-permeabilized cells were between 120 and 155 days compared to 50 and 65 h for DAAO and catalase, respectively, in untreated detergent permeabilized cells (Upadhya *et al.*, 2000).

Applications

Most of the current knowledge about enzyme reactions and regulations has been generated using enzymes in an isolated form. However, these do not reflect the enzyme's *in vivo* physiological condition. This is because many intracellular enzymes are present at much higher concentrations *in vivo* than used *in vitro*. It is, therefore, very important to study the properties of some of these enzymes at their physiological concentrations. In permeabilized cells, enzymes are restricted within the boundaries of the cell and maintaining the intracellular concentration of the enzymes and the potential influence of homologous and heterologous protein-protein interactions (Aragon and Sols, 1991). In addition, permeabilized cells can be prepared rapidly by a number of procedures. It can be used as a good tool for investigating secondary metabolites. Permeabilized cells thus are an interesting and practically significant alternative to isolated enzymes for executing a variety of bioconversion reactions.

Immobilization of Cells

In contrast to free cell catalysis, the use of immobilized cells is advantageous as it prevents the washout of the cells at a high dilution rate, permits high cell concentration in the reactor, and allows easy separation of the biocatalyst from the reactor or product containing solution. Their importance in engineering research has also been realized and has covered a wide range of aspects (e.g., immobilizing materials, mass transport effects, physical and chemical environment, kinetics and process modeling). Immobilization can be achieved by aggregation of cells or by confining or attaching them to a solid support. There are two broad types of immobilization: (i) Attachment - The microorganisms adhere to the surface or to other organisms by self-adhesion or by chemical bonding; (ii) Entrapment - The organism is caught in the interstices of fibrous or porous materials or are physically retained within or by a solid porous matrix (stabilized gel or membrane). There are also subcategories within each type of immobilization and selecting the method and the type of matrix depends on the nature of the enzyme or the cell, bioconversion reaction, and its application.

Attachment

In this method, the tendency of the organisms to adhere to other organisms or to solid surfaces has been exploited. Some organisms tend to aggregate particularly when they are in suspension culture at high concentration for extended periods. These aggregated cells have been employed in continuous tower reactors or in columnar bioreactors such as fluidized beds (Netto *et al.*, 1985).

Adsorption

The natural tendency of many types of microorganisms to adhere to solid surfaces has been exploited for many of the bioprocess applications. Sometimes cells may also be induced for adsorption by altering physical or chemical properties of the cells or cell surface by ionic attraction or chemical bonding. Adsorption of the cells on an ion-exchange resin has been extensively investigated (D'Souza, 1990). The varieties of supports used for this technique include wood chips, cotton, cloth, porous bricks, sawdust, glass, ceramics, and ion-exchange resins.

Entrapment

This is by far one of the most popular approaches used for the immobilization of cells. In this method, cells are held either within the interstices of porous materials or by physical restraints of membranes or encapsulating gel matrices. Physical entrapment is the most widely used technique for cell immobilization. The resulting material must have sufficient porosity to allow the transport of substrates in and products out while restraining the cells. Hydrocolloidal gels are frequently used which include alginate, carrageenan, hen egg white, cellulose, chitosan and synthetic polymers (polyacrylamide and photo cross-linkable resins). The gel is usually formed into useful biocatalyst beads by forcing a uniform mixture of cell suspension and aqueous solution of gelling material, dropwise through a nozzle or an orifice. The droplets are subsequently hardened via polymerization or other types of cross-linking.

Cells can also be restrained by semi-permeable membrane materials that separate the organism from the bulk liquid. The most commonly used membrane materials are polysulfones, cellulose acetate, and acrylic copolymers. The immobilization of the cells into the membrane is mainly employed for the fabrication of biosensors. Hollow-fibre bioreactors with the organisms confined to one side of the porous fibre and the soluble substrate and products on the other side seem to be most advantageous.

Conclusions and Outlook

More information about gene inventories is being revealed by several ongoing genome projects in yeast research. The process of permeabilization will be a simple and rapid approach for screening yeasts and other microorganisms for their use in potential and novel biotransformation reactions. If the genome sequence data is consistent with the initial pilot experimental result of biotransformation reaction using permeabilized cells, various conditions of permeabilization and immobilization can be optimized to increase the efficiency of the bioconversion reaction in yeasts and other microorganisms.

References

Aragon, J.J. and Sols, A. 1991. Regulation of enzyme activity in the cell: effect of enzyme concentration. *FASEB J. 5*, 2945-2950.

Argeon, J.J., Feliu, J.E., Frenkel, R.A. and Sols, A. 1980. Permeabilization of animal cells for kinetic studies of intracellular enzymes: *In situ* behavior of the glycolytic enzymes of *erythrocytes. Proc. Nat. Acad. Sci. 77*, 6324-6328.

Bisping, B. and Rehm, H.J. 1982. Glycerol production by immobilized cells of *Saccharomyces cerevisiae Appl. Microbiol. Biotechnol. 14*, 136-139.

D'Souza, S.F. 1990. Surface immobilization of food relevant microbial cells through adhesion. *Food Biotechnol. 4*, 373-398.

Felix, H.A. 1982. Permeabilized cells. *Ann. Biochem. 120*, 211-234.

Flores, M.V., Voget, C.E. and Ertola, R.J.J. 1994. Permeabilization of yeast cells (*Kluyveromyces lactis*) with organic solvents. *Enz. Microb. Technol. 16*, 340-346.

Galabova, D., Tuleva, B. and Sparova, D. 1996. Permeabilization of *Yarrowia lipolytica* cells by triton X-100. *Enz. Microb. Technol. 18*, 18-22.

Golzio, M., Rols, M.P. and Teissie, J. 2004. *In vitro* and *in vivo* electric field-mediated permeabilization, gene transfer, and expression. *Methods 33*, 126-135.

Goyal, A. and Chand, S. 1996. Isolation of NADH from *Saccharomyces cerevisiae* by ether permeabilization and its purification by affinity ultrafiltration. *Indian J. Exp. Biol. 34*, 999-1004.

Hahn-Hagerdal, B. and Mattiasson, B. 1982. Azide sterilization of fermentation media *Appl. Microbiol. Biotechnol. 14*, 140-143.

Jackson, R.W. and Demoss, J.A. 1965. Effects of Toluene on *Escherichia coli. J. Bacteriol. 90*, 1420-1425.

Joshi, M.S., Gowda, L.R. and Bhat, S.G. 1987. Permeabilization of yeast cells *(Kluyveromyces fragilis*) to lactose by *cetyltrimethylammonium* bromide. *Biotechnol. Lett. 9*, 549-554.

Joshi, M.S., Bachhawat, N. and Bhat, S.G. 1989a. Stabilization of *cetyltrimethylammonium* bromide permeabilized yeast whole cell lactase. *Biotechnol. Lett. 11*, 349-352.

Joshi, M.S., Gowda, L.R., Katwa, L.C. and Bhat, S.G. 1989b. Permeabilization of yeast cells (*Kluyveromyces fragilis*) to lactose by digitonin. *Enz. Microb. Technol. 11*, 439-443.

Kierstan, M.P.J. and Coughlan, M.P. 1991. Immobilization of proteins by noncovalent procedures: principles and applications. *Bioprocess Technol. 14*, 13-71.

Lee, Y., Chu, W. and Hsu, W. 1994. Bioconversion of *cephalosporin* C with D-amino acid oxidase from the yeast *Rhodosporidium toruloides. Biotechnol. Lett. 16*, 467-472.

Liu, Y., Fujita, Y., Kondo, A. and Fukuda, H. 2000. Preparation of high-activity whole cell biocatalysts by permeabilization of recombinant yeasts with alcohol. *J. Biosci. Bioeng. 89*, 554-558.

Netto, C.B., Destruhaut, A. and Goma, G. 1985. Ethanol fermentation by flocculating yeast: Performance and stability dependence on a critical fermentation rate *Biotechnol. Lett. 7*, 355-360.

Neufeld, R.J., Peleg, Y., Roken, J.S., Pines, O. and Goldberg, I. 1991. L-Malic acid formation by immobilized *Saccharomyces cerevisiae* amplified for fumarase *Enz. Microbiol. Technol. 13*, 991-996.

Peter, K. and Richards, F.M. 1977. Chemical xross-linking: Reagents and problems in atudies of membrane atructure. *Ann. Rev. Biochem. 46*, 523-551.

Prabhune, A.A., Rao, B.S., Pundle. A.V. and Siva Raman, H. 1992. Immobilization of permeabilized *Escherichia coli* cells with penicillin acylase activity. *Enz. Microb. Technol. 14*, 161-163.

Rodriguez, M., Guereca, L., Valla, F., Quintero, R. and Lopez, M.A. 1992. *Penicillin acylase* extraction by osmotic shock. *Process. Biochem. 27*, 217-223.

Somkuhi, G.A. and Steinberg, H.D. 1994. Permeabilization of *Streptococcus thermophilus* and the expression of beta-galactosidase. *Enz. Microb. Technol. 16*, 573-576.

Srinivasan, A.R., Nagajyothi, Gowda, L.R. and Bhat, S.G. 1994. Phenylalanine ammonia-lyase activity in permeabilized yeast cells (*Rhodotorula glutinis*). *Biotechnol. Lett. 8*, 729-732.

Upadhya, R., Nagajyothi and Bhat, S.G. 1999. D-Amino acid oxidase and catalase of detergent permeabilized *Rhodotorula gracilis* cells and its potential use for the synthesis of α-keto acids. *Process Biochem. 35*, 7-13.

Upadhya, R., Nagajyothi and Bhat, S.G. 2000. Stabilization of D-amino acid oxidase and catalase in permeabilized *Rhodotorula gracilis* cells and its application for the preparation of α-ketoacids. *Biotechnol. Bioeng. 68*, 430-436.

Vicenzi, J.T. and Hansen, J.G. 1993. Enzymatic oxidation of cephalosporin C using whole cells of the yeast *Triginopsis variabilis* within a "cross-flow filter-reactor". *Enz. Microb. Technol. 15*, 281-285.

Yamamoto, K., Tosa, T., Yamashita, K. and Chibata, I. 1976. Continuous production of L-malic acid by immobilized *Brevibacterium ammoniagenes* cells. *Eur. J. Appl. Microbiol. Biotechnol. 3*, 169-183.

Frontiers in Fungal Ecology, Diversity and Metabolites (edited by K.R. Sridhar), I.K. International Publishing House Pvt. Ltd., New Delhi, 2009.

19

Molecular Approaches in Mycology Based on Polymerase Chain Reaction

Seena Sahadevan*
Department of Biology, Mount Allison University, 63B York Street, Sackville, New Brunswick, Canada E4L 1G7
**Present address: Centre of Biology, Department of Biology, University of Minho, 4710–057 Braga, Portugal*
(E-mail: seenasahadevan@yahoo.com)

Introduction

Conventional methods like environmental sample culturing, laboratory identification by morphology, and biochemical tests are still fundamental. Traditional microscopic techniques fail to identify the vegetative hyphae, which are the major fungal biomass on decaying leaves (Nickolcheva *et al*., 2003). The nucleic acids throughout the fungal life cycle can be characterized or unveiled using molecular techniques that strengthen the demand for molecular revolution. Over the past two decades, there has been an exponential increase in the development and application of molecular techniques to study fungal phylogeny and biodiversity involving an initial polymerase chain reaction (PCR) amplification. Often the PCR-based molecular study begins with the PCR amplification and is achieved by means of primers that are specifically targeted followed by the detection of sequence variation in the PCR products by cloning and sequencing or by an analysis involving electrophoresis. The PCR technology, conventional and specific types of PCRs and PCR-based molecular tools can be applied in fungal diagnostics and diversity assessment. Introduction of molecular tools, particularly the PCR in the 1980's has made it possible to carry out detailed evaluation of the fungal biodiversity, accurate identification, molecular diagnostics and epidemiology. Since then, the PCR has been the foundation of molecular biology. The PCR is a DNA amplification technique that will eventually lead to the amplification of various parts of the genome of interest by *in vitro* DNA synthesis. The PCRs are easy to perform, can be run routinely and the analysis or interpretation of result does not demand a high level of expertise. In PCR, a pair of oligonucleotide primers is used to selectively hybridize the target DNA and will bridge our understanding of fungal ecological aspects and their interactions (fungal-plant, fungal-pest, fungal-fungal). This chapter provides an overview of PCR technology and PCR-based molecular tools in the diagnosis, and assessment of fungal diversity.

PCR Technology

The conventional PCR is not a quantitative but a qualitative technique and has been used widely to detect, monitor and identify fungi from a huge range of environmental samples. The PCR associated technology includes methods that increase the sensitivity and are more discriminative. The PCR reaction is performed in an automated thermocycler, which controls and optimizes the different temperatures that are required. The thermo-cycler is programmed to achieve certain temperatures, a high temperature to denture DNA (94-95°C, 30-60 S), a low temperature (depending on the base composition, length and concentration of primer) to anneal the primers to the template DNA and an intermediate temperature (72°C) for the extension of the primer. The entire process takes only a few minutes and is repeated many times to amplify the selected DNA by monitoring the number of cycles. The recommended number of cycles is 36 and this depends on the efficiency of primer annealing and the extension step. The *Taq* DNA polymerase, oligonucleotide primers, deoxynucleotides (dNTP), reaction buffer, template DNA and magnesium ions are the constituents of the PCR reaction. The concentration of each component that is optimized for PCR assays are mainly determined by the primers.

The selected primers should be highly specific to the target species and cross-homology to other species is not desirable. In PCR, the *in vivo* DNA replication is carried out by creating a replication start point artificially at the position of gene of interest using primers. For species specific detection, the primer pair should be very specific, 18-30 nucleotide long, amplify all strains of the species with equal sensitivity, possess similar melting temperature for each primer, low hairpin formation and offering no secondary priming site in the template. In a PCR reaction, the primer should possess sequences that are complementary to the flanking region of the desired target DNA sequences and facilitate annealing to this region. The *Taq* DNA polymerase recognizes single stranded DNA as template and replicates the gene of interest. The final PCR product obtained contains a mixture of multiple copies of the selected DNA region (Abd-Elsalam, 2003; Gorelenkov *et al.*, 2001; Rangoo *et al.*, 2000; Schmidt, 1997).

The use of standard PCR has limitations in some diagnostic assessment as it fails to deliver the sensitivity and specificity. A two-step nested PCR, can overcome the above problem. In a nested PCR, an out primer set is used for the first round of amplification and its products serve as templates in a second round of PCR by using the second set of primers, and rendering higher specificity and sensitivity. Often in an environmental sample assay, fungal universal primers are used as the first set of primers (out primers), specific primers as the second set of specific primers (inner primers) and this is advantageous in the characterization of minor quantity composite samples. However, this technique may fail to detect the species in an extremely low quantity and this can be solved by using specific primers in both rounds of the PCR. High sensitive detection has been reported by Zeng *et al.* (2005). As a remedy for reducing the contamination, one-tube nested PCR is practiced, i.e. amplification with the inner and outer primers are done in the same reaction vessel. However, the difficulty lies in designing compatible primers with different temperature profiles. The ease and short operation procedure makes it suitable for implementation in diagnostic laboratories. Multiplex PCR allows amplifying more than one target of interest simultaneously in one reaction by using more than one primer pair.⊠This method reduces the examination duration due to the detection of more than one target organism at the same time. This method is important for the identification and the detection of fungi in clinical and environmental samples (Amicucci *et al.*, 2000; Fujita *et al.*, 2001; Gariepy *et al.*, 2003; Nagao *et al.*, 2005). Nagao *et al.* (2005) used four different primer sets to detect five human pathogenic *Rhizopus* spp. This method is cost effective, labour saving, and useful in environmental sample screening (Elnifro *et al.*, 2000). The primers used in multiplex reactions must be meticulously designed, as similar annealing temperature is desired. Only a limited number of multiplex PCR assays have been made for fungi (Fujita *et al.*, 2001; Gariepy *et al.*, 2003; Nagao *et al.*, 2005). See also Chapter 20 for molecular techniques employed to identify pathogenic fungi and their significance in understanding of the evolution of species as well as differentiation of pathogens.

Target DNA

A PCR technology involves the selection of a specific target region of DNA/RNA region to identify a particular fungus, extraction of total DNA/RNA from a sample and a method to identify the presence of the target genetic material in the sample. Target DNA selection can be done by using specific sequence information from the databases and by designing primers across the conserved and variable region. The most popular DNA region that is targeted is the ribosomal DNA (rDNA). The fungal ribosomal DNA (rDNA) consist of three genes: Large Sub Unit (LSU) gene (28S), Small Sub Unit (SSU) gene (18S) and 5.8S gene separated by an internal transcribed spacer (ITS) in a unit tandemly repeated many times in the genome (Fig. 1). The LSU and SSU rDNA encompasses highly conserved and variable regions. Group-specific markers can be developed based on the conserved regions. The ITS region in the rDNA unit evolved rapidly and a notable variation among the strain within the species or genome was reported. However, the degree of variability within the ITS

the pseudo-terminal restriction fragment formations, which results in an overestimation of diversity and is unsuitable to detect complex microbial population (Egret and Friedrich, 2003; Engebretson and Moyer, 2003). The choice of primers and restriction enzyme determines the accuracy of this method in detecting microbial load (Lueders and Friedrich, 2003). The T-RFLP method is reproducible, allows both quantitative and qualitative estimations and targets 16S rRNA gene fragments. This technique has been employed to assess (both between and within) the microbial dynamics (Osborn *et al.*, 2000).

Single Strand Conformation Polymorphism (SSCP)

The SSCP method was first employed by Lee *et al.* (1996) to estimate the natural microbial diversity. It detects the sequence variations among the PCR amplified DNA fragments and is based on the principle that at low temperature a single stranded DNA adopts a three-dimensional conformation that is determined by the intramolecular interactions, which influence their electrophoretic mobility in a non-denaturing polyacrylamide gel. According to the varied electrophoretic mobilility of the PCR fragments of similar size with different nucleotide sequences are separated. The differences in mobility are detected by autoradiograms or automated DNA sequencers. Although the detection of variations in sequence using this method is good, the detection sensitivity decreases with the increase in length of the fragment. Microbial diversity in aquatic ecosystem, rhizosphere diversity of microbial community and succession during composting has been reported using this technique (Peters *et al.*, 2000; Ross *et al.*, 2001; Schmalenberger and Tebbe, 2003; Wenderoth *et al.*, 2003).

Ribosomal Intergenic Spacer Analysis (RISA) and Automated Ribosomal Intergenic Spacer Analysis (ARISA)

In this method, amplification products of the spacer region located between the small and large subunit of ribosomal RNA (rRNA) genes in the rRNA operon are separated on polyacrylamide gel on the basis of size and visualized by silver staining. The automated version of RISA and ARISA was developed by Fisher and Triplett (1999) in order to assess community diversity, rapidly and efficiently. Automated capillary electrophoresis detects the amplicons, which have been amplified using a fluorescently labeled forward primer. The total number of distinct fluorescent peaks in the ARISA data denotes an estimation of species diversity in a given sample. An investigation on the seasonal variety and species richness and diversity of microorganisms has been performed by employing ARISA (Graham *et al.*, 2004; Hewson and Fuhrman, 2004).

Random Amplified Polymorphic DNA (RAPD) and Amplified Fragment Length Polymorphism (AFLP)

The RAPD fingerprinting technique utilizes only one short primer with an arbitrary sequence, which binds many sites in the genomic DNA. The fragments of different size are separated by electrophoresis to obtain a specific banding pattern and provide a simple means to construct genetic maps and to perform DNA finger printing (Williams *et al.*, 1991). The RAPD is used to analyze intraspecific variability, to distinguish pathogens, and to determine genetic variations (Goodwin and Annis, 1991; Meyer *et al.*, 1992; Pelaez *et al.*, 1996). The AFLP, is a modified PCR technique employed initially for genetic mapping of plants (Vos *et al.*, 1995) and currently used to estimate biodiversity and fungal phylogeny. The AFLP is a PCR-based tool, amplifying the fragments from a restriction enzyme that digests the genomic DNA. It is highly reproducible compared to RAPD. It is also more efficient in the detection of polymorphism and paves the way for the detection of inter- and intra-specific variations of fungal phytopathogens (Majer *et al.*, 1996).

Inter-Simple Sequence Repeat (ISSR) and Random Amplified Microsatellite (RAMS)

The ISSR techniques are almost identical to the RAPD techniques. The ISSR primer sequences are designed from microsatellite regions and the annealing temperatures are more than those used for RAPD markers. Inter-simple sequence repeat (ISSR) markers are easily generated and were initially devised for differentiating closely related plant cultivars. Now a days, these markers are extremely useful for the studies of natural populations of plants, fungi, insects, and vertebrates (Wolfe, 2005). The ISSR markers were used by Wang *et al.* (2005) to investigate genetic diversity among 39 isolates of *Beauveria* spp. isolated from different insect hosts and geographical origins, and considerable intraspecific and interspecific variability among the isolates of *Beauveria* spp. was revealed.

The RAMS is a PCR fingerprinting technique, which utilizes single primers originally designed as hybridization probes to detect the mini- or microsatellite DNA in conventional DNA fingerprinting for the amplification of different eukaryotic DNA sectors. The identification and assessment of the geographical variant species, and different strains has been achieved by the amplification of a microsatellite region (Hantula and Müller, 1997; Longato and Bonfante, 1997; Marmeise *et al.*, 1992).

Quantitative Real-Time PCR (qRT-PCR)

The conventional PCR method permits only qualitative but not quantitative analyses. In qRT-PCR, a flourimeter automatically measures the PCR product during each cycle. The monitoring of the accumulation of products in qRT-PCR is enabled by the labeling of primers and oligonulcleotide probes or amplification products with fluorescent dyes. The signal strength is determined by the amount of amplification products during each cycle and increases with an increase in specific amplification products. The initial target DNA amount in the reaction can be related to a cycle threshold (c_t). The c_t is the cycle number at which the fluorescence signal surpasses the background during the exponential phase of amplification. The quantification of target DNA is attained by correlating to a standard curve that relates c_t to known amounts of template DNA. The monitoring of the accumulation of products in qRT-PCR⊠s enabled using fluorescent DNA intercalating dyes such as SYBR Green 1 (Manerkar *et al*., 2007), sequence specific probe-based assays using TaqMan probes (Wittwer *et al*., 1997), molecular beacons (Tyagi *et al*., 1998) and dual hybridization probes (Bernard and Wittwer, 2000). The SYBR green 1 and TaqMan probe-based assays are becoming attractive in detection and quantification of fungal contaminants in clinical specimens and environmental samples. The multiplex RT-PCR can be used effectively for the detection and quantification of more than one target DNA molecule from a single reaction tube. However, the number of successful multiplex RT-PCR system for fungal detection is limited (Bluhm *et. al*., 2004; Bu *et al*., 2005; Wittwer *et al*., 2001). Further development and optimization of multiplex RT-PCR systems will lead to the expansion of this powerful tool in fungal diagnostics and biodiversity assessment in future.

Conclusions and Outlook

Rapid strides in molecular techniques have made a huge impact on mycology. Many studies indicate that the molecular methods are more sensitive than the traditional culture methods for the detection of fungi in environmental and clinical specimens (Kim *et. al.,* 2005; Rimek *et al*., 1998; Rintala, 2003; Shin *et al.,* 1997; Willinger *et. al.,* 2003; Baerlocher, 2007). Not all fungi have an ideal medium and therefore, the non-culturable and non-viable spores can be detected by molecular methods. However, PCR inhibitors are the main problem when environmental samples are examined, which results in false or negative results and need careful confirmation and interpretations. The PCR inhibitors can be eliminated by the dilution of the template DNA and further purification by agarose gel electrophoresis, various columns and resins (Rintala, 2003). Emphasis should always be directed on extensive sequence analysis and careful primer design, and also on PCR conditions as all these affect the specificity of a PCR assay. The fingerprint approaches based on PCR amplification, which do not demand any establishment of clone libraries, are currently the most popularly used tools in research involving the assessment of biodiversity in aquatic microbial communities (Dorigo *et al.,* 2005). The RT-PCR now makes the quantitative determination of DNA and RNA more accurately and reproducible than any other PCR-based tool (Schmittgen, 2001). The scope for PCR technologies becomes greater with the increased knowledge on fungal genomics and gene function. All the PCR techniques are still laboratory-based. Therefore, efforts should be channelized in developing PCR/probe-based detection kits that can be taken into the field to perform various assessments.

References

Abd-Elsalam, K.A. 2003. Bioinformatic tools and guideline for PCR primer design. *Afr. J. Biotech. 2*, 91-95.

Amicucci, A., Guidi, C., Zambonelli, A., Potenza, L. and Stocchi, V. 2000. Multiplex PCR for the identification of white *tuber* species. *FEMS Microbiol. Lett. 189*, 265-269.

Baerlocher, F. 2007. Molecular approaches applied to aquatic hyphomycetes. *Fungal Biology Reviews 21*: 19-24.

Baldauf, S.L. and Palmer, J.D. 1993. Animals and fungi are each other's closest relatives: Congruent evidence from multiple proteins. *Proc. Nat. Acad. Sci. 90*, 11558-11562.

Berbee, M.L., Yoshimura, A., Sugiyama, J. and Taylor, J.W. 1995. Is *Penicillium* monophyletic? An evaluation of phylogeny in the family Trichocomaceae from 18S, 5.8S and ITS ribosomal DNA sequence data. *Mycologia 87*, 210-222.

Bernard, P.S. and Wittwer, C.T. 2000. Homogeneous amplification and variant detection by fluorescent hybridization probes. *Clin. Chem. 46*, 147-148.

Bluhm, B.H., Cousin, M.A. and Woloshuk, C.P. 2004. Multiplex real-time PCR detection of fumonisin-producing and trichothecene-producing groups of *Fusarium* species. *J. Food Prot. 67*, 536-543.

Bruns, T.D., Vilgalys, R., Barns, S.M., Gonzalez, D., Hibbett, D.S., Lane, D.J., Simon, L., Stickel, S., Szaro, T.M., Weisberg, W.G. and Sogin, M.L. 1992. Evolutionary relationships within the fungi: Analysis of nulclear small subunit rRNA sequences. *Mol. Phylogenet. Evol. 1*, 231-234.

Bu, R., Sathiapalan, R.K., Ibrahim, M.M., Al-Mohensen, I., Almodavar, E., Gutierrez, M.I. and Bhatia, K. 2005. Monochrome light cycler PCR assay for detection five common species of *Candida* and *Aspergillus. J. Med. Microbiol. 54*, 243-248.

Carbone, I. and Kohn, L.M. 1993. Ribosomal DNA sequence divergence within internal transcribed spacer I of the Sclerotiniaceae. *Mycologia 85*, 415-427.and quantification of

Dorigo, U., Volatier, L. and Jean-Francois, H. 2005. Molecular approaches to the assessment of biodiversity in aquatic microbial communities. *Water Res. 39*, 2207-2218.

Edlind, T.D., Li, J., Visvesvara, G.S., Vodkin, M.H., McLaughlin, G.L. and Katiyar, S.K. 1996. Phylogenetic analysis of b-tubulin sequences from a mitochondrial protozoa. *Mol. Phylogenet. Evol. 5*, 359-367.

Egret, A. and Friedrich, M.W.⊠2003. Formation of pseudo terminal restriction fragments, a PCR-related bias affecting terminal restriction fragment length polymorphism analysis of microbial community structure. *Appl. Environ. Microbiol. 69*, 2555-2562.

Elnifro, E.M., Ashshi, A.M., Cooper, R.J. and Klapper, P.E. 2000. Multiplex PCR: optimization and application in diagnostic virology. *Clin. Microbiol. Rev. 13*, 559-570.

Engebretson, J.J. and Moyer, C.L. 2003. Fidelity of selected restriction endonucleases in determining microbial diversity by terminal-restriction fragment length polymorphism. *Appl. Environ. Microbiol. 69*, 4823-4829.

Feibelman, T.P., Doudrick, R.L., Cibula, W.G. and Bennett, J.W. 1997. Phylogenetic relationships within the *Cantherellaceae* inferred from sequence analysis of the nuclear large subunit rDNA. *Mycol. Res. 101*, 1423-1430.

Fisher, M. and Triplett, E.W. 1999. Automated approach for ribosomal intergenic spacer analysis of microbial diversity and its application to freshwater bacterial communities. *Appl. Environ. Microbiol. 65*, 4630-4636.

Fletcher, L.D., McDowell, J.M., Tidwell, R.R., Meager, R.B. and Dykstra, C.C. 1994. Structure, expression and phylogenetic analysis of the gene encoding actin I in *Pneumocystis carinni. Genetics 137*, 743-750.

Fujita, S.I., Senda, Y., Nakaguchi, S. and Hashimoto, T. 2001. Multiplex PCR using internal transcribed spacer 1 and 2 regions for rapid detection and identification of yeast strains. *J. Clin. Microbiol. 39*, 3617-3622.

Gargas, A. and Taylor, J.W. (1992). Polymerase chain reaction (PCR) primers for amplifying and sequencing nuclear 18S rDNA from lichenized fungi. *Mycologia 84*, 589-592.

Gariepy, T.D., Levesque. C.A., de Jong, S.N. and Rhe, J.E. 2003. Species specific identification of the *Neofabraea* pathogen complex associated with pome fruits using PCR and multiplex DNA amplification. *Mycol. Res. 107*, 528-536.

Giovannoni, S.J. Britschgi, T.B., Moyer, C.L. and Field, K.G. 1990. Genetic diversity in Sargasso Sea bacterioplankton. *Nature 345*, 60-63.

Glenn, A.E., Bacon, C.W., Price, R. and Hanlin, R.T. 1996. Molecular phylogeny of *Acremonium* and its taxonomic implications. *Mycologia 88*, 369-383.

Goodwin, P.H. and Annis, S.L. 1991. Rapid identification of genetic variation and pathotype of *Leptosphaeria maculans* by Random Amplified Polymorphic DNA assay. *Appl. Environ. Microbiol. 57*, 2482-2486.

Gorelenkov, V., Antipov, A., Lejnine, S., Draselia, N. and Yuryev, A. 2001. Set of novel tools for PCR primer design. *Biotechniques 31*, 1326-1330.

Graham, J.M., Kent, A.D., Lauster, G.H., Yannarell, A.C., Graham, L.E., Triplett, E.W., 2004. Seasonal dynamics of phytoplankton and planktonic protozoan communities in a Northern temperature humic lake: Diversity in a dinoflagellate dominated system. *Microb. Ecol. 48*, 528-540.

Hantula, J. and Müller, M.M. 1997. Variation within *Gremmeniella abietina* in Finland and other countries as determined by Random Amplified Microsatellite (RAMS). *Mycol. Res. 101*, 169-175.

Hasegawa, M., Hashimoto, T., Adachi, J., Iwabe, N. and Miyata, T. 1993. Early branchings in the evolution of eukaryotes: Ancient divergence of *Entamoeba* that lacks mitochondria revealed by protein sequence data. *J. Mol. Evol 36*, 380-388.

Hayes, V.W., Y.Wu, Y., Osinga, J., Mulder, I.M., van der Vlies, P., Elfferich, P., Buys, C.H. and Hofstra, R.M. 1999. Improvements in gel composition and electrophoretic conditions for broad-range mutation analysis for denaturing gradient gel electrophoresis. *Nucl. Acids Res.* 27, 29.

Hewson, I. and Fuhrmam, J.A. 2004. Richness and diversity of bacterioplankton species along an estuarine gradient in Moreton Bay, Australia. *Appl. Environ. Microbiol. 70*, 3425-3433.

Jasalavich, C.A., Morales, V.M., Pelcher, L.E. and Seguin-awaetz, G. 1995. Comparison of nuclear ribosomal DNA sequences from *Alternaria* species pathogenic to crucifers. *Mycol. Res. 99*, 604-614.

Kim, S.T., Choi, J.H., Jeon, H.G., Cha, H.E., Hwang, Y.J. and Chung, Y.S. 2005. Comparison between polymerase chain reaction and fungal culture for the detection of fungi in patients with chronic sinusitis and normal controls. *Acta Otolaryngol. 125*, 72-75.

Lee, D.H., Zo, Y.G. and Kim, S.J. 1996. Nonradioactive method to study genetic profiles of natural bacterial communities by PCR-single-strand-conformation polymorphism. *Appl. Environ. Microbiol. 62*, 3112-3120.

LoBuglio, K.F., Pitt, J.I. and Taylor, J.W. 1993. Phylogenetic analysis of two ribosomal DNA regions indicates multiple independent losses of a sexual *Talaromyces* state among asexual *Penicillium* species in subgenus *Biverticilium. Mycologia 85*, 592-604.

Longato, S. and Bonfante, P. 1997. Molecular identification of mycorrhizal fungi by direct amplification of microsatellite regions. *Mycol. Res. 101*, 425-432.

Lueders, T. and Friedrich, M.W. 2003. Evaluation of PCR amplification bias by terminal restriction fragment length polymorphism analysis of small–subunit rRNA and mcrA genes by using defined template mixtures of methanogenic pure cultures and soil DNA extracts. *Appl. Environ. Microbiol. 69*, 320-326.

Majer, D., Mitchen, R., Lewis, B.G., Vos, P. and Oliver, R.P. 1996. The use of AFLP fingerprinting for the detection of genetic variation in fungi. *Mycol. Res. 100*, 1107-1111.

Manerkar, M.A., Seena, S. and Baerlocher, F. 2008. Q-RT-PCR for assessing archaea, bacteria, and fungi during leaf decomposition in a stream. *Microbial Ecology 55*, (in press).

Marmeise, R., Debaud, J.C. and Casselton, L.A. 1992. DNA probes for species and strain identification in the ectomycorrhizal fungus *Hebeloma. Mycol. Res. 96*, 161-165.

Marsh, T.L. 1999. Terminal restriction fragment length polymorphism (T-RFLP): an emerging method of characterizing diversity among homlogous populations of amplification products. *Curr. Opin. Microbiol. 2*, 323-327.

Martin, W., Brinkmann, H., Savonna, C. and Cerff, R. 1993. Evidence for a chimeric nature of nuclear genomes: Eubacterial origin of eukaryotic glyceraldhyde-3-phosphate dehydrogense genes. *Proc. Nat. Acad. Sci. USA 90*, 8692-8696.

Meyer, W., Lieckfeldt, E., Wostemeyer, J. and Borner, T.H. 1992. DNA fingerprinting for differentiating aggressivity groups of the rape seed pathogen *Leptosphaeria maculans*. *Mycol. Res.* 96, 651-657.

Muyzer, G., 1999. DGGE/TGGE a method for identifying genes from natural ecosystems. *Curr. Opin. Microbiol. 2*, 317-322.

Nagao, K., Ota, T., Tanikawa, A., Takae, Y., Mori, T., Udagawa, S.-I. and Nishikawa, T. 2005. Genetic identification and detection of human pathogenic *Rhizopus* species, a major mucormycosis agent by multiples PCR-based on internal transcribed spacer region of rRNA gene. *J. Dermatol. Sci. 39*, 23-31.

Nickolcheva, L.G., Cockshutt, A.M. and Bärlocher, F. 2003. Determining diversity of freshwater fungi on decaying leaves: Comparison of traditional and molecular approaches. *Appl. Environ . Microbiol. 69*, 2548-2554.

Niessen, L. and Vogel, R.F. 1998. Group specific PCR- detection of potential trichothecene-producing *Fusarium* species in pure cultures and cereal samples. *Syst. Appl. Microbiol. 21*, 618-631.

Olsen, G.J., Lane, D.J., Giovannoni, S.J., Pace, N.R.1986. Microbial ecology and evolution: a ribosomal RNA approach. *Ann. Rev. Microbiol. 40*, 337-365.

Osborn, A.M., Moore, E.R.B. and Timmis, K.N. 2000.An evaluation of terminal-restriction fragment length polymorphism (T-RFLP) analysis for the study of microbial community structure and dynamics. *Environ. Microbiol. 2*, 39-50.

Paleaz, F., Platas, G., Collado, J. and Diez, M.T. 1996. Intraspecific variation in two species of aquatic hyphomycetes assessed by RAPD analysis. *Mycol. Res. 100*, 831-837.

Peters, S., Koschinsky, S., Schwieger, F.and Tebbe, C.C. 2000. Succession of microbial communities during hot composting as detected by PCR-single-strand-conformation polymorphism-based genetic profiles of small-subunit rRNA genes. *Appl. Environ. Microbiol. 66*, 930-936.

Rangoo, V.M., Hyde, K.D. and Liew, C.Y.E. 2000. Molecular techniques. In: *Marine Mycology - A Practical Approach* (Ed. Hyde, K.D. and Pointing, S.B.). Fungal Diversity Research Series1, Fungal Diversity Press, Hong Kong, 308-336.

Riemann, L. and Winding, A., 2001. Community dynamics of free-living and particle-associated bacterial assemblages during a freshwater phytoplankton bloom. *Microb. Ecol. 42*, 219-232.

Rimek, D., Garg, A.P., Kappe, R. and Sonntag, H.G. 1998. Fungal nucleic acid detection for invasive aspergillosis. *Mycoses 41*, 65-68.

Rintala, H. 2003. *Streptomycetes in Indoor Environments-PCR Based Detection and Diversity*. Ph.D thesis, University of Kuopio, Kuopio, Finland.

Ross, N., Villemur, R., Marcandella, E. and Deschenes, L. 2001. Assessment of changes in biodiversity when a community of ultramicrobacteria isolated from ground water is stimulated to form a biofilm. *Microb. Ecol. 42*, 56-58.

Spatafora, J.W. and Blackwell, M. 1993. Molecular systematics of unitunicate paerithecial ascomycetes: the claviceptales-hypocorales connection. *Mycologia 85*, 912-922.

Schmalenberger, A. and Tebbe, C.C. 2003. Bacterial diversity in maize rhizospheres: Conclusions on the use of genetic profiles based on PCR-amplified partial small subunit rRNA genes in ecological studies. *Mol. Ecol. 12*, 251-262.

Schmidt, B.L. 1997. PCR in laboratory diagnosis of human *Borrelia burgdorferi* infections. *Clin. Microbiol. Rev. 10*, 185-201.

Schmittgen, T.D. 2001. Real-time quantitative PCR. *Methods 25*, 383-385.

Shin, J.H., Nolte, F.S. and Morrison, C.J. 1997. Rapid identification of *Candida* species in blood cultures by a clinically useful PCR method. *J. Clin. Microbiol. 35*, 1454-1459.

Tyagi, S., Bratu, D.P. and Krammer, F.R. 1998. Multicolor molecular beacons for allele discrimination. *Nature Biotechnol. 16*, 44-53.

Vallaeys,T., Topp, E., Muyzer, G., Macheret, V., Laguerrie, G., Rigaud, A. and Soulas, G. 1997. Evaluation of denaturing gel gradient electrophoresis in the detection of 16SrDNA sequence variation in rhizobia and methanotrops. *FEMS Microb. Ecol. 24*, 279-285.

Vos, P., Hogers, R., Bleeker, M., Reijans, M., Van de Lee, T., Homes. M., Frijters, A., Pot, J., Peleman, J., Kuiper, M. and Zabeau, M. 1995. AFLP: A new concept for DNA fingerprinting. *Nucleic Acids Res. 23*, 4407-4414.

Wang, S., Miao, X., Zhao, W., Huang B., Fan. M., Li, Z. and Huang Y. 2005. Genetic diversity and population structure among strains of the entomopathogenic fungus, *Beauveria bassiana*, as revealed by inter-simple sequence repeats (ISSR). *Mycol. Res. 109*, 1364-1372.

Wenderoth, D.F., Rosenbrock, P., Abraham, W.R., Pieper. D.H. and Hofle, M.G. 2003. Bacterial community dynamics during biostimulation and bioaugmentation experiments aiming at chlorobenzene degradation in ground water. *Microb. Ecol. 46*, 161-176.

Williams, M.N.V., Pande, N., Nair, M., Mohan, M. and Bennet, J. 1991. Restriction fragment length polymorphism analysis of polymerase chain reaction products amplified from mapped loci of rice (*Oryza sativa* L.) genomic DNA. *Theor. Appl. Genet. 82*, 489-498.

Wittwer, C.T., Hermann, M.G., Moss, A.A. and Rasmussen, R.P. 1997. Continuous fluorescence monitoring of rapid cycle DNA amplification. *Bio Techniques 22*, 130-138.

Wittwer, C.T., Herrmann, M.G., Gundry, C.N. and Elenitoba-Johnson, K.S. 2001. Real-time multiplex PCR assays. *Methods 25*, 430-442.

Willinger, B., Obradovic, A., Selitsch, B., Beck-Mannagetta, J., Buzina, W., Braun, H., Apfalter, P., Hirschl, A.M, A.M., Makristathis, A. and Rotter, M. 2003. Detection and identification of fungi from fungus balls of the maxillary sinus by molecular techniques. *J. Clin. Microbiol. 41*,581-585.

Woese, C.R. 1987. Bacterial evolution. *Microbiol. Rev. 51*, 221-271.

Wolfe, A.D. 2005. ISSR techniques for evolutionary biology. *Methods Enzymol. 395*,134-144.

Zeng, Q-Y., Per, H. and Wang, X-R. 2005. Specific and sensitive detection of the conifer pathogen *Gremmeniella abietina* by nested PCR. *BMC Microbiol. 5*, 65.

Zhou, M.E. and Kleinhofs, A. 1996. Molecular evolution of nitrate reductate genes. *J. Mol. Evol. 42*, 432-442.

Frontiers in Fungal Ecology, Diversity and Metabolites (edited by K.R. Sridhar), I.K. International Publishing House Pvt. Ltd., New Delhi, 2009.

20

Phylogenetic Tools for the Identification of Fungi

M.V. Tejesvi and H.S. Prakash*
Department of Applied Botany and Biotechnology, University of Mysore, Manasagangotri, Mysore 570 006, Karnataka, India
*(*E-mail: hsp@appbot.uni-mysore.ac.in)*

Introduction

The precise identity of pathogens is a necessary step to decide their control method and trade. Traditional identification techniques such as morphological and biological methods are of poor diagnostic value. In recent years, nucleic acid-based techniques have become most potential diagnostic tools (Baerlocher, 2007). The DNA sequence database of fungi helps in sequence analysis and the comparison of closely related genera and species for precise taxonomic placement. A phylogenetic approach based on multiple gene sequences is found to be well suited to distinguish fungal species. Different molecular tools like ITS, IGS, β-tubulin and histone gene sequencing, ITS-RFLP and various primers have been tried for the detection of pathogenic fungi. The PCR-based molecular approach is important in the diagnosis of invasive mycotoxin producing fungi. Molecular biological techniques are advantageous for the study of the properties of pathogenic fungi and are essential for the quick and early diagnosis of fungal infections.

The identification and taxonomic positioning of fungi, bacteria, viruses, and parasites are very important for biologists who are interested in their diversity or their conservation as well as for others to help in distinguishing the organism of their interest (e.g., plant pathologists, quarantine personnel, medical personnel, manufacturers of pharmaceutics). Most of the biological identifications are still carried out using traditional paper-based expertise, which has to be followed manually, although molecular methods are becoming more common in recent years (Clarke, 2003). While using this taxonomic key, the user makes a series of choices from successive groups of contrasting statements and thereby culminating in the identification of species. This identification is only as good as the observations of an expert who has compiled the key and its correct interpretation by the user. The taxonomic identification at the morphological level is however, time consuming, labor-intensive, requires specific skill and above all, possesses the inherent possibility of mis-identification or mis-interpretation. Further, the morphological characters could be highly variable depending on the media and culture conditions (Konietzny and Greiner, 2003). Traditionally, fungal species have been separated and described on the basis of morphology, cultural characteristics (e.g., pigment production, the presence of microconidia and chlamydospores) and the new species may still be described on such basis (Leslie *et al.*, 2005; Rheeder *et al.*, 1996). However, the number of these markers available is generally low, which makes classification difficult and thus, the identification of the related species (Magnani *et al.*, 2005). In recent years, multidisciplinary approaches have been launched and the rapid development in molecular biology has a major impact on fungal systematics. Indeed, many recently described *Fusarium* spp. have initially been discovered by molecular tools used in phylogenetic studies followed by a formal description of the species (Benyon *et al.*, 2000; Tan and Niessen, 2003).

Since the invention of PCR, several tests have been described that enable DNA amplification from most plant pathogenic species. However, these assays aim at conserved segments of the ribosomal genes that are present at several hundred to thousand copies per cell in an organism. It is now possible to characterize in a more objective manner and identify fungal species because of inexpensive DNA sequencing and a relatively large database of ribosomal DNA sequences. Such targets are readily detectable but reflect the status of non-ribosomal nuclear genes in tissues distantly (Bataille *et al.*, 1999). The development of additional molecular tools enabling the study of nuclear DNA in phylogenetically distant genus is still necessary. To assess the quality of pathogenic samples after genomic DNA extraction more accurately, PCR assay targeting a nuclear gene will simplify the identification process. Nuclear DNA would supply a much higher set of characters for phylogenetic analysis and open up the possibility of studying genetic loci that are involved in determining genotypic traits in fungal species.

The need for more objective and rapid methods to identify fungi and mycotoxin producing fungi in cereals and human foods are of prime importance. Recently, molecular techniques have been introduced as powerful tools for the detection and the identification of fungi. The development of molecular biological techniques for genetic differentiation of species due to their sensitivity and specificity has resulted in substantial advances in taxonomy. An amplification of internal transcribed spacer (ITS1-5.8S-ITS2) of ribosomal DNA (rDNA) by the polymerase chain reaction (PCR) (Mullis and Faloona, 1987) combined with the sequencing of the amplicon and analysis of similarity between the sequences obtained and those already deposited in the gene bank has been frequently employed for the identification of fungal species. The current chapter mainly emphasizes various molecular techniques that are employed to identify pathogenic fungi and their importance in understanding of the evolution of species and pathogen differentiation.

Mycotoxins

Mycotoxins are secondary metabolites that are produced by many phytopathogenic and food spoiling fungi including *Aspergillus, Fusarium* and *Penicillium*. The presence of mycotoxins in cereals is considered a global problem in public health. About 25-50% of the crops harvested worldwide are contaminated with mycotoxins. The percentage is highest in the tropical regions, where up to 80% of the crops are reported to contain significant amounts of mycotoxins (Konietzny and Greiner, 2003). The toxicity and carcinogenicity of mycotoxins and their potential to contaminate foods and feedstock is a cause of serious global concern from the point of food safety and trade. The varieties of mycotoxin producing fungi and their metabolites are presented in the Table 1. *Fusarium* toxins are produced by numerous fusaria, which occur widely in cereal grains and animal feeds and cause spontaneous outbreaks of mycotoxicoses in humans and livestock in Africa, Asia, Europe, New Zealand and South America (Cetin and Bullerman, 2005; Placinta *et al*., 1999). The most important *Fusarium* toxins that potentially affect human and animal health and productivity are deoxynivalenol (DON), zearalenone (ZEN), fumonisin (FB1) and moniliformin (MON) (D'Mello *et al*., 1999). The DON or vomitoxin, is a type B trichothecene associated primarily with *Fusarium graminearum* (perfect state: *Gibberella zeae*) and *Fusarium culmorum*. Both are important plant pathogens, which cause *Fusarium* head blight in wheat and *Gibberella* ear rot in maize (McLean, 1996). The DON reduces growth and feed consumption (anorexia) at low concentrations in the diet, thus induces vomiting (emesis) when consumed in acute doses (Rotter *et al*., 1996). Thus, the rapid identification of mycotoxigenic fungi would be desirable for early intervention steps that can be applied to limit the amounts of

Table 1. Mycotoxins produced by various fungi

Mycotoxin	Fungus	Contaminated commodity	Reference
Deoxynivalenol	*Fusarium*	cereals	Tanaka *et al*., 1988
Nivalenol	*Fusarium*	cereals	
Zearalenone	*Fusarium*	cereals	
T2 toxin	*Fusarium*	cereals	Sweeney *et al*., 1998
Fumonisins	*Fusarium*	cereals, legumes	
Trichothecens	*Fusarium*	paddy, corn	
Alternariol	*Alternaria*	cereals, oilseeds	Paterson, 2000
Patulin	*Penicillium*	cereals	
Penicillic acid	*Penicillium*	cereals	
Ochratoxin A	*Penicillium*	cereals, coffee, tree nuts, spices	Oslen *et al*., 2006
Aflatoxins	*Aspergillus*	cereals, coffee, tree nuts, spices, peanuts	Yu *et al*., 2005

contaminated materials (e.g., cereals, cereal-based products) gaining access to the human food chain. With this aim, a number of PCR-based methods have been developed for the identification of mycotoxin biosynthetic genes in different fungal genera for the identification of specific toxigenic fungi (Table 2, 3).

DNA Extraction Methods

The PCR methods have been described in more detail by Hoelzel and Green (1998). Saiki *et al.* (1988) published the first experimental data on PCR and since then, the PCR technique (Mullis and Faloona, 1987) has tremendously influenced research in diverse areas of biological sciences leading to an understanding of microorganisms more explicitly. On using PCR, it is now possible to make virtually unlimited copies of a fragment of DNA (Field and Wills, 1998). The organism of interest can be detected directly through PCR assays in a much shorter duration than by employing conventional culture methods. An important issue that must be considered while determining the sensitivity and precision of PCR-based methods for the detection of microorganisms is the efficiency of the cell lysis and nucleic acid purification methods that are used to recover DNA or RNA templates (Wilson, 1997). The well known resistance of fungal spores and conidia to lysis as well as the potential presence of inhibitory compounds in these cells (Martin and Haider, 1971) could significantly impair the sensitivity of the PCR method designed for their detection. An ideal technique for DNA extraction should minimize the number of times a tissue sample is handled from collection to analysis, optimize the yield of DNA extracted from a sample, be applicable to diverse organisms, be suited to mass handling of samples to minimize labor and materials costs and do not generate hazardous waste that negatively influence the environment (Steiner *et al.*, 1995). Several studies have been devoted to the development and or comparison of different procedures for the extraction of DNA from fungal and bacterial spores (Gang and Weber, 1995; Kuske *et al.*, 1998; Moller *et al.*, 1992).

Ribosomal Gene Complex

The cellular factory that is responsible for the production of protein is the ribosome. As the proper functioning of ribosomes are critical for survival, their physical parameters have been conserved in all forms of life from bacteria to humans. However, some components within the ribosomes have altered during the evolutionary process. These similarities as well as the changes within the genetic material can be used as tools for the identification of microorganisms including fungi. The sequence homology within the rDNA genes of fungi (18S, 5.8S and 28S) and differences within the spacer regions (ITS1 and ITS2) are the genetic basis of organization of the fungi into various taxonomic groups (White *et al.*, 1990). The ribosomal DNA (rDNA) genes are found in all microorganisms and are known to accumulate mutations at a slow constant

Table 2. Species-specific primers used for identification of *Fusarium* spp.

Target species	Primer	Sequence	Amplicon size (bp)	Reference
Fusarium spp.	ITS-Fu-f	5'-CAACTCCCAAACCCCTGTGA-3'	389	Abd-Elsalam *et al.*, 2003
	ITS-Fu-r	5'-GCGACGATTACCAGTAACGA-3'		
F. graminearum	Fg16F	5'-CTCCGGATATGTTGCGTCAA-3'	450	Nicholson *et al.*, 1998
	Fg16R	5'-GGTAGGTATCCGACATGGCAA-3'		
F. pseudograminearum	Fp1-1	5'-CGG GGTAGTTTCACATTTCCG-3'	523	Aoki and O'Donnell, 1999
	Fp1-2	5'-GAGAATGTGATGACGACAATA-3'		
F. avenaceum	FAC R	5'-GGGATATCGGCAAGATCG-3'	920	Doohan *et al.*, 1999
	Fa F	5'-CAAGCATTGTCGCCACTCTC-3'		
F. culmorum	Fc01F[2]	5'-ATGGTGAACTCGTCGTGGC-3'	570	Nicholson *et al.*, 1998
	Fc01R[2]	5'-CCCTTCTTACGCCAATCTCG-3'		
F. poae	Fp82F[5]	5'-CAAGCAAACAGGCCTCTTGACC-3'	220	Parry and Nicholson, 1996
	Fp82R[5]	5'-TGTTCCACCTCAGTGACAGGTT-3'		
F. acuminatum	FAC F	5'-GGGATATCGGGCCTCA-3'	600	Williams *et al.*, 2002
	FAC R	5'-GGGATATCGGCAAGATCG-3'		
F. oxysporum	PFO3[7]	5'-CGGGGGATAAATGCGG-3'	70	Edel *et al.*, 2000
	PFO2[7]	5'-CCCAGGGTATTACACGGT-3'		
F. crookwellense	CroA F	5'-CTCAGTGTCCACCGCGTTGCGTAG-3'	842	Yoder and Christianson, 1998
	CroA R	5'-CTCAGTGTCCCAATCAAATAGTCC-3'		
F. equiseti	FEF1	5'-CATACCTATACGTTGCCTCG-3'	400	Mishra *et al.*, 2003
	FER1	5'-TTACCAGTAACGAGGTGTATG-3'		
F. sporotrichioides	AF330109CF	5'-AAAAGCCCAAATTGCTGATG-3'	332	Demeke *et al.*, 2005
	AF330109CR	5'-TGGCATGTTCATTGTCACCT-3'		
F. solani	Fusofor	5'-CCAATGCCCTCCGGGGCTAAC-3'	565	Jaeger *et al.*, 2000
	Fusorev	5'-GCATAGGCCTGCCTGGCG-3'		
A. fumigatus	Asfufor	5'-CCAATGCCCTTCGGGGCTCCT-3'	520	Jaeger *et al.*, 2000
	Asfurev	5'-CCTGGTTCCCCCCACAG-3'		
F. oxysporum f. sp. *niveum*	Fn-1	5'-TACCACTTGTTGCCTCGGC-3'	327	Zhang *et al.*, 2005
	Fn-2	5'-TTGAGGAACGCGAATTAAC-3'		

Table 3. Group-specific primers for the identification of toxin producing fungi

Target species	Mycotoxin	Primer	Sequence	Amplicon size (bp)	Reference
Fusarium spp.	Trichothecene	Tox5/1	5'-GCTGCTCATCACTTTGCTCAG-3'	685	Niessen and Vogel, 1998
		Tox5/2	5'-CTGATCTGGTCACGCTCATC-3'		
F. oxysporum	fumonisin B1 and B2	Clox1	5'-CAGCAAAGCATCAGACCACTATAACTC-3'	534	Mule *et al.*, 2004
		Clox2	5'-CTTGTCAGTAACTGGACGTTGGTACT-3'		
F. proliferatum	fumonisin B1 and B2	Clpro1	5'-TGCATCAGACCACTCAAATCCT-3'	526	Mule *et al.*, 2004
		Clpro2	5'-TGTCAGTAACTCGACGTTGTTGTT-3'		
Fusarium spp.	Fumonisin B1	Fum5F	5'-GTCGAGTTGTTGACCACTGCG-3'	845	Bluhm *et al.*, 2002
		Fum5R	5'-CGTATCGTCAGCATGATGTAGC-3'		
Fusarium spp.	DON	N1-2	5'-CTTGTTAAGCTAAGCGTTTT-3'	650	Baken *et al.*, 2002
		N1-2r	5'-AACCCCTTTCCACCGAGTATTTC-3'		
Aspergillus spp.	Aflatoxin	aflR660	5'-CGCGCTCCCAAGTCCCCTTCATT-3'	630	Sweeney *et al.*, 1998
		aflR1249	5'-CTTGTTCCCCGAGATGACCA-3'		
Pencillium spp.	Patulin	IDH-1 IDH-2	5'-CAATGTGTCGTACTGTGCCC-3' 5'-ACCTTCAGTCGCTGTTCCTC-3'	600	Paterson, 2000

ribosome (28S rDNA) of each of the targeted groups from GenBank (National Center for Biotechnology Information) for designing the primers. They also employed Multalin for aligning the sequences (Corpet, 1988). The ITS5 was used as forward primers for the amplification of the target groups (Nikolcheva and Bärlocher, 2004). See Chapter 19 for the PCR-based molecular tools in the diagnosis and assessment of fungal diversity.

Table 7. Primers specific to fungi (Ascomycota, Basidiomycota, Zygomycota, Chytridiomycota) and Stramenopila (Oomycota)

Target group*	Primer	Sequence	Reference
Ascomycota	ITS4Asco	5'-CGTTACTRRGGCAATCCCTGTTG-3'	Nikolcheva and Bärlocher, 2004
Basidiomycota	ITS4Basidio	5'-GCRCGGAARACGCTTCTC-3'	
Zygomycota	ITS4Zygo	5'-AAAACGTWTCTTCAAA-3'	
Chytridiomycota	ITS4Chytrid	5'-TTTTCCCGTTTCATTCGCCA-3'	
Oomycota	ITS4Oo	5'-ATAGACTACAATTCGCC 3'	
Fungi	ITS 5	5'-GGAAGTAAAAGTCGTAACAAGG-3'	White *et al.*, 1990

*Used in combination with the forward primer ITS5

Novelties of DNA-Based Markers

The novelties of DNA-based markers are many:

- They are rapid and require a tiny sample
- It does not require species of *Fusarium*, *Aspergillus* and other fungi to form specialized identifying structures such as conidia
- DNA probes are very specific and could be easily and reproducibly synthesized and stored ready for use
- Species-specific data (such as sequences) can be obtained without the need to culture or even isolate a species
- Detection system has the potential to be easily automated

Conclusions and Outlook

The molecular tools employed to study pathogenic fungi are becoming increasingly sophisticated. The identification of fungal pathogens is essential for all aspects of plant pathology (basic research on the biology of pathogens to the control of diseases). Molecular methods involving PCR can be routinely used in the diagnosis of plant diseases and to detect and identify plant pathogens through various primers that are specific to coding or noncoding genes. Molecular methods have been used to identify several pathogens simultaneously using multiplex PCR. Nucleic acid sequence analysis has proven to be a valuable tool for the identification of fungi in a number of applications. Some of the most interesting applications of this technology include: identification of variant strains of known species, identification of non-cultivatable fungi in infected samples and recognition of new species. As a number of fungal genomes have now been sequenced, it has provided a framework to a better understanding of the relationships between different orders not merely at the gene level, but also at genomic level. Establishing the degree of synteny between species of closely related fungi will allow us better comprehension of species relationships as well as enable a better understanding of pathogen differentiation and the evolution of species.

References

Abd-Elsalam, K.A., Aly, I.N., Abdel-Satar, M.A., Khalil, M.S. and Verreet, J.A. 2003. PCR identification of *Fusarium* genus based on nuclear ribosomal-DNA sequence data. *Afr. J. Biotechnol. 2*, 82-85.

Aoki, T. and O'Donnell, K. 1999. Morphological and molecular characterization of *Fusarium pseudograminearum* sp. nov., formerly recognized as the Group 1 population of *F. graminearum. Mycologia 91*, 597-609.

Appel, D.J. and Gordon, T.R. 1996. Relationships among pathogenic and nonpathogenic isolates of *Fusarium oxysporum* based on the partial sequence of the intergenic spacer region of the ribosomal DNA. *Mol. Plant Microbe Inter. 9*, 125-138.

Baerlocher, F. 2007. Molecular approaches applied to aquatic hyphomycetes. *Fungal Biology Reviews 21*: 19-24.

Bakan, B., Giraud-Delville, C., Pinson, L., Richard-Molard, D., Fournier, E. and Brygoo, Y. 2002. Identification of PCR of *Fusarium culmorum* strains producing large and small amounts of deoxynivalenol. *Appl. Environ. Microbiol. 68*, 5472-5479.

Bataille, M., Crainic, K. Leterreux, M., Durigon, M. and de Mazancourt, P. 1999. Multiplex amplification of mitochondrial DNA for human and species identification in forensic evaluation. *Foren. Sci. Int. 99*, 165-170

Benyon, F.H.L., Burgess, L.W. and Sharp, P.J. 2000. Molecular genetic investigations and reclassification of *Fusarium* species in sections *Fusarium* and *Roseum. Mycol. Res. 104*, 1164-1174.

Bluhm, B.H., Flaherty, J.E., Cousin, M.A. and Woloshuk, C.P. 2002. Multiplex polymerase chain reaction assay for the differential detection of trichothecene and fumonisin-producing species of *Fusarium* in cornmeal. *J. Food Protec. 65*, 1955-1961.

Bruns, T.D. 2001. ITS reality. *Inoculum 52*, 2-3.

Cetin, Y. and Bullerman, L.B. 2005. Cytotoxicity of *Fusarium* mycotoxins to mammalian cell cultures as determined by the MTT bioassay. *Food Chem. Toxicol. 43*, 755-764

Clark, J.Y. 2003. Artificial neural networks for species identification by taxonomists *BioSystems 72*, 131-147

Corpet, J. 1988. Multiple sequence alignment with hierarchical clustering; *Nucleic Acids Res. 16*, 10881-10890.

Demeke, T.T., Clear, R.M., Patrick, S.K. and Gaba, D. 2005. Species-specific PCR-based assays for the detection of *Fusarium* species and a comparison with the whole seed agar plate method and trichothecene analysis. *Int. J. Food Microbiol. 103*, 271-284.

D'Mello, J.P.F., Placinta, C.M. and Macdonald, A.M.C. 1999. *Fusarium* mycotoxins: a review of global implications for animal health, welfare and productivity. *Anim. Feed Sci. Technol. 80*, 183-205.

Doohan, F.M., Weston, G., Rezanoor, H.N., Parry, D.W. and Nicholson, P. 1999. Development and use of reverse transcription- PCR assay to study expression of Tri5 by *Fusarium* species *in vitro* and in planta. *Appl. Environ. Microbiol. 65*, 3850-3854.

Edel, V., Steinberg, C., Gautheron, N. and Alabouvette, C. 2000. Ribosomal DNA-targeted oligonucleotide probe and PCR assay specific for *Fusarium oxysporum. Mycol. Res. 104*, 518-526.

Edlind, T.D., Li, J., Visvesvara, G.S., Vodkin, M.H., McLaughlin, G.L. and Katiyor, S.K. 1996. Phylogenetic analysis of β-tubulin sequences from a mitochondrial protozoa. *Mol. Phylogen. Evol. 5*, 359-367.

Field, D. and Wills, C. 1998. Abundant microsatellite polymorphism in *Saccharomyces cerevisiae*, and the different distributions of microsatellites in eight prokaryotes and *S. cerevisiae*, result from strong mutation pressures and a variety of selective forces. *Proc. Nat. Acad. Sci. 95*, 1647-1652.

Gang, D.R. and Weber, D.J. 1995. Preparation of genomic DNA for RAPD analysis from thick-walled dormant teliospores of *Tilletia* species. *BioTechniques 19*, 92–97.

Gantley, A.R.D. and Scott, B. 1998 Extraordinary ribosomal spacer length heterogeneity in a *neotyphodium endophyte* hybrid: implications for concerted evolution. *Genetics 150*, 1625-1637.

Gardes, M. and Bruns, T.D. 1993. ITS primers with enhanced specificity for *basidiomycetes* – application to the identification of mycorrhizae and rusts. *Mol. Ecol. 2*, 113-118.

Gaskell, G.J., Carter, D.A. Britton, W.J. Tovey, E.R. Benyon, F.H. and Lovborg, U. 1997. Analysis of the internal transcribed spacer regions of ribosomal DNA in common airborne allergenic fungi. *Electrophoresis 18*, 1567-1569.

Geiser, D.M., Frisvad, J.C. and Taylor, J.W. 1998. Evolutionary relationships in *Aspergillus* section *Fumigati* inferred from partial β-tubulin and hydrophobin DNA sequences. *Mycologia 90*, 831-845.

Geiser, D.M., Dorner, J.W., Horn, B.W. and Taylor, J.W. 2000. The phylogenetics of mycotoxin and sclerotium production in *Aspergillus flavus* and *Aspergillus oryzae*. *Fungal Genet. Biol. 31*, 169-179.

Glass, N.L. and Donaldson, G.C. 1995. Development of Primer Sets Designed for Use with the PCR To Amplify Conserved Genes from Filamentous *Ascomycetes*. *Appl. Environ. Microbiol. 61*, 1323-1330.

Guarro, J., Gene, J. and Stchigel, A.M. 1999. Developments in fungal taxonomy. *Clin. Microbiol. Rev. 12*, 454-500.

Hennequin, C., Abachin, E. and Symoens, F. 1999. Identification of *Fusarium* species involved in human infections by 28S rRNA gene sequencing. *J. Clin. Microbiol. 37*, 3586-3589.

Hoelzel, A. and Green, A. 1998. PCR protocols and population analysis by direct DNA sequencing and PCR-based DNA fingerprinting In: *Molecular Genetic Analysis of Populations: A Practical Approach* (Ed. Hoelzel, A.R.). IRL Press, Oxford, 201-233.

Iwen, P.C., Hinrichs, S.H. and Rupp, M.E. 2002. Utilization of the internal transcribed spacer regions as molecular targets to detect and identify human fungal pathogens. *Medical Mycology 40*, 87-109.

Jaeger, E.M.E., Carroll, N.M., Choudhury, S., Dunlop, A.A.S., Towler, H.M.A., Matheson, M.A., Adamson, P., Okhravi, N. and Lightman, S. 2000. Rapid detection and identification of *Candida*, *Aspergillus*, and *Fusarium* species in ocular samples using nested PCR. *J. Clin. Microbiol. 38*, 2902-2908.

Karvonen, P. and Savolainen, O. 1993. Variation and inheritance of ribosomal DNA in *Pinus sylvestris* L. (Scots pine). *Hereditas 71*, 614-622.

Konietzny, U. and Greiner, R. 2003. The application of PCR in the detection of mycotoxigenic fungi in foods. *Brazilian J. Microbiol. 34*, 283-300.

Kuske, C.R., Banton, K.L., Adorada, D.L., Stark, P.C., Hill, K.K. and Jackson, P.L. 1998. Small-scale DNA sample preparation method for field PCR detection of microbial cells and spores in soil. *Appl. Environ. Microbiol. 64*, 2463-2472.

Lee S.B. and Taylor J.W. 1992. Phylogeny of the five fungus-like protoctistan *Phytophthora* species, inferred from the internal transcribed spacers of ribosomal DNA. *Mol. Biol. Evol. 9,* 636-653.

Leslie, J.F., Zeller, K.A., Lamprecht, S.C., Rheeder, J. P. and Marasas, W.F.O. 2005. Toxicity, pathogenicity, and genetic differentiation of five species of *Fusarium* from sorghum and millet. *Phytopathology 95*, 275-283.

Lott, L.F., Kuykendall, R.J. and Reiss E. 1993. Nucleotide sequence analysis of the 5·8S rDNA and adjacent ITS2 region of *Candida albicans* and related species. *Yeast 9*, 1199-1206

Magnani, M., Fernandes, T., Prete, C.E.C., Homechim, M., Ono, U.Y.S., Vilas-Boas, L.A., Sartori, D., Cristina, M., Furlaneto, S. and Fungaro, M.H.P. 2005. Molecular identification Of *Aspergillus* spp. isolated from coffee beans. *Sci. Agric. 62*, 45-49.

Martin, J.P. and Haider, K. 1971. Microbial activity in relation to soil in humic formation. *Soil Sci. 111*, 54-63.

McLean, M., 1996. The phytotoxicity of *Fusarium* metabolites, an update since 1989. *Mycopathol. 133*, 163-179.

Mishra P.K., Fox R.T.V. and Culham A. 2003. Development of A PCR-Based assay for rapid and reliable identification of pathogenic *fusaria. FEMS Microbiol. Lett. 218*, 329-332.

Moller, E.M., Bahnweg, G., Sandermann, H. and Geiger, H.H. 1992. A simple and efficient protocol for isolation of high molecular weight DNA from filamentous fungi, fruit bodies, and infected plant tissues. *Nucl. Acids Res. 20*, 6115-6116.

Mule, G., Susca, A., Stea, G. and Moretti, A. 2004. Specific detection of the toxigenic species *Fusarium proliferatum* and *F. oxysporum* from asparagus plants using primers based on calmodulin gene sequences *FEMS Microbiol. Lett. 230*, 235-240.

Mullis, K. B. and Faloona, F. 1987. Specific synthesis of DNA *in vitro* via a polymerase catalysed chain reaction. *Methods Enzymol. 155*, 335-350.

Nicholson, P., Simpson, D.R., Weston, G., Rezanoor, H.N., Lees A.K. Parry, D.W. and Joyce, D. 1998. Detection and quantification of *Fusarium culmorum* and *Fusarium graminearum* in cereals using PCR assays. *Physiol. Mol. Plant Pathol. 53*, 17-37.

Niessen, L. and Vogel, R.F. 1998. Group specific PCR-detection of potential *trichothecene*-producing *Fusarium*-specis in pure cultures and cereal samples. *Syst. Appl. Microbiol. 21*, 618-631.

Nikolcheva, L.G. and Bärlocher, F. 2004. Taxon-specific fungal primers reveal unexpectedly high diversity during leaf decomposition in a stream. *Mycol. Prog. 3*, 41-50.

Olsen, M., Jonsson, N., Magan, N., Banks, J., Fanelli, C., Rizzo, A., Haikara, A., Dobson, A., Frisvad, J., Holmes, S., Olkku, J., Persson, S. J. and Borjesson, T. 2006. Prevention of ochratoxin A in cereals in Europe. *Adv. Exp. Med. Biol. 571*, 317-342.

Parry, D.W. and Nicholson, P. 1996. Development of a PCR assay to detect *Fusarium poae* in wheat. *Plant Pathol. 45*, 383-391.

Paule, M.R. and Lofquist, A.K. 1996. Organization and expression of eukaryotic ribosomal RNA genes. In: *Ribosomal RNA, Structure, Evolution, Processing and Function in Protein Biosynthesis* (Ed. Zimmerman, R.A. and Dahlberg, A.E.). CRC Press, New York, 395-420.

Peterson, S.W. 2000. Phylogenetic relationships in *Aspergillus* based on rDNA sequence analysis. In: *Integration of Modern Taxonomic Methods for Penicillium* and *Aspergillus Classification* (Ed. Samson R.A. and Pitt, J.I.). Harwood Academic Publishers, Amsterdam, The Netherlands, 323-355.

Placinta, C.M., D'Mello, J.P.F. and Macdonald, A.M.C. 1999. A review of worldwide contamination of cereal grains and animal feed with *Fusarium* mycotoxins. *Anim. Feed Sci. Technol. 78*, 21-37.

Pryce, T.M., Pallidino, S., Kay, D. and Coombs, G.W. 2003. Rapid identification of fungi by sequencing the ITS1 and ITS2 regions using an automated capillary electrophoresis system. *Med. Mycol. 41*, 369-381.

Rheeder, J.P., Marasas, W.F.O. and Nelson, P.E. 1996. *Fusarium globosum*, a new species from corn in southern Africa. *Mycologia 88*, 509-513.

Rotter, B.A., Prelusky, D.B. and Pestka, J.J., 1996. Toxicology of *deoxynivalenol* (vomitoxin). *J. Toxicol. Environ. Health 48*, 1-34.

Saiki, R.K., Gelfand, D.H., Stoffel, S., Scharf, S.J., Higuchi, R., Horn, G.T., Mullis, K.B. and Erlich, H.A., 1988. Primer-directed enzymatic amplification of DNA with a thermostable DNA polymerase. *Science 239*, 487-491.

Steenkamp, E.T., Wingfield, B.D., Coutinho, T.A., Wingfield, M.J., and Marasas, W.F.O. 1999. Differentiation of *Fusarium subglutinans* f. sp. *pini* by histone gene sequence data. *Appl. Environ. Microbiol. 65*, 3401-3406.

Steiner, J.J., Poklemba, C.J., Fjellstrom, R.G. and Elliott, L.F. 1995. A rapid one-tube genomic DNA extraction process for PCR and RAPD analyses. *Nucl. Acids Res. 23*, 69-2570.

Sweeney, M.J. and Dobson, A.D.W. 1998. Mycotoxin production by *Aspergillus, Fusarium* and *Penicillium* species. *Int. J. Food Microbiol. 43*, 141-158.

Tan, M.K. and Niessen, L.M. 2003. Analysis of rDNA ITS sequences to determine genetic relationships among, and provide a basis for simplified diagnosis of, *Fusarium* species causing crown rot and head blight of cereals. *Mycol. Res. 107*, 811-821

Tanaka, T., Hasegawa, A., Yamamoto, S., Lee, U.S., Sugiura, Y. and Ueno, Y. 1988. Worldwide Contamination of Cereals by the *Fusarium* Mycotoxins Nivalenol, Deoxynivalenol, and Zearalenone survey of 19 Countries. *J. Agric. Food Chem. 36*, 979-983

Thomson, J.D., Gibson, T.J., Plewniak, F., Jeanmougin, F. and Higgins, D.G., 1997. The Clustal-X windows interface: flexible strategies for multiple sequence alignment aided by quality analysis tools. *Nucl. Acids Res. 25*, 4876-4882.

Thon, M.R. and Royse, D.J. 1999. Partial β-tubulin gene sequences for evolutionary studies in the Basidiomycotina. *Mycologia 91*, 468-474.

Varga, J., Toth, B., Rigo, K., Debets, F. and Kozakiewicz, Z. 2000. Genetic variability within the *Aspergillus viridinutans* species. *Folia Microbiol. 45*, 423-428.

Varga, J., Rigo, K., Molnar, J., Toth, B., Szencz, S., Teren, J. and Kozakiewicz, Z. 2003. Mycotoxin production and evolutionary relationships among species of *Aspergillus* section Clavati. *Ant. Leeuwen. 83*, 191-200.

Vilgalys, R. and Gonzalez. D. 1990. Organization of ribosomal DNA in the basidiomycete *Thanatephorus praticola. Cur. Gen. 18*, 277-280.

Vilgalys, R., Hopple, J.S.Jr. and Hibbett, D.S. 1994. Phylogenetic implications of generic concepts in fungal taxonomy: The impact of molecular systematic studies. *Mycol. Helvet. 6*, 73-91.

White, T.J., Bruns, T., Lee, S. and Taylor, J. 1990. Amplification and direct sequencing of fungal ribosomal RNA genes for phylogenetics. In: *PCR Protocols: A Guide to Methods and Applications* (Ed. Innis, M.A., Gelfand, D.H. and Sninsky, J.J.). Academic Press, New York, 315-322.

Williams, K.J., Dennis, J.J., Smyl, C. and Wallwork, H. 2002. The application of species-specific assays based on the polymerase chain reaction to analyze *Fusarium* crown rot of durum wheat. *Aust. Plant Pathol. 31*, 119-127.

Wilson, I.G., 1997. Inhibition and facilitation of nucleic acid amplification. *Appl. Environ. Microbiol. 63*, 3741-3751.

Wu, Z., Tsumura, Y., Blomquist, G. and Wang, X.R. 2003. 18S rRNA gene variation among common airborne fungi and development of specific *oligonucleotide* probes for the detection of fungal isolates. *Appl. Environ. Microbiol. 69*, 5389-5397.

Yoder, O.C. and Turgeon, B.G. 2001. Fungal genomics and pathogenicity. *Curr. Opinion Plant Biol. 4*, 315-321.

Yoder, W.T. and Christianson, L.M. 1998. Species-specific primers resolve members of *Fusarium* section *Fusarium*: taxonomic status of the edible “Quorn” fungus reevaluated. *Fungal Genet. Biol. 23*, 68-80.

Yu, J., Cleveland, T.E., Nierman, W.C. and Bennett, J.W. 2005. *Aspergillus flavus* genomics: gateway to human and animal health, food safety, and crop resistance to diseases. *Rev. Iberoamer. Micol. 22*, 194-202.

Zhang, Z., Zhang, J., Wang, Y. and Zheng, X. 2005. Molecular detection of *Fusarium oxysporum* f. sp. *niveum* and *Mycosphaerella melonis* in infected plant tissues and soil. *FEMS Microbiol. Lett. 249*, 39-47.

other fungi. The genus *Trichoderma* was first identified in 1794 having four species and subsequent investigations showed that *Trichoderma* is the asexual stage of *Hypocrea* (cf. Druzinina *et al.*, 2006). Antibiotics producing property of these fungi was detected (Weindling, 1934). Taxonomic status and number of species present in the genus *Trichoderma* is ever changing with several hundreds of strains being isolated and investigated (Bissett *et al.*, 2003; Wuczskowski *et al.*, 2003; Zhang *et al.*, 2005; Druzinina and Kubicek, 2005). *Trichoderma* isolates are found in most of the soils, decaying wood and in live plant tissues as endophytes (Danielson and Davey, 1973; Widden and Abitbol, 1980; Samuels *et al.*, 2002; Kubicek *et al.*, 2003; Harman *et al.*, 2004). It is evident from literature that different isolates from distinct geographical locations have showed significant differences in colony morphology, microscopic features and mechanism of antagonism (Lieckfeldt *et al.*, 1999; Samuels *et al.*, 2002; Kubicek *et al.*, 2003; Kucuk and Kivanc, 2003; Sariah *et al.*, 2005; Verma *et al.*, 2007; Vinale *et al.*, 2008). More than 100 species have been identified based on molecular data along with phylogenetic analysis (Kullnig-Gradinger *et al.*, 2002; Druzinina *et al.*, 2006). In an excellent review, Druzinina *et al.* (2006) discussed developments of *Trichoderma* taxonomy based on sequence analysis with current trends in molecular fungal taxonomy. However, morphological and microscopic examinations continue to provide insight into new species/strains of *Trichoderma* in spite of its taxonomic complexity (Bisset *et al.*, 2003; Chaverri *et al.*, 2003; Wuczskowski *et al.*, 2003; Zhang *et al.*, 2005). Oligonucleotide barcode system of *Tricoderma* species identification has also been developed by Druzhinina *et al.* (2005).

The greatest significance of *Trichoderma* spp. lies in their ability of mycoparasitism. Many of the *Trichoderma* isolates were found to be opportunistic and avirulent plant symbionts (Harman *et al.*, 2004). Species of *Trichoderma* have shown unique ability in terms of soil/rhizosphere competence for survival through hyperparasitism and antagonism against other fungi. They are also well known producers of enzymes of industrial importance. Series of antimicrobial compounds have been found to be associated with these species. Remarkable capacity of *Trichoderma* for rapid growth under various soil conditions, efficient degradation of diverse substrates, resistant to harmful chemicals and pathogenic on commercially cultivated mushrooms is well known (Samuels *et al.*, 2002; Kubicek *et al.*, 2003; Harman *et al.*, 2004; Druzinina *et al.* 2006). These unique properties of *Trichoderma* led extensive use of different isolates as biological control agents (BCAs) at laboratory/green house conditions as well as at filed level against phytopathogenic fungi of agricultural crops. Commercial formulations of *Trichoderma* in the trade names (e.g. Trichodex™, Trichodowels, Binab T) are currently available (Fravel, 2005; Deacon, 2006).

The mechanisms of antagonism of *Trichoderma* against other fungi are interesting. Many species/strains of *Trichoderma* can directly parasitize several fungi through hyphal coils around hyphae of other fungi, eventually penetrating and dissolving the target fungi (Howell, 2003; Deacon, 2006). *Trichoderma* spp. are known to produce considerable quantity of chitinases and glucanases to digest and kill other fungi (Howell, 2003; Benitez *et al.*, 2004). Different strains of *Trichoderma* have been shown to produce a series of volatile and non-volatile antibiotics. Interestingly, *Trichoderma* spp. have also been recognized as inducers of plant defense mechanisms and growth promoter leading to elevated yield in several agricultural crops. Excellent information is available on the mechanisms involved in induction of systemic acquired resistance (SAR) in plants through *Trichoderma* spp. (Harman *et al.*, 2004; Brunner *et al.*, 2005; Vinale *et al.,* 2008). Efficient antagonism of *Trichoderma* appears to be due to synergistic action of enzymes and antimicrobial compounds apart from its rhizosphere competence and induction of plant resistance (Schirmbock *et al.*, 1994; Vassueur *et al.*, 1995; Howell, 2003; Kubicek *et al.*, 2003; Benitez *et al.*, 2004; Kucuk and Kivanc, 2004; Brunner *et al.*, 2005; Deacon, 2006; Verma *et al.*, 2007). Signal transduction pathways involving protein kinases in *Trichoderma* during mycoparasitism has been reviewed (Zeilinger and Omann, 2007; Reithner *et al.*, 2007). Molecular techniques for elucidating the mechanism of action of fungal biocontrol agents have been recently reported by Massart and Jijakli (2007). Fravel (2005) discussed in detail about commercialization of biocontrol agents for plant disease control. As the mechanisms involved in antagonism of *Trichoderma* is beyond the scope of this Chapter, a major emphasis has been given to peptaibols, one of the groups of antimicrobial compounds of *Trichoderma*.

Peptaibols

Peptaibols are linear peptides produced predominantly by *Trichoderma*. Origin of the name is due to **Pept**ide nature, presence of high number of non-proteinogenic amino acid **Aib** (α-aminoisobutyric acid) (Fig. 2) and conversion of C-terminal amino acid residue as aminoalcoh**ol**. These peptides belong to the larger group peptaibiotics which are characterized by having 5-21 amino acid residues, molecular weight of 500 to 2200 Da, high amount of Aib and other non-proteinogenic amino acids and acylated N-terminus. The C-terminal residue in these peptides shows various modifications (Degenkolb *et al.*, 2007). In case of peptaibols, the N-terminal is acetylated and C-terminal residue is

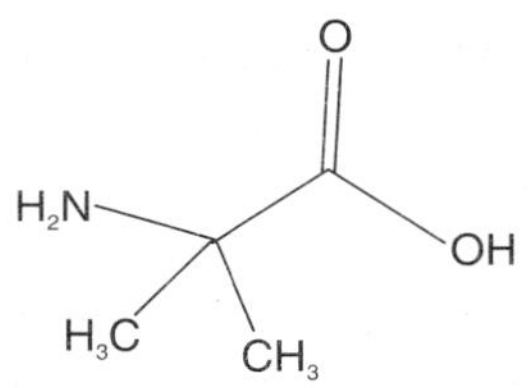

Fig. 2. α-aminoisobutyric acid (Aib) – major non-standard amino acid found in peptaibols.

represented by 1,2-amino alcohol (e.g. valinol, leucinol, isoleucinol or tryptophanol) (Chugh and Wallace, 2001; Schiell *et al.*, 2001; Degenkolb *et al.*, 2003; Szekeres *et al.*, 2005; Ruiz *et al.*, 2007). Unique features of peptaibols have been listed in Box 1. The other related molecules with different chemical modification such as lipoamino peptides (aminolipopetides) are also reported by Degenkolb *et al.* (2007).

Box 1. Unique features of peptaibols

- **Pept**ides containing high content of **Aib** (α-aminoisobutyric acid) residues, acetylated N-terminal and C-terminal alcoh**ol**.
- Antibiotic peptides produced from *Trichoderma* and related fungi.
- Molecular weight between 500 and 2200 Da.
- Linear peptides containing 5 to 21 amino acid residues.
- Characterized by the presence of the unusual amino acids like, α-aminoisobutyric acid (Aib).
- C-terminus alcohol moiety (phenylalaninol, leucinol, isoleucinol, valinol/ tryphtophanol).
- Microheterogenous polypeptides which differ in mass by units of 14 Da or 16 Da.

Extensive literature is available on peptaibiotics regarding their structure, sequence, mode of action and diversity. Originally, the term peptaibol was proposed independently by Benedetti *et al.* and Bruckner *et al.* (cf. Degenkolb *et al.*, 2003). Previously, these molecules have been referred to as peptiabophols by Pandey *et al.* (1977) considering the presence of phenylalaninol at C-terminus. Later, reviews of Degenkolb *et al.* (2003, 2007) reconsidered the terminology by clearly distinguishing peptaibols from other peptaibiotics based on C-terminal amino alcohol. Generally, all these peptides with Aib have been considered under **P**eptaibols and **r**elated **P**eptaibiotic**s** (**PrPs**) (Szekeres *et al.*, 2005). It is important to understand the chemical nature and modifications of different amino acid residues in these molecules.

Chugh and Wallace (2001), in their review elegantly classified peptaibols into nine subfamilies based on the sequences, size and structures. These subfamilies also provide basis for modeling peptaibols for the membrane channel activity. Most of the peptaibols reported from *Trichoderma* belong to subfamilies: SF1, SF4, SF5 and SF9 (Szekeres *et al.*, 2005). All the peptaibols reported so far have been considered to be made up of 5-21 amino acid residues (Degenkolb *et al.*, 2007). In our studies, different size peptaibols have been detected in *Trichoderma* isolates indicating the extent of diversity within the same organism to produce various peptaibols (Shishupala, 2001; Shishupala *et al.*, 2003). These peptaibols have been attributed to exhibit different biological activities (Box 2). These activities have been attributed due to nature of amino acid sequence and structures (Chugh and Wallace, 2001).

Box 2. Biological activity of peptaibols.

- Antibacterial activity
- Antifungal activity
- Antiviral activity
- Antimalarial activity
- Antitumor activity
- Alteration of membrane permeability
- Induce pores in lipid membranes leading to ion channels
- Leakage of cytoplasmic material can occur through such channels, leading to cell death

Fungal Source of Peptaibols

Trichoderma spp. are the major producers of petaibols and peptaibiotics (Daniel and Filho, 2007) and a few other fungi have also been reported to produce these molecules (Table 1). These fungi commonly exhibit antagonism in soil and some have been shown to possess entamopathogenic activity. All the peptaibols are synthesized by large peptide synthetases found in such fungi as well as bacteria. Several species of fungi belonging to *Acremonium, Emericellopsis, Paecilomyces, Sepedonium* and *Trichoderma* are known to produce peptaibols (Degenkolb *et al.,* 2003; Neuhof *et al.*, 2007a). It is interesting to study the biosynthetic pathways of these fungi to assess the phylogenetic relationships. Several reports on peptaibol producing fungi may be misleading due to complexity of *Trichoderma* taxonomy. Druzhunina *et al.* (2006) reviewed the taxonomic status of *Trichodema* and opined that abundant homoplasy in morphological and phenetic characters results difficulty in identification as well as classification. Owing to chemotaxonomy, production of specific type of peptaibol may be considered as one of the important criteria to distinguish *Trichoderma.*

Table 1. Fungal producers of peptaibiotics other than Trichoderma

Fungal Species	Peptaibiotic
Acremonium lindtneri	Liphexin
Culicinomyces clavisporus	Culicinin
Hypocrea muroiana	Ampullosporin
Sepedonium chrysospermum	Chrysospermin
Stilbella erythrocephala	Antiamoebin
Tolypocladium inflatum	Efrapeptin

Peptaibol Sequences

Amino acid sequences of most of the characterized peptibols are available in peptaibol database. The amino acid sequences of representative types of petaibols from different species of *Trichoderma* are given in Table 2. Even with *T. harzianum*, different types of petaibols with significant variation in the sequences have been characterized (Table 3). Comparison and analysis of such sequences reveals extensive diversity as well as common features among peptaibols. The major constituent of these peptaibols is Aib (Fig. 2). The other non-proteinogenic amino acids found in such peptaibiotics are Iva (L- and D-Isovaline), Hyp (*cis*- and *trans*-4-L-Hydroxyproline), MePro (cis-4-L-Methylproline), Hyleu (β-Hydroxy-L-leucine), β-Ala (β-Alanine), Pip (L-Pipecolic acid), Etnor (α-Ethyl-norvaline) and AHMOD (2-Amino-4-methyl-6-hydroxy-8-oxodecanoic acid) (Degenkolb *et al.*, 2003). Sequence alignment of petaibols as studied by Chugh and Wallace (2001) has resulted in precise classification of peptaibols. Although molecular weights of some of the peptaibols are similar, the differences in the sequences are evident (Szekeres *et al.,* 2005; He *et al.*, 2006; Sabareesh and Balaram, 2006; Degenkolb *et al.*, 2007). The sequence data is upgraded as more and more sequences being determined using novel peptide sequencing methods.

Table 2. Amino acid sequence of Peptaibols from different species of Trichoderma (Source: Petaibol database from www.cryst.bbk.ac.uk/peptaibol)

Trichoderma atroviride	
Atroviridin A	Ac U P U A U A Q U V U G L U P V U U Q Q F OH
Atroviridin B	Ac U P U A U A Q U V U G L U P V U J Q Q F OH
Trichoderma harzianum	
Trichorozin I	Ac U N I L U P I L U P V OH
Trichokindin Ia	Ac U S A U U Q J L U A U U P L U U Q I OH
Trichoderma saturnisporum	
Paracelsin E	Ac U A U A U A Q U L U G U A P V U U Q Q F OH
Saturnisporin SA I	Ac U A U A U A Q U L U G U U P V U U Q Q F OH
Trichoderma viride	
Alamethicin F-30	Ac U P U A U A Q U V U G L U P V U U E Q F OH
Suzukacillin	Ac U A U A U A Q U U U G L U P V U U Q Q F OH
Trichoderma atroviride	
Atroviridin A	Ac U P U A U A Q U V U G L U P V U U Q Q F OH
Atroviridin B	Ac U P U A U A Q U V U G L U P V U J Q Q F OH
Trichoderma pseudokoningii	
Harzianin HK VI	Ac U N I I U P L L U P L OH
Trichoderma koningii	
Trichokonin Ia	Ac U A U A U A Q U V U G L A P V U U Q Q F OH
Trichoderma polysporum	
Polysporin A	Ac U P U A U U Q U V U G V U P V U U Q Q F OH
Trichoderma longhibrachitum	
Trichobrachin A -I	Ac U N L L U P L U U P L OH
Longibrachin A I	Ac U A U A U A Q U V U G L U P V U U Q Q F OH

Note: Non-standard amino acids found in peptaibols are Aminoisobutyric Acid (Aib; U); Ethylnorvaline (EtN or EtNor; Z); Isovaline (Iva; J); Hydroxyproline (Hyp; O); Ac- acylation of N-terminal, OH- C-terminal amino alcohol

Table 3. Representative sequences of Peptaibols from *Trichoderma harzianum* (Source: www.cryst.bbk.ac.uk/peptaibol)

a)	**11 residue peptaibols**	
		Amino acid sequence
	Harzianin HB I	Ac U **N** L I U P **J** L U P **L** OH
	Trichorozin I	Ac U **N** I L U P **I** L U P **V** OH
	Trichorozin II	Ac U **Q** I L U P **I** L U P **V** OH
	Trichorozin III	Ac U **N** I L U P **I** L U P **L** OH
	Trichorozin IV	Ac U **Q** I L U P **I** L U P **L** OH
b)	**14 residue peptaibols**	
		Amino acid sequence
	Harzianin HC XIV	Ac U N L U P **A** **I** U P **U** L U P L OH
	Harzianin HC VIII	Ac U N L U P **A** **V** U P **J** L U P L OH
	Harzianin HC I	Ac U N L U P **S** **V** U P **U** L U P L OH
	Harzianin HC III	Ac U N L U P **S** **V** U P **J** L U P L OH
	Harzianin HC XII	Ac U N L U P **S** **I** U P **J** L U P L OH
c)	**18 residue peptaibols**	
		Amino acid sequence
	Trichorzin HA V	Ac U **G** A U J Q **U** **V** U **G** **L** U P L U **J** Q L OH
	Trichorzin HA VII	Ac U **G** A U J Q **V** **V** U **G** **L** U P L U **J** Q L OH
	Trichokindin Vb	Ac U **S** A U J Q **U** **L** U **A** **J** U P L U **U** Q **I** OH
	Trichokindin VI	Ac U **S** A U J Q **J** **L** U **A** **J** U P L U **U** Q L OH
	Trichorzin PA IX	Ac U **S** A J J Q **U** **V** U **G** **L** U P L U **U** Q **F** OH
d)	**19 residue peptaibols**	
		Amino acid sequence
	Trichorzianine TB VII	Ac U A A U **J** Q U U U S L U P **V** U I Q **E** **F** OH
	Trichorzianine TB Vb	Ac U A A U **U** Q U U U S L U P **L** U I Q **E** **F** OH
	Trichorzianine TB VIa	Ac U A A U **J** Q U U U S L U P **L** U I Q **E** **F** OH
	Trichorzianine TA IIIc	Ac U A A U **U** Q U U U S L U P **V** U I Q **Q** **W** OH
	Trichorzianine TB IVb	Ac U A A U **J** Q U U U S L U P **V** U I Q **E** **W** OH

Note: Microheterogeneity of amino acid sequence in peptaibols is indicated by bold face for the amino acid residue in the changed position. Non-standard amino acids found in peptaibols are a minoisobutyric Acid (Aib; U); Ethylnorvaline (EtN or EtNor; Z); Isovaline (Iva; J); Hydroxyproline (Hyp; O); Ac - acylation of N-terminal; OH - C-terminal amino alcohol

Extensive sequence diversity has been noticed in peptaibols. It is mainly attributed to modular nature and multifunctional activity of peptide synthetases (Konz and Marahiel, 1999; Wiest *et al.,* 2002; Vizcaino *et al.*, 2005). The sequence analysis provided information on composition of these peptides. The non-standard amino acid, Aib represents 40% of the residues in all the known sequences. Glutamine is the only significant polar residue in peptaibols, and the position and number of these residues appear to be related to their functional properties as ion channels (Whitmore and Wallace, 2004a). Mass spectrometric studies have revealed extensive information of peptaibol sequences (Rebuffat *et al.*, 1989, 1995; Wada *et al.*, 1994; Hlimi *et al.*, 1995; Pocsfalvi *et al.*, 1997; Ishiyama *et al.*, 2000; Jaworski and Bruckner, 2000; Psurek *et al.*, 2005; Mohmed-Benkada *et al.*, 2006; Wada and Tanaka, 2007; Degenkolb *et al.*, 2007). Karle and Balaram (1990) showed alpha-helical nature of these molecules. Microhetergeneity of peptaibol is attributed to amino acid exchanges or replacements between different types as evidenced by the sequence comparison (see Table 2, 3). The types of amino acid replacement (Box 3) may result in sequence diversity among peptaibols. Such property of peptaibols is due to multifunctional nature of peptaibol synthetase (Wiest *et al.*, 2002). The sequence information may be useful in assessment of metabolic diversity among *Tricoderma* species and strains.

Box 3. Possible amino acid exchanges involved in sequence diversity of peptaibols.

Aib	↔	Ala
Aib	↔	Val
Leu	↔	Val
Gly	↔	Ala
Gln	↔	Glu

Biological Activity of Peptaibols

Various biological activities of peptaibols have been given in Box 2. Antibacterial activity of petabiols has been documented in literature. The major biological activity of peptaibols is in their capacity to form voltage dependent ion channels in lipid bilayer membrane owing to amphipathic nature (Chugh and Wallace, 2001). The channel-forming activities of Trichotoxin_A50E, an 18 residue peptaibol has been analyzed through crystal structure. This peptide induces single-channel conductance in contrast to multi-channel conductance induced by alamethicin (Duclohier *et al.*, 2004). Different peptaibiols have inhibitory effect on a wide range of bacteria: *Bacillus subtilis, B. cereus, Brucella bronchiseptica, Micrococcus luteus, Mycobacterium phlei, Staphylococcus aureus, Streptococcus feacalis, S. lactis,* and *S. thermophilus* (Szkeres *et al.*, 2005).

Most of the antagonistic property of *Trichoderma* against phytopathogenic fungi (e.g. *Pythium, Phytophthora, Rhizoctonia, Sclerotinia*) is found to be due to production of antifungal compounds (Schirmbock, *et al.*, 1994; Benitez *et al.*, 2004; Howell, 2003; Kucuk and Kivanc, 2004; Deacon, 2006). Host-sensing pathways and signal transduction mechanisms have also been demonstrated in *Trichoderma* (Zeilinger and Omann, 2007). Antifungal activity of petaibols is one of the major reasons for the exploitation of *Trichoderma* spp. for biocontrol programme.

Antiviral property of peptaivirins A and B belonging to peptaibols group has been reported against tobacco mosaic virus infection to tobacco plants (Yun *et al.*, 2000). Antimalarial activities of such metabolites have also been recorded by Nagaraja *et al.* (2001). Peptaibols inducing plant defence reactions through salicylate signal pathway leading to systemic acquired resistance is an interesting feature (Engelberth *et al.*, 2001; Howell, 2003; Harman *et al.*, 2004; Brunner *et al.*, 2005). Such potential of peptaibols of *Trichoderma* qualifies its exploitation as an important plant protectant.

Peptabiobls were found to damage the cell membrane barrier function of boar spermatozoa and dissipated mitochondrial membrane potential of human lung epithelial carcinoma cells (Peltola *et al.*, 2004). This feature is a serious concern for mass production of *Trichoderma* as a biocontrol agent. Antitumor activity of peptaibols of *Culcinomyces clavisporus* was studied by He *et al.* (2006). Some of the peptaibols were also shown to activate membrane-bound adenylate and guanylate cyclases (Szkeres *et al.*, 2005). Majority of these properties are due to the pore-forming ability of peptaibols in the cell membrane resulting in leakage and cell death. Wide range of activities shown by peptaibols certainly qualifies their application in chemotherapy.

Biosynthesis of Peptaibols

Analysis of amino acid sequences of peptaibols revealed extensive diversity. Such remarkable sequence diversity and presence of unusual amino acids makes peptaibols as interesting molecular tool to understand biosynthetic pathways. Peptaibols are synthesized without involvement of ribosomes. Their biosynthesis involves the activity of multi-enzyme complex modular peptide synthetases. These large and multifuncational enzymes assemble peptides from remarkable range of precursors including non-protein amino acids and hydroxy or carboxy acids (Wiest *et al.,* 2002).

Non-ribosomal peptides of microbial origin have an outstanding structural diversity, which has been partly attributed to the incorporation of many unusual non-protein amino acid residues (Kleinkauf and von Dohren, 1996; Konz and Marahiel, 1999). Non-ribosomal peptide synthetases are organized into repetitive synthetase units or modules. Each of these modules is a semiautonomous unit and has the function necessary to complete a different single amino acid elongation step during peptide synthesis. Each module recognizes, activates and modifies a single residue. Each module is further partitioned into adenylation (A), thiolation (T) and condensation (C) domains, which together represent a minimal repeating unit of the enzyme (Kleinkauf and von Dohren, 1987; Wiest *et al*., 2002; Chutrakul and Peberdy, 2005; Vizcaino *et al*., 2005). The mechanism involved in such peptide synthesis using peptide synthetases is referred as multi-enzyme thio-template mechanism, since all activated intermediates formed are thioesters (Kleinkauf and von Dohren, 1987). In this mechanism, the amino- and hydroxyl acids that have to be sequentially connected by corresponding peptide synthetase are first activated as acyladenylates (unstable intermediates) using ATP hydrolysis. The enzyme covalently link activated amino acid as a carboxy thioester (Marahiel, 1992). For an incorporated amino acid residue, the

of genetic coding of the sequences, peptaibols are not generally included in the protein sequence databases. Considerable data is available on peptaibols, specifically on amino acid sequences and structure determination. This led to the creation of the Peptaibol Database in 1997 at the Department of Crystallography, Birkbeck College, University of London, United Kingdom (Whitmore and Wallace, 2004b). The peptaibol database is available at: http://www.cryst.bbk.ac.uk/peptaibol. The major features of peptaibol database are given in Box 4. Currently, about 317 sequences of peptaibols are available and their number is increasing owing to the development of novel techniques for peptide sequencing with software for sequence analysis. The peptaibol database is a very important source of information for sequence and structural elucidation of fungal peptides. This database includes only naturally occurring peptaibols from literature. The database has been arranged in an easily accessible form comprising two major sections: 1) Sequences and 2) Structures. Complete information is available on the source of fungal species, sequence, nomenclature, definitions of non-standard amino acid residues and classification of peptaibols (Chugh and Wallace, 2001; Whitmore *et al.*, 2003; Whitmore and Wallace, 2004b). The 'Sequence' section provides information on peptaibols through search navigation based on peptaibol name, group, sequence motify and fungal source. Links for full record of information of each of the peptaibols is also available. As relatively a few structures of peptaibols have been elucidated, 'Structure' section of the database provides table entry of downloadable crystal structure or nuclear magnetic resonance (NMR) structure data of peptaibols. Overall, the peptaibol database is an excellent document and serves as a ready reference for biochemists as well as mycologists.

Box 4. Peptaibol database

- Available at www.cryst.bbk.ac.uk/peptaibol
- Developed by Department of Crystallography, Birkbeck College, London, UK
- Sequence and structure resource
- 317 sequence entries available till date
- Sequences section of the database allows searches by name, group, sequence motif and biological source

Conclusions and Outlook

Significance of antimicrobial compounds in terms of petaibols is evident in antagonistic property of *Trichoderma* towards other fungi. Extensive diversity among species aggregate and strains of *Trichoderma* have been indicated through peptaibol diversity (Box 5). Microheterogeneity of petaibols is a direct indication of efficiency of multifunctional non-ribosomal peptide synthetic machinery with peptide synthetases. Remarkable functional ability of these enzymes results in the production of such distinct peptaibols with unique properties. Biosynthesis of peptaibols by these enzymes involves ATP hydrolysis. The fungi producing such peptides need to spend lot of cellular energy for biosynthesis. Hence, considering the process as 'secondary' seems to be inappropriate. The ecological role of these molecules makes them as 'special molecules' in offering competitive ability to *Trichoderma.* The fungus has been widely considered as biocontrol agent and thus peptaibol producing capacity has to be exploited. The literature on peptaibol would be immensely useful to mycologists to elucidate biochemical complexity as well as chemotaxonomy of fungi. Extensive literature on the chemistry and structural features of peptaibols will be highly valuable to mycologists for future landmark.

Box 5. Peptaibols from *Trichoderma harzianum*

- Total of 55 peptaibols identified.
- 05 peptaibols are 11 residues long.
- 14 peptaibols are 14 residues long.
- 28 peptaibols are 18 residues long.
- 08 peptaibols are 19 residues long.
- N-terminus acetylated Aib.
- Microheterogeneity in terms of amino acid sequence.

Bacterial multi-enzyme polyketide syntheases were found to be involved in the production of diverse compounds with antibiotic and anticancer activity (Kolter and Moreno, 1992; Walsh, 2004). The modular genetic architecture will be valuable for production of novel candidates of drug through combinatorial biosynthesis and enzyme engineering (Weissman and Leadlay, 2005). Such studies are also necessary for peptaibols of *Trichoderma.* Cloning and expression of peptaibol synthetases of *Trichoderma* was already achieved (Wiest *et al.*, 2002, Vazcaino *et al.*, 2005; Chutrakul and Peberdy, 2005; Wei *et al.*, 2005). Challenges in understanding the biosynthetic pathways of peptabiols in *Trichoderma* have been discussed by Kubicek *et al.* (2007). Further research is urgently warranted to understand the biosynthesis of peptaibols in total and the necessity of peptiabol synthetase in ample sequence diversity. Biosynthetic pathways in different peptaibiotics producing fungi need to be understood in order to elucidate phylogenetic relationship of peptaibol producing fungi.

The secondary metabolites of fungi are of great significance as natural products for medical and or agricultural value (Carlile *et al.*, 2001; Shishupala, 2003; Deacon, 2006; Mukherjee *et al.*, 2006). Although many such metabolites are deleterious (e.g. mycotoxins), several of them have been considered useful. Calvo *et al.* (2002) discussed the role of secondary metabolism in fungal development. In olden days, the products of secondary metabolism were considered to be of negligible value for the producing organisms. Biosynthetic pathways of fungi and their biotechnological exploitation have shown the significance of such metabolites (Conesa *et al.*, 2001; Smedsgaard and Nielsen, 2005; Isaka *et al.*, 2005; Keller *et al.*, 2005; Jewett *et al.*, 2006). Monaghan *et al.* (1995) considered searching for useful secondary metabolites as random walk in random forest. Recent developments in understanding of biosynthetic machinery of these metabolites along with genome exploration in fungi should pave way for better utilization of peptaibols. Proteome analysis of *Trichoderma* has been already performed by Grinyer *et al.* (2004a; 2004b). The genome and proteome understanding of *Trichoderma* may unravel unknown potentials of peptaibols.

Fungi are known to possess several modes of genetic recombination (Kister and Miao, 1992; Kohn, 2005). Technological innovations for production of secondary metabolites of fungi are available (Robinson *et al.*, 2001). Fishing of new antimicrobials is essential to combat growing incidences of microbial infection and development of antibiotic resistance in bacterial pathogens (Hancock and Chappel, 1999; Mukhopadhyay and Peterson, 2006). High-throughput technologies for cellular analysis are accessible (Rausch, 2006). Peptaibols have been thought to be alternative for classical antibiotics (Duclohier, 2007). Remarkable capacity of peptaibols may make them highly suitable candidates as therapeutic agents. Montesinos (2007) reviewed the application of antimicrobial peptides in plant disease control. A wide range of antimicrobial property of peptaibols makes them highly suitable for biopesticides. The peptide sequence and structure may also make peptaibols as lead molecules for utilization in plant disease management. Additionally, they are known to induce systemic acquired resistance in plants making them special molecules of choice in integrated disease management. Archer (2000) considered the use of filamentous fungi as microbial cell factories for food applications. Metabolic diversity and exclusive capacity of *Trichoderma* spp. make them as potential biosynthetic machinery of peptaibol synthesis. Extraordinary diversity of peptaibols, significant mode of action and copious production of peptaibols by different *Trichoderma* is of great significance of academic interest and also for commercial exploitation. Peptaibols are found in the cells of producing fungi as well as secreted out. It is essential to understand the biological role of peptaibols besides providing competitive advantage for the producer organism. Antifungal peptides have been considered significant with the backdrop of increasing fungal infection. Such peptides with strong antifungal activities are being potential therapeutic agents against emerging pathogens. These antifungal peptides must be exploited for clinical use after ascertaining their interaction with normal human system. Overall, peptaibol diversity may provide an excellent opportunity for the assessment of fungal diversity precisely at biochemical level apart from providing potential molecules of therapeutic, medical and agricultural value.

Acknowledgements

The author is grateful to Prof. P. Balaram, Director, Indian Institute of Science, Bangalore for having introduced to the field of peptaibols, constructive discussion and providing laboratory facilities. Thanks are also due to Mr. S. Prakash for his technical assistance in procuring mass spectra. The author acknowledges the Teacher Research Fellowship awarded by the Indian Academy of Sciences, Bangalore and Visiting Fellowship awarded by the Jawaharalal Nehru Centre for Advanced Scientific Research, Bangalore.

REFERENCES

Alexopoulos, C.J., Mims, C.W. and Blackwell, M. 1996. *Introductory Mycology*, 4th Edition. John Wiley and Sons, Inc., New York.

Archer, D.B. 2000. Filamentous fungi as microbial cell factories for food use. *Curr. Opin. Biotechnol. 11*, 478-483.

Augeven-Bour, I., Rebuffat, S., Auvin,C., Prigent, Y. and Bodo, B. 1997. Harzianin HB I, an 11-residue peptaibol from *Trichoderma harzianum*: isolation, sequence, solution synthesis and membrane activity. *J. Chem. Soc. Perkin Trans. 1*, 1587-1592.

Benitez, A., Rincon, A.M., Limon, M.C. and Codon, A.C. 2004. Biocontrol mechanisms of *Trichoderma* strains. *Int. Microbiol. 7*, 249-260.

Bentley, R. and Bennett, J.W. 1999. Constructing polyketides: from collie to combinatorial biosynthesis. *Ann. Rev. Microbiol. 53*, 411-446.

Bissett, J., Szakacs, G., Nolan, C.A., Druzhinina, I.S., Kullnig-Gradinger, C.M. Kubicek, C.P. 2003. Seven new taxa of *Trichoderma* from Asia. *Can. J. Bot. 81*, 570-586.

Brunner, K., Zeilinger, S., Ciliento, R., Woo, S.L., Lorito, M., Kubicek, C.P. and Mach, R.L. 2005. Improvement of the fungal biocontrol agent *Trichoderma atroviride* to enhance both antagonism and induction of plant systemic disease resistance. *Appl. Environ. Microbiol. 71*, 3959-3965.

Calvo, A.M., Wilson, R.A., Bok, J.W. and Keller, N.P. 2002. Relationship between secondary metabolism and fungal development. *Microbiol. Mol. Biol. Rev. 66*, 447-459.

Carlile, M.J., Watkinson, S.C. and Gooday, G.W. 2001. *The Fungi*, 2nd Edition. Academic Press, San Diego.

Chaverri, P., Castlebury, L.A., Overton, B.E. Samuels, G.J. 2003. *Hypocrea/Trichoderma*: species with conidiophore elongations and green conidia. *Mycologia 95*, 1100-1140.

Chugh, J.K. and Wallace, B.A. 2001. Peptaibols: models for ion channels. *Biochem. Soc. Trans.* 29, 565-570.

Chutrakul, C. and Peberdy, J.F. 2005. Isolation and characterisation of a partial peptide synthetase gene from *Trichoderma asperellum. FEMS Microbiol. Lett. 252*, 257-265.

Conesa, A., Punt, P.J., van Luijk, N. and van den Hondel, C.A.M.J.J. 2001. The secretion pathway in filamentous fungi: A biotechnological view. *Fungal Genet. Biol. 33*, 155-171.

Daniel, J.F. and Filho, E.R. 2007. Peptaibols of *Trichoderma. Nat. Prod. Rep. 24*, 1128-1141.

Danielson, R. and Devey, C. 1973. The abundance of *Trichoderma* propagules and the distribution of species in forest soils. *Soil Biol. Biochem. 5*, 485-494.

Deacon, J. 2006. *Fungal biology*. 4th Edition, Blackwell Publishing, UK.

Degenkolb, T., Berg, A., Gams, W., Schlegel, B. and Grafe, U. 2003. The occurrence of petaibols and structurally related peptaibiotics in fungi and their mass spectrometric identification via diagnostic fragment ions. *J. Peptide Sci. 9*, 666-678.

Degenkolb, T., Grafenhan, T., Berg, A., Nirenber, H.I., Gams, W. and and Bruckner, H. 2006. Peptaibiomics: screening for polypeptide antibiotics (peptaibiotics) from plant-protective *Trichoderma* species. *Chem. Biodivers. 3*, 593-609.

Degenkolb, T., Kirschbaum, J. and Bruckner, H. 2007. New sequences, constituents, and producers of peptaibiotics: An updated review. *Chem. Biodivers. 4*, 1052-1067.

De Paoli, P. 2005. Biobanking in microbiology: From sample collection to epidemiology, diagnosis and research. *FEMS Microbiol. Rev. 29*, 897-910.

De Lucca. A.J. and Walsh, T.J. 1999. Antifungal peptides: Novel therapeutic compounds against emerging pathogens. *Antimicrob. Agents Chemotherap. 43*, 1-11.

De Lucca. A.J. and Walsh, T.J. 2000. Antifungal peptides: origin, activity and therapeutic potential. *Rev. Iberoam. Micol. 17*, 116-120.

Druzhinina, I.S. and Kubicek, C.P. 2005. Species concepts and biodiversity in *Trichoderma* and *Hypocrea*: from aggregate species to species cluster? *J. Zhejiang. Univ. Sci. 6*, 100-112.

Druzhinina, I.S., Kophinskiy, A.G., Komon, M., Bissett, J., Szakacs, G. and Kubicek, C.P. 2005. An oligonucleotide barcode for species identification in *Trichoderma* and Hypocrea. *Fungal Genet. Biol. 42*, 813-828.

Druzhinina, I.S., Kophinskiy, A.G. and Kubicek, C.P. 2006. The first 100 *Trichoderma* species characterized by molecular data. *Mycoscience 47*, 55-64.

Duclohier, H. 2007. Peptaibiotics and peptaibols: An alternative to classical antibiotics? *Chem. Biodivers. 4*, 1023-1026.

Duclohier, H., Alder, G.M., Bashford, C.L., Bruckner, h., Chugh, J.K. and Wallace, B.A. 2004. Conductance studies on Trichotoxin_A50E and implications for channel structure. *Biophys. J. 87*, 1705-1710.

Engelberth, J., Koch, T., Schuler, G., Bachmann, N., Rechtenback, J. and Boland, W. 2001. Ion channel-forming alamethicin is a potent elicitor of volatile biosynthesis and tendril coiling: Cross talk between jasmonate and salicylate signaling in lima bean. *Plant Physiol. 125*, 369-377.

Fravel, D.R. 2005. Commercialization and implementation of biocontrol. *Annu. Rev. Phytopathol. 43*, 337-359.

Gaston, K.J. and Spicer, J.I. 2004. *Biodiversity: An Introduction.* Blackwell Publishing, Oxford, UK.

Giuliani, A., Pirri, G. and Nicoletto, S.F. 2007. Antimicrobial peptides: an overview of a promising class of therapeutics. *Cent. Eur. J. Biol. 2*, 1-33.

Goulard, C., Hlimi, S., Rebuffat, S. and Bodo, B. 1995. Trichorzins HA and MA, antibiotic peptides from *Trichoderma harzianum* I: fermentation, isolation and biological properties. *J. Antibiotics 48*, 1248-1253.

Grinyer, J., McKay, M., Herbert, B.R. and Nevalainen, H. 2004a. Fungal proteomics: mitochondrial proteins of a *Trichoderma harzianum* strain applied for biological control. *Curr. Gen. 45*, 170-175.

Grinyer, J., McKay, M., Nevalainen, H. and Herbert, B.R. 2004b. Fungal proteomics: initial mapping of biological control strain *Trichoderma harzianum. Curr. Gen. 45*, 163-169.

Hancock, R.E.W. and Chappel, D.S. 1999. Peptide antibiotics. *Antimicrob. Agents Chemother. 43*, 1317-1323.

Harman, G.E., Howell, C.R., Viterbo, A., Chet., I. and Lorito, M. 2004. *Trichoderma* species – opportunistic, avirulent plant symbionts. *Nat. Rev. Microbiol. 2*, 43-56.

He, H., Janso, J.E., Yang, H.Y., Bernan, V.S., Lin, S.L. and Yu, K. 2006. Culicinin D, an antitumor peptaibol produced by the fungus *Culicinomyces clavisporus*, strain LL-121252. *J. Nat. Prod. 69*, 736-741.

Hlimi, S., Rebuffat, S., Goulard, C., Duchamp, S. and Bodo, B. 1995. Trichorzins HA and MA, antibiotic peptides from *Trichoderma harzianum* II. Sequence determination. *J. Antibiotics 48*, 1254-1261.

Howell, C.R. 2003. Mechanisms employed by *Trichoderma* species in the biological control of plant disease: The history and evolution of current concepts. *Plant Dis. 87*, 4-10.

Isaka, M., Kittakoop, P., Kirtikara, K., Hywel-Jones, N.L. and Thebtaranonth, Y. 2005. Bioactive substances from insect pathogenic fungi. *Acc. Chem. Res. 38*, 813-823.

Ishiyama, D., satou, T., Senda, H., Fujimaki, T., Honda, R. And Kanazawa, S. 2000. Heptaibin, a novel antifungal peptaibol antimbiotic from *Emericellopsis* sp. BAUA8289. *J. Antibiotics. 53*, 728-732.

Jaworski, A. and Bruckner, H. 1999. Detection of new sequences of peptaibol antibiotics trichotoxins A-40 by on-line liquid chromatography – electrospray ionization mass spectrometry. *J. Chromatogr. A. 862*, 179-189.

Jaworski, A. and Bruckner, H. 2000. New sequences and new fungal producers of peptaibol antibiotics antiamoebins. *J. Pept. Sci. 6*, 149-167.

Jewett, M.C., Hofmann, G. and Nielsen, J. 2006. Fungal metabolite analysis in genomics and phenomics. *Curr. Opin. Biotechnol. 17*, 191-197.

Karle, I.L. and Balaram, P. 1990. Structural characteristic of alpha-helical peptide molecules containing Aib residues. *Biochemistry 29*, 6747-6756.

Keller, N.P., Turner, G. and Bennett, J.W. 2005. Fungal secondary metabolism - from biochemistry to genomics. *Nat. Rev. Microbiol. 3*, 937-947.

Kesseler, N., Schuhmann, H., Morneweg, S., Linne, V. and Marahiel, M.A. 2004. The linear pentadecapeptide gramicidin is assembled by four multimodular non-ribosomal peptide synthetases that comprise 16 modules with 56 catalytic domains. *J. Biol. Chem. 279*, 7413-7419.

Kister, H.C. and Miao, V.P.W. 1992. New modes of genetic change in filamentous fungi. *Ann. Rev. Phytopathol. 30*, 131-152.

Kleinkauf, H and von Dohren, H. 1987. Biosynthesis of peptide antibiotics. *Ann. Rev. Microbiol. 41*, 259-289.

Kleinkauf, H and von Dohren, H. 1996. A non-ribosomal system of peptide synthesis. *Eur. J. Biochem. 236*, 335-351.

Kohn, L.M. 2005. Mechanisms of fungal speciation. *Ann. Rev. Phytopathol. 43*, 279-308.

Kolter, R. and Moreno, F. 1992. Genetics of ribosomally synthesized peptide antibiotics. *Ann. Rev. Microbiol. 46*, 141-163.

Konz, D. and Marahiel, M.A. 1999. How do peptide synthetases generate structural diversity? *Chem. Biol. 6*, R39-R48.

Krause, C., Kirschbaum, J. and Brückner, H. 2006. Peptaibiomics: an advanced, rapid and selective analysis of peptaibiotics/peptaibols by SPE/LC-ES-MS. *Amino Acids 30*, 435-443.

Kubicek, C.P., Bissett, J., Druzhinin, I., Kullnig-Gradinger, C. and Szakacs, G. 2003. Genetic and metabolic diversity of *Trichoderma*: a case study on south-east Asian isolates. *Fungal Genet. Biol.. 38*, 310-319.

Kubicek, C.P., Komon-Zelazowska, M., Sandor, E. and Druzhinina, I.S. 2007. Facts and challenges in the understanding of the biosynthesis of peptaibols of *Trichoderma*. *Chem. Biodivers. 4*, 1068-1082.

Kucuk, C. and Kivanc, M. 2003. Isolation of *Trichoderma* spp. and determination of their antifungal, biochemical and physiological features. *Turk J. Biol. 27*, 247-253.

Kucuk, C. and Kivanc, M. 2004. *In vitro* antifungal activity of strains of *Trichoderma harzianum*. *Turk J. Biol. 28*, 111-115.

Kullnig-Gradinger, C.M., Szkacs, G. and Kubicek, C.P. 2002. Phylogeny and evolution of the fungal genus *Trichoderma*: a multigene approach. *Mycol. Res. 106*, 757-767.

Landreau, A., Pouchus, Y.F., Sallenave-Namont, C., Biard, J., Boumard, M., du Pont, T.R., Mondeguer, F., Goulard, C. and Verbist, J. 2002. Combined use of LC/MS and a biological test for rapid identification of marine mycotoxins produced by *Trichoderma koningii*. *J. Microbiol. Methods 48*, 181-194.

Lieckfeldt, E., Samuels, G.J., Nirenberg, H.I. and Petrini, O. 1999. A morphological and molecular perspective of *Trichoderma virde*: Is it one or two species? *Appl. Environ. Microbiol. 65*, 2418-2428.

Marahiel, M.A. 1992. Multidomain enzymes involved in peptide synthesis. *FEBS Lett. 307*, 40-43.

Massart, S. and Jijakli. H.M. 2007. Use of molecular techniques to elucidate the mechanisms of action of fungal biocontrol agents: A review. *J. Microbiol. Methods 69*, 229-241.

Mohamed-Benkada, M., Montagu, M., Biard, J., Mondeguer, F., Verite, P., Dalgalarrondo, M., Bissett, J. and Pouchus, Y.F. 2006. New short peptaibols from a marine *Trichoderma* strain. *Rapid Commun. Mass Spectrom. 20*, 1176-1180.

Monaghan, R.L., Polishook, J.D., Pecore, V.J., Billis, G.F., Nallin-Omsted, M. and Streicher, S.L. 1995. Discovery of novel secondary metabolites from fungi – is it really a random walk through a random forest? *Can. J. Bot. 73*, S925-S931.

Montesinos, E. 2007. Antimicrobial peptides and plant disease control. *FEMS Microbiol. Lett.* 270, 1-11.

Mukherjee, M., Horwitz, B.A., Sherkhane, P.D., Hadar, R. and Mukherjee, P.K. 2006. A secondary metabolite biosynthesis cluster in *Trichoderma virens*: evidence from analysis of genes underexpressed in a mutant defective in morphogenesis and antibiotic production. *Curr. Genet. 50*, 193-202.

Mukhopadhyay, A. and Petrson, R.T. 2006. Fishing for new antimicrobials. *Curr. Opin. Chem. Biol. 10*, 327-333.

Vizcaino, J.A., Sanz, L., Cardoza, R.E., Monte, E. and Gutierrez, S. 2005. Detection of putative peptide synthetase genes in *Trichoderma* species: application of this method to the cloning of a gene from *T. harzianum* CECT 2413. *FEMS Microbiol. Lett. 244*, 139-148.

Wada, S., Nishimura, T., Iida, A., Toyama, N. and Fujita, T. 1994. Primary structures of antibiotic peptides, Trichocellins-A and –B from *Trichoderma viride. Tetrhedron Lett. 35*, 3095-3098.

Wada, S. and Tanaka, R. 2007. A functionalized 20-residue peptaibol derivative for nucleic acid delivery. *Chem. Biodivers. 4*, 991-997.

Walsh, C.T. 2004. Polyketide and non-ribosomal peptide antibiotics: modularity and versatility. *Science 303*, 1805-1810.

Wei, X., Yang, F. and Straney, D.C. 2005. Multiple non-ribosomal peptide synthetase genes determine peptabiol synthesis in *Trichoderma virens. Can. J. Microbiol. 51*, 423-429.

Weindling, R. 1934. Studies on lethal principle effective in the parasitic action of *Trichoderma lignorum* on *Rhizoctonia solani* and other soil fungi. *Phytopathol. 24*, 1153-1179.

Weissman, K.J. and Leadlay, P.F. 2005. Combinatorial biosynthesis of reduced polyketides. *Nature Rev. Microbiol. 3*, 925-936.

Whitmore, L., Chugh, J.K., Snook, C.F. and Wallace, B.A. 2003. The peptaibol database: a sequence and structure resource. *J. Pept. Sci. 9*, 663-665.

Whitmore, L. and Wallace, B.A. 2004a. Analysis of peptaibol sequence composition: implications for in vivo synthesis and channel formation. *Eur. Biophys. J. 33*, 233-237.

Whitmore, L. and Wallace, B.A. 2004b. The Peptaibol Database: a database for sequences and structures of naturally occurring peptaibols. *Nucleic Acids Res. 32*, D593-594.

Widden, P. and Abitbol, J.J. 1980. Seasonality of *Trichoderma* species in a spruce-forest soil. *Mycologia 72*, 775-784.

Wiest, A., Grzegorski, D., Xu, B., Goulard, C., Rebuffat, S., Ebbole, D.J., Bodo, B. and Kenerley, C. 2002. Identification of peptaibols from *Trichoderma virens* and cloning of a peptaibol synthetase. *J. Biol. Chem. 277*, 20862-20868.

Wilhite, S.E., Lumsden, R.D. and Straney, D.C. 2001. Peptide synthetase gene in *Trichoderma virens. Appl. Environ. Microbiol. 67*, 5055-5062.

Wuczskowski, M., Druzhinina, I.S., Gherbawy, Y., Klug, B., Prillinger, H.J. and Kubicek, C.P. 2003. Taxon pattern and genetic diversity of *Trichoderma* in a mid-European, primeval floodplain forest. *Microbiol. Res. 158*, 125-134.

Yun, B., Yoo, I., Kim, Y.H., Kim, Y., Lee, S., Kim, K. and Yeo, W. 2000. Peptaivirins A and B, two new antiviral petaibols against TMV infection. *Tetrahedron Lett. 41*, 1429-1431.

Zeilinger, S. and Omann, M. 2007. *Trichoderma* biocontrol: signal transduction pathways involved in host sensing and mycoparasitism. *Gene Regn. Systems Biol. 1*, 227-234.

Zhang, C.L., Druzhinina, I.S., Kubicek, C.P. and Xu, T. 2005. Biodiversity of *Trichoderma* in China: evidence for a North to South difference of species distribution in East Asia. *FEMS Microbiol. Lett. 251*, 251-257.

Frontiers in Fungal Ecology, Diversity and Metabolites (edited by K.R. Sridhar),
I.K. International Publishing House Pvt. Ltd., New Delhi, 2009.

22

Emerging Issues in Mycology

K.R. Sridhar
Microbiology and Biotechnology, Department of Biosciences,
Mangalore University, Mangalagangotri, Mangalore 574 199, Karnataka, India
(E-mail: sirikr@yahoo.com)

"Serendipity is the hallmark of pharmacological research. A chance discovery can lead not only to a successful drug but to advances in fundamental science. Cyclosporin, derived from a Norwegian mountain fungus and epibatidine, an opium-like toxin secreted by dendrobatid frogs, are two examples".

— Edward O. Wilson

Introduction

Mycology has become a crucial issue in almost all facets of human, livestock and ecosystem health. Fungi are the non-photosynthetic lower eukaryotes comprising single celled and multicellular filamentous forms. They have been placed in the '*Fifth Kingdom*' based on unique absorptive mode of nutrition. Efforts to probe the origin of species in recent past, a team of scientists has reconstructed the early evolutionary history of fungi and linked to the animals as closest relatives (James *et al*., 2006). Although microfungi are invisible, they are the largest living organisms (Smith *et al*., 1992). Among eukaryotes, fungi are the only organisms those have the unique ability to thrive at a temperature 45-60°C (Maheshwari, 1998). Besides that, the fungi are the inhabitants of diverse habitats, they live mutualistically within the plant tissues (endophytes), symbiotically with plants (mycorrhizas), algae (lichens) and animals (trichomycetes). Majority of fungi involve in decomposition of dead materials and recycle the essential mineral nutrients such as nitrogen, phosphorus and potassium. Fungal distribution in a wide variety of habitats and their interaction with biotic and abiotic system emerged '*fungal ecology*' as one of the frontier areas of research. The Global Taxonomy Initiative (GTI) and Convention on Biological Diversity (CBD, 2002) projected the significance and magnitude of '*fungal diversity*' and revealed our ignorance on fungal resources. The '*fungal metabolites*' possess life-saving qualities include enzymes, therapeutic proteins and antibiotics (http://www.bio.mq.edu.au/dept/centres/edge/fungalbt.html, www.genengnews.com). The current chapter bridges the contributions of this book with a brief commentary of the emerging issues of fungal ecology, diversity and metabolites.

Fungal Ecology

One of the major challenges of contemporary ecology is linking species and ecosystem functioning (Cardinale *et al*., 2000). Due to relevance of fungal interaction with biotic and abiotic systems, fungal ecology has become an emerging area of mycology, which deals with distribution, interaction and ecosystem functions. Wicklow and Carroll (1981) stated that the fungal ecology suffers due to lack of generalizations on the significance of results of mycological observations. Cooke and Rayner (1984) also addressed on the lack of adequate knowledge on fungi, which retarded the emergence of clear ecological principles. Baerlocher (1992) indicted that fungal ecologist facing problems in identifying the members of the community as they occur as mycelia, fructifications or spores and thus it is necessary to identify an active mycelial state of fungi. However, production of fungal spores, fungal-specific signature molecules (e.g., ergosterol) and immunological/fluorescence/radioactive techniques demonstrate the activity of fungal mycelia (Graça *et al*., 2005). Several chapters of this book address the fungal ecology with major emphasis on aquatic and mycorrhizal studies.

Chapter 1 projects structural and functional significance of freshwater hyphomycetes in aquatic ecosystem. Freshwater hyphomycetes constitutes a vital link and mediates the energy flow between dead plant materials to higher trophic levels. They are most dominant among microbial communities on submerged decaying leaf litter in streams and contribute 63-100% of total microbial biomass (Baldy *et al*., 2002; Gulis and Suberkropp, 2003). Their biomass will build up to 18% of total detritus mass and allocate 80% to conidial production (Suberkropp, 1991, 1995). They transform their mycelial biomass into conidial biomass, thus conidial estimate gives a close picture of fungal activity in a substrate (e.g., leaf, wood) exists in a specific ecological niche. Freshwater hyphomycetes has become a model community to evaluate the structure and function in streams. Conidia of freshwater hyphomycetes are similar to plankton as they are designed for floatation, transport and attachment to substrate. Many techniques applied to study terrestrial fungi can also be adapted to understand their ecological functions in aquatic habitats (Graça *et al*., 2005). Human interference has a major influence on the assemblage, structure and functions of these fungi (e.g., pollution, change in water course). Baerlocher (2007), reviewed molecular approaches applied to aquatic hyphomycetes and indicated that these methods do not rely on the presence of reproductive stages to identify taxa. Therefore, molecular approaches refine our knowledge on many aspects of the taxonomy and ecology of aquatic hyphomycetes. However, for investigation of aquatic fungal properties at higher levels of integration (e.g., spore shape, population survival during resource scarcity) has to rely on more conventional techniques (Baerlocher, 2007).

In aquatic mycology, the role of marine and mangrove fungi has attracted the attention of several mycologists. Mangrove fungi constitute the second largest group next to fungi on marine intertidal woody litter (Hyde, 1990). Similar to the dominance of mitosporic fungi in freshwater streams, ascomycetes dominate in mangrove and marine habitats. The ascospores show specific adaptation to the marine ecosystem in appendages and these appendages help in buoyancy, entrapment and adherence to the surfaces. Chapter 3 discusses the current status of mangrove filamentous fungi of the Indian Peninsula with an emphasis on the pattern of colonization, litter decomposition, endophytic fungi and terrestrial fungi in Indian mangroves. Indian mangroves is estimated to be about 7% of the world mangroves extended over 7000 km^2 with highest distribution in the East Coast (70%). A variety of ecological niches available for fungi in mangrove habitats (e.g., intertidal, canopy, sediment), which facilitates mass accumulation strategy of filamentous fungi. Similarly, animal remains of mangroves also support a variety of filamentous fungi. *Aigialus mangrovei, Cirrenalia pygmea, C. tropicalis, Halosarpheia cincinnatula, Lulworthia grandispora, Passeriniella mangrovei, Verruculina enalia* and *Zalerion maritimum* are typical mangrove fungi, but they are also core-group fungi on introduced wood in monsoon season indicates their high colonization activity on wood even under low salinity. On the contrary, some terrestrial anamorphic fungi (e.g., *Alternaria, Arthrobotrys, Aspergillus, Penicillium, Phoma, Tetracrium*) were common during monsoon season and their activity usually diminishes in high saline conditions. A wide range of endophytic fungi harbour mangrove plant tissues and many of them are typical terrestrial fungi. Several terrestrial fungi have been isolated from mangrove and marine habitats, which are of immense value as they have adapted to saline habitats.

Mycorrhizal symbiosis is universal in terrestrial plant communities. However, the role of endomycorrhizal symbiosis in plant communities has been ignored. Chapter 10 emphasizes the significance of arbuscular mycorrhizal (AM) fungi on growth and mineral acquisition in crop plants growing under salinity stress. The AM fungal colonization of plants under stressed or disturbed habitats serves alternative mechanisms to satisfy plant's nutritive requirements, which differs among AM fungi. *Glomus*-colonized lettuce grew less and accumulates less N and P at the highest salt levels (0.75 g NaCl/kg dry soil). Thus, suitable AM fungal inoculation can be proposed to improve salt tolerance of plant species. Host-AM fungus specificity, interaction and competitiveness under different environmental stresses such as salinity, drought, N and P deficiency are the major factors to be considered in selection of AM fungi for application in

The diversity of 17 AM fungi in Khazan ecosystem and their ecological importance has been emphasized in Chapter 9. The diversity of lichens of Schirmacher Oasis of Antarctica has been dealt in Chapter 7. Out of total 48 lichen species that occur in Schirmacher Oasis, 19 are endemic to Antarctica, 15 are bipolar and 14 are cosmopolitan in distribution. There is an urgent need to intensify investigation on the diversity of fungi associated with lichens of Antarctica.

Conventional methods such as culturing and identification by morphology and biochemical tests are the basic requirements. Traditional microscopic techniques fail to identify specific fungal vegetative hyphae. Chapter 19 emphasizes the significance of PCR-based molecular tools in diagnosis and fungal diversity assessment. As not all fungi have ideal medium for grow, many are uncultivable and such fungi or their activities can be evaluated by molecular methods. The PCR-based finger printing approaches are currently most popularly used tool in research involving assessment of biodiversity of microbial communities (Dorigo *et al.,* 2005), fungal genomics and gene function. Identification and taxonomic positioning of fungi is important for biologists interested in diversity, conservation and for others aims (e.g., plant pathology, quarantine, medical, pharmaceutics). In Chapter 20, authors have emphasized on various molecular methods employed to identify the pathogenic fungi and their significance in understanding species evolution and pathogen differentiation. Nucleic acid sequence analysis has proved to be a viable tool for the identification of fungi in a number of applications such as identification of fungal variants, non-cultivatable fungi in samples and recognition of new species.

Fungal Metabolites

Fungi have been broadly classified as 'creative' and 'non-creative' species depending on the ability to generate biologically viable secondary metabolites (see Miller, 2000). Basic studies on fungal ecology paved way to understand the importance of fungi in the production of secondary metabolites. Over the last six decades, although considerable efforts have been channelled towards isolation and characterization of fungal metabolites, it is still in its infant stage to make a breakthrough. However, new ideas towards the utilization of fungi for production of novel metabolites are emerging due to significant efforts on fungal ecology and biodiversity inventions. For instance, research on endophytic and extremophilic fungi has generated ample interest in fungal biologists and chemists to look for novel metabolites. Similarly, mangrove and marine fungi are the targets for production of value-added metabolites. Research on fungal natural product suffers due to two major reasons: firstly, low initiative in ecology and biodiversity research; secondly, lack of application of sophisticated techniques or uncertainty of production of novel metabolites.

Several chapters of this book address the importance and novelties of fungal metabolites. Marine fungi have been the target of new bioactive compounds and enzymes. In Chapter 4, emphasis has been given on marine fungal value-added metabolic products. A variety of novel metabolites have been produced and evaluated from fungi of marine origin for antibacterial, antifungal, antiviral, antimalarial, anti-inflammatory, cytotoxic and antineoplastic properties. Chapter 13 emphasizes on *Fusarium* spp. reported to colonize *Sorghum* worldwide causing serious problems of seed rot, seedling blight, root rot, stalk rot, Pokkah boeng, pre-emergence ear rot, head blight and grain mold. Many fusaria heavily colonize post harvest agricultural produce and generate mycotoxins. A commentary has been given on different groups of fusarial toxins (polyketides, terpenoides, amino acid derivatives and shikimic acid derivatives) and their impact on humans and livestock. A new class of fusarial toxins (fumonisins) of agricultural produce and their toxicity, risk assessment and strategies necessary for its reduction and management are given in Chapter 14. A variety of mechanisms have been shown to be involved in antagonism of *Trichoderma* against plant pathogenic fungi and thus *Trichoderma* has become one of the most powerful biocontrol agents of plant diseases. Shishupala in Chapter 21, describes the biological control potential of *Trichoderma* spp. Antagonistic properties of *Trichoderma* is due to their ability to produce multifunctional linear peptides called 'peptaibols' which have the capacity to induce membrane channels in the target organisms. Besides *Trichoderma*, a few other fungi also known to produce peptaibols. Peptaibols showed remarkable sequence diversity by the presence of unusual amino acids. Their synthesis is devoid of involvement of ribosomes and possible by the activity of multi-enzyme complex modular peptide synthetases. These complexes assemble peptides from a remarkable range of precursors such as non-protein amino acids and hydroxy or carboxy acids. The study of peptaibol diversity helps to assess fungal diversity explicitly at biochemical level and paptaibols are potential molecules of therapeutic, medical and agricultural value. The author emphasizes that the term 'secondary metabolites' for peptaibols seems to be inappropriate as they are 'special molecules' offering competitive ability to *Trichoderma.*

The quality of coffee, safety, trade, implications of mould proliferation, mycotoxins and mould prevention at different stages including positive implications of coffee moulds in natural fermentation are discussed in Chapter 15.

Symbiotic or mutualistic fungi are also the potential source of value-added metabolites, thus, two main aspects have been covered in this book (endophytic fungi and ectomycorrhizae). Endophytic fungi are largely untapped resource having economically viable applications (Chapter 12). Endophytic fungal metabolites are the major players in agriculture, pharmaceutical industry and plant breeding. In addition, endophytes are important due to their ability to control pests, production of novel bioactive metabolites and bioprospecting avenues. A brief account of metabolites of endophytic fungi of mangroves is also available in Chapter 3. Production of IAA by ectomycorrhizal fungi and their role in the early initiation of colonization with tree roots and its beneficial exploitation in micropropagation of commercially important tree species have been evaluated in Chapter 11.

Fungi become primarily or opportunistically pathogenic depending on the host immune response. The primary pathogens cause disease in humans with perfect immune system, while opportunistics in the immuno-compromised humans (e.g. HIV patients). The fungal infections may be superficial, subcutaneous and systemic. Chapter 16 addresses on causative fungi, clinical symptoms, pathogenesis, diagnosis and the line of treatment. Over 30 years, amphotericin B (which causes nephro-toxicity) was the sole drug available to control fungal infections. Subsequent invention of imidazoles and triazoles, the treatment of fungal infections has become effective and safe. As a few drugs are available for fungal control in humans, there is ample scope to advent fungal natural products as appropriate therapeutic agents for fungal diseases.

Future Perspectives

More efforts need to be channeled to understand the diverse fungal resources, their functions in the ecosystem and judicious harnessing of their potentialities and metabolites. Hawksworth (2003) drafted explicitly how to safeguard the global fungal resources through personal, national and international collaborative efforts, while Hyde (2003) commented on the current status and future of mycological research in Asia. Both contributions concerned with enormous fungal diversity in tropical region of the world and its assessment, conservation and exploitation. Barnett (1998) linked contemporary biotechnological methods to harness filamentous fungi for value-added products in his review.

As eukaryotes, fungi are the ideal organisms to generate suitable macromolecules to serve basic needs of human and livestock. Two chapters in this book deal with contemporary advances in yeast research (Chapters 17, 18). Industrially biocatalyzed processes generally utilize extracellular enzymes such as proteolytic, lipolytic and amylolytic enzymes. But, the applications of intracellular enzymes are greatly hampered by the laborious expensive isolation procedures as well as enzyme instability. Thus, cell permeabilization is gaining importance in biocatalysis. Chapter 18 explores the whole cell permeabilization, permeabilization methods and applications in yeasts. It also deals with limitations of the whole cell permeabilization (permeability barrier and unwanted side reactions by the other enzymes) with suggestions to overcome such hurdles. Undissociated form of short-chain carboxylic acids can permeate microbial plasma membranes by simple diffusion and thus serve as food preservatives. Interestingly, about one half of the total yeast species listed has strains able to use carboxylic acids as carbon and energy source and thus seriously threaten the food industry. On the contrary, the ability of yeasts to grow on organic acids facilitates their exploitation as sources of proteins, enzymes and vitamins. Chapter 17 provides an overview of developments on the biology of monocarboxylate transporters in yeasts by summarizing the transport physiological data (from whole cells or cell-free systems), identification of the genes/proteins associated with monocarboxylate transport and structural details on plasma membrane carboxylate permeases. The mammalian monocarboxylate transporters have been compared with those from yeast cells, as yeasts share basic structural and functional organization with higher eukaryotes and ease of manipulation and making them potential models to study important mechanisms for mammalian cells and cancer therapy.

Research on plant diversity and production demonstrates an asymptotic relationship between diversity and ecosystem function. The ecosystem function collapses when species number drops, which applies to fungal dependent ecosystems too. Mycological research gained momentum by two contemporary developments in biology (Baerlocher, 2003): firstly, Convention on Biological Diversity (CBD) renewed research efforts on fungal diversity due to threat of human interference in fungal extinction and thus their ecological services; secondly, application of sophisticated techniques to understand the fungal diversity and their ecological functions. A number of avenues have been opened up by the basic and applied investigations in recent past favoring future studies and outlook on fungal diversity, ecology and metabolites.

"The fungi are progressive, ever changing and evolving rapidly in their own way, so that they are capable of becoming adapted to every condition of life. We may rest assured that as green plants and animals disappear one by one from the face of the globe, some of the fungi will always be present to dispose of the last remains".

– B.O. Dodge

References

Baerlocher, F. 1992. Community organization. In: *The Ecology of Aquatic Hyphomycetes* (Ed. Baerlocher, F.). Springer-Verlag, Berlin, 38-76.

Baerlocher, F. 2003. Aspects of biodiversity in relation to ecological functions. *The Campus Scholar (Mount Allison's newsletter of research and creative activity) 1*, 5.

Baerlocher, F. 2007. Molecular approaches applied to aquatic hyphomycetes. *Fungal Biology Reviews 21*: 19-24.

Baldy, V., Chauvet, E., Charcosset, J.Y. and Gessner, M.O. 2002. Microbial dynamics associated with leaves decomposing in the mainstream and floodplain pond of a large river. *Aq. Microb. Ecol. 28*, 25-36.

Barnett, J.W. 1998. Mycotechnology: The role of fungi in biotechnology. *J. Biotechnol. 66*, 101-107.

Cannon, P.F. 1997. Strategies for rapid assessment of fungal diversity. *Biodivesity and conservation 6*, 669-680.

Cardinale, B.J., Nelson, K. and Palmer, M.A. 2000. Linking species diversity to the functioning of ecosystem: on the importance of environmental context. *Oikos 91*, 175-183.

CBD 2002. Global Taxonomy Initiative. http://www.biodiv.org/programmes/cross-cutting/taxonomy/default.asp.

Cooke, R.C. and Rayner, A.D.M. 1984. *Ecology of Saprophytic Fungi*. Longman, London.

Dorigo, U., Volatier, L. and Jean-Francois, H. 2005. Molecular approaches to the assessment of biodiversity in aquatic microbial communities. *Water Res. 39*, 2207-2218.

Fröhlich, J. and Hyde, K.D. 1999. Biodiversity of palm of fungi in the tropics: are global fungal diversity estimates realistic? *Biodiver. Conser. 8*, 977-1004.

Graça, M.A.S., Baerlocher, F. and Gessner, M.O. 2005. *Methods to Study Litter Decomposition: A Practical Guide.* Springer, Netherlands.

Gulis, V. and Suberkropp, K. 2003. Effect of inorganic nutrients on relative contribution of fungi and bacteria to carbon flow from submerged decomposing leaf litter. *Microb. Ecol. 45*, 11-19.

Hart, M.M. and Reader, R.J. 2004. Do arbuscular mycorrhizal fungi recover from soil disturbance differently? *Tropical Ecol. 45*, 97-111.

Hawksworth, D.L. 1991. The fungal dimension of biodiversity: Magnitude, significance and conservation. *Mycol. Res. 95*, 641-655.

Hawksworth, D. L. 2001. The magnitude of fungal diversity: the 1.5 million species estimate revisited. *Mycol. Res. 105*, 1422-1432.

Hawksworth, D.L. 2003. Monitoring and safeguarding fungal resources worldwide: the need for an international collaborative MycoAction Plan. *Fungal Diversity 13*, 29-45.

Hyde, K.D. 1990. A study of the vertical zonation of intertidal fungi on *Rhizophora apiculata* at Kampong Kapok mangrove, Brunei. *Aquat. Bot. 36*, 255-262.

Hyde, K.D. 1995. Measuring biodiversity of microfungi in the wet tropics of North Queensland. In: *Measuring and Monitoring Biodiversity of Tropical and Temperate Forests* (Ed. Boyle, T.J.B. and Boontawee, B.). CIFOR, Indonesia, Bogor, Indonesia, 271-286.

Hyde, K.D. 2003. Mycology and its future in Asia region. *Fungal Diversity 13*, 59-68.

James, T.Y., Kauff, F., Schoch, C.L., Matheny, P.B., Hofstetter, V., Cox, C.J., Celio, G., Gueidan, C., Fraker, E., Miadlikowska, J., Lumbsch, H.T., Rauhut, A., Reeb, V., Arnold, A.E., Amtoft, A., Stajich, J.E., Hosaka, K., Sung, G.-H., Johnson, D., O'Rourke, B., Crockett, M., Binder, M., Curtis, J.M., Slot, J.C., Wang, Z., Wilson, A.W., Schüβler, A., Longcore, J.E., O'Donnell, K., Mozley-Standridge, S., Porter, D., Letcher, P.M., Powell, M.J., Taylor, J.W., White, M.M., Griffith, G.W., Davies, D.R., Humber, R.A., Morton, J.B., Sugiyama, J., Rossman, A.Y., Rogers, J.D., Pfister, D.H., Hewitt, D., Hansen, K., Hambleton, S., Shoemaker, R.A., Kohlmeyer, J., Volkmann-Kohlmeyer, B., Spotts, R.A., Serdani, M., Crous, P.W., Hughes, K.W., Matsuura, K., Langer, E., Langer, G., Untereiner, W.A., Lücking, R., Büdel, B., Geiser, D.M., Aptroot, A., Diederich, P., Schmitt, I., Schultz, M., Yahr, R., Hibbett, D.S., Lutzoni, F., McLaughlin, D.J., Spatafora, J.W., and Vilgalys, R. 2006. Reconstructing the early evolution of fungi using a six-gene phylogeny. *Nature 443*, 818-822.

Johnston, N.C. and Pfleger, F.L. 1992. Vesicular-arbuscular mycorrhizae and cultural stresses. In: *Mycorrhizae in Sustainable Agriculture* (Ed. Bethlenfalvay, G.J. and Linderman, R.G.). ASA Special Publication # 54, Madison, WI, USA, 71-79.

Kim, S.T., Choi, J.H., Jeon, H.G., Cha, H.E., Hwang, Y.J. and Chung, Y.S. 2005. Comparison between polymerase chain reaction and fungal culture for the detection of fungi in patients with chronic sinusitis and normal controls. *Acta Otolaryngol. 125*, 72-75.

Maheshwari, R. 1998. Upper temperature limit of fungi. *Current Science 72*, 537.

Maheshwari, R. 2005. Fungal biology in 21st century. *Current science 88*, 1406-1418.

May, R.M. 2000. The dimensions of life on Earth. In: *Nature and Human Society: The Quest for a Sustainable World* (Ed. Raven, P.H. and Williams, T.). National Academy Press, Washington, DC, 13-20.

Miller, J.D. 2000. Screening for secondary metabolites. In: *Marine Mycology – A Practical Approach* (Ed. Hyde, K.D. and Pointing, S.B.). Fungal Diversity Series # 1, Fungal Diversity Press, Hong Kong, 158-171.

Smith, M., Bruhn, J.N. and Anderson, J.B. 1992. The fungus *Armillaria bulbosa* is among the largest and oldest living organisms. *Nature 356*, 428–431.

Suberkropp, K. 1991. Relationships between growth and sporulation of aquatic hyphomycetes on decomposing leaf litter. *Mycol Res. 95*, 843-850.

Suberkropp, K. 1995. The influence of nutrients on fungal growth, productivity and sporulation during leaf breakdown in streams. *Can. J. Bot. 73*, S1361-S1369.

Wicklow, D.T. and Carroll, G.C. 1981. Introduction. In: *The Fungal Community* (Ed. Wicklow, D.T. and Carroll, G.C.). Marcel Dekker, New York, 1-8.

Willinger, B., Obradovic, A., Selitsch, B., Beck-Mannagetta, J., Buzina, W., Braun, H., Apfalter, P., Hirschl, A.M, A.M., Makristathis, A. and Rotter, M. 2003. Detection and identification of fungi from fungus balls of the maxillary sinus by molecular techniques. *J. Clin. Microbiol. 41*, 581-585.

Index